Cohomology of Groups

Pure and Applied Mathematics

A Series of Monographs and Textbooks

Editors **Samuel Eilenberg and Hyman Bass**

Columbia University, New York

RECENT TITLES

Ronald G. Douglas. Banach Algebra Techniques in Operator Theory

Willard Miller, Jr. Symmetry Groups and Their Applications

Arthur A. Sagle and Ralph E. Walde. Introduction to Lie Groups and Lie Algebras

T. Benny Rushing. Topological Embeddings

James W. Vick. Homology Theory: An Introduction to Algebraic Topology

E. R. Kolchin. Differential Algebra and Algebraic Groups

Gerald J. Janusz. Algebraic Number Fields

A. S. B. Holland. Introduction to the Theory of Entire Functions

Wayne Roberts and Dale Varberg. Convex Functions

A. M. Ostrowski. Solution of Equations in Euclidean and Banach Spaces, Third Edition of Solution of Equations and Systems of Equations

H. M. Edwards. Riemann's Zeta Function

Samuel Eilenberg. Automata, Languages, and Machines: Volume A. *In preparation:* Volume B

Morris Hirsch and Stephen Smale. Differential Equations, Dynamical Systems, and Linear Algebra

Wilhelm Magnus. Noneuclidean Tesselations and Their Groups

François Treves. Basic Linear Partial Differential Equations

William M. Boothby. An Introduction to Differentiable Manifolds and Riemannian Geometry

Brayton Gray. Homotopy Theory: An Introduction to Algebraic Topology

Robert A. Adams. Sobolev Spaces

John J. Benedetto. Spectral Synthesis

D. V. Widder. The Heat Equation

Irving Ezra Segal. Mathematical Cosmology and Extragalactic Astronomy

J. Dieudonné. Treatise on Analysis: Volume II, enlarged and corrected printing; Volume IV

Werner Greub, Stephen Halperin, and Ray Vanstone. Connections, Curvature, and Cohomology: Volume III, Cohomology of Principal Bundles and Homogeneous Spaces

In preparation

I. Martin Isaacs. Character Theory of Finite Groups

K. D. Stroyan and W. A. J. Luxemburg. Introduction to the Theory of Infinitesimals

James R. Brown. Ergodic Theory and Topological Dynamics

Clifford A. Truesdell. A First Course in Rational Continuum Mechanics: Volume 1, General Concepts

Cohomology of Groups

EDWIN WEISS

Department of Mathematics
Boston University
Boston, Massachusetts

1969

ACADEMIC PRESS New York San Francisco London

A Subsidiary of Harcourt Brace Jovanovich, Publishers

ACADEMIC PRESS, INC.
111 Fifth Avenue, New York, New York 10003

United Kingdom Edition published by
ACADEMIC PRESS, INC. (LONDON) LTD.
24/28 Oval Road, London NW1

LIBRARY OF CONGRESS CATALOG CARD NUMBER: 78-84239
AMS 1968 SUBJECT CLASSIFICATIONS 1820, 1068

PRINTED IN THE UNITED STATES OF AMERICA

To My Parents

עֲטֶרֶת זְקֵנִים בְּנֵי בָנִים וְתִפְאֶרֶת בָּנִים אֲבוֹתָם ..... משלי יז

Preface

The prerequisites for the study of the "modern version" of class field theory, as formulated by Artin and Tate, are of two kinds. On the one hand, the student needs to know certain basic arithmetic results. This subject matter, which is commonly referred to as algebraic number theory, is now readily available in several books, including one by the author. On the other hand, the student must be well versed in certain topics from homological algebra, centering upon the cohomology of groups. It is the purpose of this book to provide a reasonably mature student of mathematics with direct access to this somewhat specialized material. Thus, in virtue of its focus and content, the most appropriate title for this book is the overly long ... Cohomology of Groups in Preparation for Class Field Theory.

We have taken a fairly classical approach to cohomology theory. Among the reasons underlying this choice of approach, the following may be listed: it is historically valid; it is, hopefully, relatively easy to understand; it probably constitutes a shorter path to the desired results than other, more abstract, approaches. The presentation of the material is intended to be detailed, expansive, and essentially self-contained. A conscious attempt has been made not to lose sight of the concrete since explicit computations are important for future class field theoretic applications.

The structure of the book is apparent from the table of contents and from the introductory remarks at the beginning of chapters and sections. There is little to be gained by a detailed analysis of the structure at this point. The problems and exercises are of varying degrees of difficulty or significance, but as a rule they are not required for later use in the text. A number of topics have been omitted—for example, spectral sequences, of the Brauer group (including its connections with central simple algebras). The fact that our attitude is not encyclopedic, but rather expository, is reflected in the bibliography, which is far from complete.

My sincere thanks are due to a number of individuals—especially to Sandra Spinacci for her skill and efficiency in typing the manuscript, to my family and to S. Shufro for relieving me of assorted burdens, and for assistance in both tangible and intangible form.

May, 1969 E. WEISS

Brookline, Massachusetts

Contents

Cohomology of Groups

CHAPTER

I

Cohomology Groups of G in A

In this chapter, and throughout this book, $G = \{\sigma, \tau, \rho, \ldots\}$ denotes a finite multiplicative group (unless there is an explicit statement made to the contrary).

The main objectives of this chapter include : introduction of the notion of a G-complex, detailed construction of the standard G-complex, use of an arbitrary G-complex to define the cohomology groups of G in a G-module A (the question of uniqueness of cohomology groups will be treated in Chapter II), and the interpretation (which depends on the standard G-complex) of the lower-dimensional cohomology groups.

Differential groups serve as a convenient starting point, as an introduction to some of the basic techniques, and as a tool for arriving at the infinite cohomology sequence (which is discussed in Chapter II); in addition, they are used to give a cohomological approach to the Herbrand quotient and its properties.

1-1. DIFFERENTIAL GROUPS

A **differential group** is a pair (A, d) where A is an abelian group (which we shall usually write additively) and d (the

differential operator) is an endomorphism of A such that $d^2 = d \circ d = 0$. Since image $d \subset$ kernel d, we may form the group $H(A) = \ker d/\operatorname{im} d$ and call it the **derived group** of (A, d). If (A_1, d_1) and (A_2, d_2) are differential groups, then a homomorphism $f : A_1 \longrightarrow A_2$ is said to be **admissible** when $d_2 \circ f = f \circ d_1$. In this section, all differential operators (and, in particular, d_1 and d_2 above) will be denoted by d; this should cause no confusion.

1-1-1. Proposition. An admissible map $f : (A, d) \longrightarrow (B, d)$ of differential groups induces a natural homomorphism

$$f_* : H(A) \longrightarrow H(B)$$

given by

$$f_*(a + dA) = f(a) + dB \qquad \text{where} \quad da = 0$$

If $f, g : (A, d) \longrightarrow (B, d)$ are both admissible, then $f \pm g$ is admissible and

$$(f \pm g)_* = f_* \pm g_*$$

If $f : (A, d) \longrightarrow (B, d)$ and $g : (B, d) \longrightarrow (C, d)$ are both admissible, then $g \circ f$ is admissible and

$$(g \circ f)_* = g_* \circ f_*$$

Proof: f_* is well-defined—for if a is replaced by another representative of the same class, say $a + da_1$ with $a_1 \in A$, then $f(a + da_1) = f(a) + f(da_1) = f(a) + df(a_1)$ and this belongs to $f(a) + dB$. The remaining assertions are clear. ▮

1-1-2. Corollary. We have

$$0_* = 0 \qquad \text{and} \qquad 1_* = 1$$

In more detail, if 0 is the trivial map of (A, d) into (B, d) then 0_* is the trivial map of $H(A)$ into $H(B)$, and if $1 : A \longrightarrow A$ is the identity map then $1_* : H(A) \longrightarrow H(A)$ is the identity.

1-1-3. Theorem. Suppose that

$$0 \longrightarrow A \xrightarrow{i} B \xrightarrow{j} C \longrightarrow 0$$

is an exact sequence (sometimes it will be referred to as a **short exact sequence**) of differential groups with admissible maps, then there exists a homomorphism $d_* : H(C) \to H(A)$ such that the following triangle is exact :

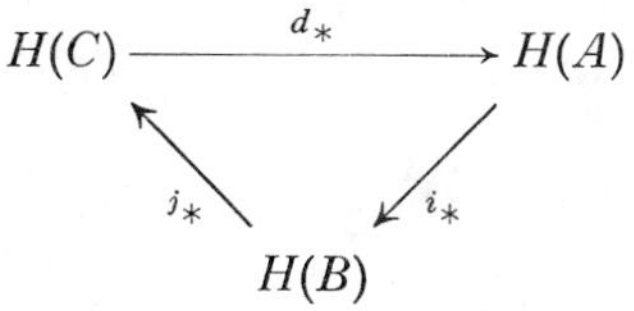

Moreover, if

$$\begin{array}{ccccccccc} 0 & \longrightarrow & A & \xrightarrow{i} & B & \xrightarrow{j} & C & \longrightarrow & 0 \\ & & \downarrow{\scriptstyle f} & & \downarrow{\scriptstyle g} & & \downarrow{\scriptstyle h} & & \\ 0 & \longrightarrow & A' & \xrightarrow{i'} & B' & \xrightarrow{j'} & C' & \longrightarrow & 0 \end{array}$$

is a commutative diagram of differential groups with exact rows and all maps admissible, then the following prism has exact triangles and commutative faces :

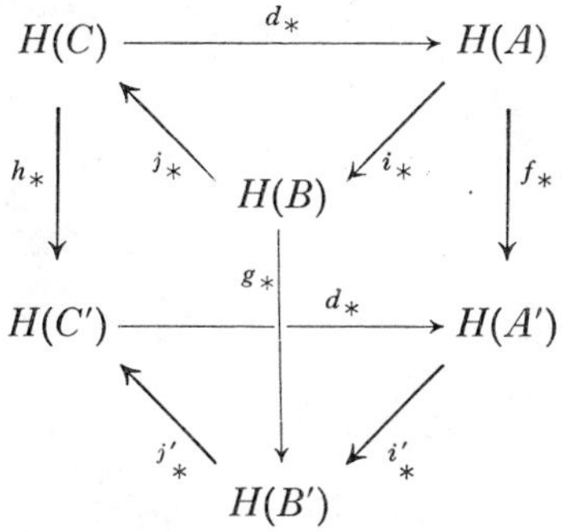

Proof: (The first formulation of this result probably appears in [50].)

definition of d_*: An element $\gamma \in H(C)$ is represented by an element $c \in C$ with $dc = 0$; that is, $\gamma = c + dC$. Since j is onto, there exists $b \in B$ with $jb = c$. Because $j(db) = d(jb) = dc = 0$, there exists $a \in A$ such that $ia = db$. Now, i is a monomorphism and $i(da) = d(ia) = d(db) = 0$, so that $da = 0$. Thus, a determines an element $\alpha = a + dA \in H(A)$, and we put $d_*(\gamma) = \alpha$. In other words,

$$d_*(c + dC) = a + dA \qquad \text{where } jb = c,\ ia = db$$

or symbolically : $d_*(c + dC) = (i^{-1}dj^{-1})(c) + dA$.

d_* is well-defined: First let us show that if $\gamma = 0 \in H(C)$ then $\alpha = d_*\gamma = 0 \in H(A)$—that is, if $c = dc_1$, $c \in C$ then $a = da_1$. Now, if $c = dc_1$ there exists $b_1 \in B$ such that $jb_1 = c_1$ so that $j(b - db_1) = jb - jdb_1 = c - djb_1 = 0$. Hence, there exists $a_1 \in A$ such that $ia_1 = b - db_1$. Therefore,

$$ia = db = d(ia_1 + db_1) = ida_1 \qquad \text{and} \qquad a = da_1.$$

Furthermore, it is immediate from its definition that d_* is additive, so it follows that d_* is indeed a well-defined homomorphism of $H(C) \longrightarrow H(A)$.

The next step is to prove the six kernel–image relations which imply that the triangle is exact.

im i_* ⊂ ker j_* : $j_* \circ i_* = (j \circ i)_* = 0_* = 0$.

im j_* ⊂ ker d_* : An element of $H(B)$ is of form $b + dB$ with $db = 0$, and $d_*[j_*(b + dB)] = d_*(jb + dC) = a + dA$ where $ia = db = 0$. Hence, $a = 0$ and $a + dA = 0 \in H(A)$.

im d_* ⊂ ker i_* :

$$i_*[d_*(c + dC)] = i_*(a + dA) = ia + dB = db + dB = dB = 0 \in H(B).$$

ker j_* ⊂ im i_* : Suppose that

$$db = 0 \qquad \text{and} \qquad j_*(b + dB) = jb + dC = 0 \in H(C).$$

There exists $c \in C$ such that $jb = dc$. Choose $b_1 \in B$ such that $jb_1 = c$. Since $j(b - db_1) = 0$ there exists $a \in A$ with $ia = b - db_1$.

Now, $ida = dia = d(b - db_1) = 0$, so that

$$da = 0 \quad \text{and} \quad i_*(a + dA) = b - db_1 + dB = b + dB.$$

ker $d_* \subset$ im j_* : Suppose that

$$dc = 0 \quad \text{and} \quad d_*(c + dC) = a + dA = 0 \in H(A),$$

so that $a = da_1$, $a_1 \in A$. Put $b_1 = b - ia_1$ (where $c = jb$, $ia = db$). Then $jb_1 = jb = c$ and $db_1 = db - ida_1 = 0$, and we have $j_*(b_1 + dB) = jb_1 + dC = c + dC$.

ker $i_* \subset$ im d_* : Suppose that

$$da = 0 \quad \text{and} \quad i_*(a + dA) = 0 \in H(B).$$

This means that $ia = db$ for some $b \in B$. Putting $c = jb$, we have $dc = 0$ and $d_*(c + dC) = a + dA$.

Finally, the relations $g_* \circ i_* = i'_* \circ f_*$ and $h_* \circ j_* = j'_* \circ g_*$ follow from (1-1-1), while $g_* \circ d_* = d_* \circ h_*$ may be checked as follows: $f_* d_*(c + dC) = f_*(a + dA) = f(a) + dA'$ where $dc = 0$, $jb = c$, $ia = db$, and $d_* h_*(c + dC) = d_*(hc + dC') = f(a) + dA'$ since $hc = hjb = j'gb$, $dgb = gdb = gia = i'(fa)$. This completes the proof. ∎

An abelian group A is said to be **graded** when it is of the form $A = \sum_{-\infty}^{+\infty} \oplus A_n$ (weak direct sum). Suppose that the graded group A is also a differential group with differential operator d, then d is said to be **compatible** with the grading if for $r = +1$ or -1 we have $d : A_n \longrightarrow A_{n+r}$ for each n. In such a situation (A, d, r) is said to be a **differential graded group**. Note that any abelian group A can be made into a differential graded group in trivial fashion by putting $A_0 = A$, $A_n = (0)$ for $n \neq 0$, $d = 0$. (It is clear that the notion of a differential graded group could be defined for any integer r and the subsequent theory developed on this basis, but these matters are not of interest to us.)

Consider a differential graded group (A, d, r); then for each n, $d_n = d \mid A_n$ is a homomorphism of $A_n \longrightarrow A_{n+r}$ and $d_{n+r} \circ d_n = 0$. Thus, (A, d, r) may be written in a more suggestive fashion as a chain or sequence

$$\cdots \longrightarrow A_{n-r} \xrightarrow{d_{n-r}} A_n \xrightarrow{d_n} A_{n+r} \xrightarrow{d_{n+r}} A_{n+2r} \longrightarrow \cdots \qquad (*)$$

Conversely, suppose we are given a chain $(*)$ of abelian groups and homomorphisms d_n such that $d_{n+r} \circ d_n = 0$ (and $r = \pm 1$), then upon writing $A = \sum_{-\infty}^{+\infty} \oplus A_n$, $d = \sum_{-\infty}^{+\infty} \oplus d_n$ it is clear that (A, d, r) is a differential graded group. These two attitudes toward a differential graded group will be used interchangeably.

The case $r = +1$ is usually called **cohomology** (and d is then denoted by δ and called the **coboundary** operator), while the case $r = -1$ is called **homology** (and d is then denoted by ∂ and called the **boundary** operator). This distinction is highly artificial in that if $(A, d, +1)$ is a differential graded group of cohomology type we may put $A' = \sum_{-\infty}^{+\infty} \oplus A'_n$ where $A'_n = A_{-n}$ and $d' = \sum_{-\infty}^{+\infty} \oplus d'_n$ where $d'_n = d_{-n}$, and then $(A', d', -1)$ is a differential graded group of homology type having all properties possessed by $(A, d, +1)$—and conversely. Because of this, we shall usually treat one of the cases $r = \pm 1$ and leave it to the reader to supply the details for the other case, when necessary.

Consider the differential graded group $(A, d, +1)$. For each n, the group A_n (which we may view as a subgroup of A) is called the **group of *n*-cochains**. The operator d determines for each n the groups : $\mathscr{Z}_n = A_n \cap \ker d = A_n \cap \ker d_n = A \cap \ker d_n$ (the **group of *n*-cocycles**) and

$$\mathscr{B}_n = A_n \cap d_{n-1}A_{n-1} = A_n \cap dA = A \cap d_{n-1}A$$

(the **group of *n*-coboundaries**). The ***n*th derived group** $H_n(A) = \mathscr{Z}_n/\mathscr{B}_n$ is also known as the ***n*th cohomology group of A**. Since $\ker d = \sum_{-\infty}^{+\infty} \oplus \mathscr{Z}_n$ and $\operatorname{im} d = \sum_{-\infty}^{+\infty} \oplus \mathscr{B}_n$ it follows that

$$H(A) = \frac{\sum \oplus \mathscr{Z}_n}{\sum \oplus \mathscr{B}_n} \approx \sum \oplus \frac{\mathscr{Z}_n}{\mathscr{B}_n} = \sum_{-\infty}^{+\infty} \oplus H_n(A)$$

so that we may identify, and write $H(A) = \sum_{-\infty}^{+\infty} \oplus H_n(A)$.

A homomorphism $f : A \longrightarrow B$ is said to be an **admissible map** for the differential graded groups $(A, d, +1)$ and $(B, d, +1)$ when $f \circ d = d \circ f$ and f maps A_n into B_n for every n. It is then clear that for the induced map $f_* : H(A) \longrightarrow H(B)$ we have $f_* : H_n(A) \longrightarrow H_n(B)$ for every n. It may also be noted that if we put $f_n = f \mid A_n$, then $f_n : A_n \longrightarrow B_n$, $f_{n+1} \circ d = d \circ f_n$, and $f = \sum_{-\infty}^{+\infty} \oplus f_n$; conversely, if for each n we have a homomorphism

$f_n : A_n \longrightarrow B_n$ such that $f_{n+1} \circ d = d \circ f_n$, then $f = \sum_{-\infty}^{+\infty} \oplus f_n$ is an admissible map for the differential graded groups.

1-1-4. Proposition. Suppose that we are given the following commutative diagram of differential graded groups, all of cohomology type, with all maps admissible and the rows exact :

$$\begin{array}{ccccccccc}
0 & \longrightarrow & A & \xrightarrow{i} & B & \xrightarrow{j} & C & \longrightarrow & 0 \\
 & & \downarrow{\scriptstyle f} & & \downarrow{\scriptstyle g} & & \downarrow{\scriptstyle h} & & \\
0 & \longrightarrow & A' & \xrightarrow{i'} & B' & \xrightarrow{j'} & C' & \longrightarrow & 0
\end{array}$$

Then, the following diagram is commutative and has exact rows :

$$\begin{array}{ccccccccccccc}
\cdots & \longrightarrow & H_{n-1}(C) & \xrightarrow{d_*} & H_n(A) & \xrightarrow{i_*} & H_n(B) & \xrightarrow{j_*} & H_n(C) & \xrightarrow{d_*} & H_{n+1}(A) & \longrightarrow & \cdots \\
 & & \downarrow{\scriptstyle h_*} & & \downarrow{\scriptstyle f_*} & & \downarrow{\scriptstyle g_*} & & \downarrow{\scriptstyle h_*} & & \downarrow{\scriptstyle f_*} & & \\
\cdots & \longrightarrow & H_{n-1}(C') & \xrightarrow{d_*} & H_n(A') & \xrightarrow{i'_*} & H_n(B') & \xrightarrow{j'_*} & H_n(C') & \xrightarrow{d_*} & H_{n+1}(A') & \longrightarrow & \cdots
\end{array}$$

Proof: Immediate from (1-1-3) as soon as one checks that d_* raises dimension by 1. It should be noted that the earlier discussion of differential groups with the consequent result (1-1-3) on the derived groups of an exact triangle was, in large part, a matter of technical convenience—namely, to minimize the need for subscripts in the proof of (1-1-4). ▮

1-1-5. Exercise. If the following diagram of differential groups and admissible maps has exact rows and columns

$$\begin{array}{ccccccccc}
 & & 0 & & 0 & & 0 & & \\
 & & \downarrow & & \downarrow & & \downarrow & & \\
0 & \longrightarrow & A' & \xrightarrow{f'} & A & \xrightarrow{f} & A'' & \longrightarrow & 0 \\
 & & \downarrow{\scriptstyle i'} & & \downarrow{\scriptstyle i} & & \downarrow{\scriptstyle i''} & & \\
0 & \longrightarrow & B' & \xrightarrow{g'} & B & \xrightarrow{g} & B'' & \longrightarrow & 0 \\
 & & \downarrow{\scriptstyle j'} & & \downarrow{\scriptstyle j} & & \downarrow{\scriptstyle j''} & & \\
0 & \longrightarrow & C' & \xrightarrow{h'} & C & \xrightarrow{h} & C'' & \longrightarrow & 0 \\
 & & \downarrow & & \downarrow & & \downarrow & & \\
 & & 0 & & 0 & & 0 & &
\end{array}$$

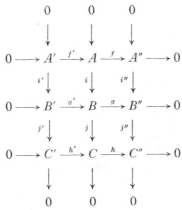

then the following diagram is anticommutative—that is, $d_* \circ d_* = -d_* \circ d_*$:

$$\begin{array}{ccc} H(C'') & \xrightarrow{d_*} & H(A'') \\ {\scriptstyle d_*}\downarrow & (-1) & \downarrow{\scriptstyle d_*} \\ H(C') & \xrightarrow{d_*} & H(A') \end{array}$$

1-2. HERBRAND'S LEMMA

This section is in the nature of a digression from the main theme of the chapter. We apply the results of Section 1-1 on derived groups to interpret cohomologically a classical index computation (see, for example, [16, p. 375]) which is useful in class field theory.

1-2-1. Herbrand's Lemma. Suppose that A is an additive abelian group and that σ and τ are endomorphisms of A such that $\sigma \circ \tau = \tau \circ \sigma = 0$. Suppose that B is a subgroup of A of finite index which is stable under both σ and τ (that is, $\sigma B \subset B$ and $\tau B \subset B$). Let A_σ denote the kernel of σ on A and $A^\sigma = \sigma A$ denote the image of A under σ. If $(B_\sigma : B^\tau)$ and $(B_\tau : B^\sigma)$ are both finite, then so are $(A_\sigma : A^\tau)$ and $(A_\tau : A^\sigma)$, and then

$$\frac{(A_\sigma : A^\tau)}{(A_\tau : A^\sigma)} = \frac{(B_\sigma : B^\tau)}{(B_\tau : B^\sigma)}$$

Proof: Since the natural map of $A \longrightarrow A^\tau/B^\tau$ is onto with kernel BA_τ, and since $BA_\tau/B \approx A_\tau/B \cap A_\tau = A_\tau/B_\tau$ it follows that

$$\begin{aligned}(A : B) &= (A : BA_\tau)(BA_\tau : B) \\ &= (A^\tau : B^\tau)(A_\tau : B_\tau) \\ &= (A^\tau : B^\tau)\frac{(A_\tau : B^\sigma)}{(B_\tau : B^\sigma)} \\ &= (A^\tau : B^\tau)\frac{(A_\tau : A^\sigma)(A^\sigma : B^\sigma)}{(B_\tau : B^\sigma)}\end{aligned}$$

where all the factors are finite. By interchanging the roles of σ and

τ we get another formula for $(A : B)$; comparing the two gives the desired result. ▮

It may be noted that the proof applies when A is not abelian; of course, B must then be a normal subgroup.

One may also give a cohomological treatment of Herbrand's lemma. Given A, σ, τ as before, we construct a differential graded group $(\bar{A} = \sum_{-\infty}^{+\infty} \oplus A_n, d = \sum_{-\infty}^{+\infty} \oplus d_n, +1)$ by putting $A_n = A$ for all n, $d_n = \sigma$ for even n, $d_n = \tau$ for odd n. Thus $(\bar{A}, d, 1)$ may be viewed as a sequence

$$\cdots \longrightarrow A_{-1} = A \xrightarrow{d_1=\tau} A_0 = A \xrightarrow{d_0=\sigma} A_1 = A \xrightarrow{d_1=\tau} \cdots$$

and because this sequence has period 2, it may be expressed in closed form as

$$A_0 \xrightarrow{d_0} A_1, \qquad d_1 = d_{-1}$$

The derived groups of $(\bar{A}, d, 1)$ are

$$H_{\text{even}}(\bar{A}) = H_0(\bar{A}) = \frac{\ker d_0}{\operatorname{im} d_1} = \frac{A_\sigma}{A^\tau}$$

$$H_{\text{odd}}(\bar{A}) = H_1(\bar{A}) = \frac{\ker d_1}{\operatorname{im} d_0} = \frac{A_\tau}{A^\sigma}$$

If both $H_0(\bar{A})$ and $H_1(\bar{A})$ are finite groups we put

$$Q(A) = \frac{\#(H_0(\bar{A}))}{\#(H_1(\bar{A}))} = \frac{(A_\sigma : A^\tau)}{(A_\tau : A^\sigma)}$$

and call it the **Herbrand quotient** of A with respect to σ and τ. Note that $Q(A)$ depends on the ordering of the endomorphisms σ and τ. If we wish to keep track of σ and τ we write $Q_{\sigma,\tau}$ for Q; of course, $Q_{\tau,\sigma}(A) = 1/Q_{\sigma,\tau}(A)$.

1-2-2. Proposition. If A is finite, then $Q(A) = 1$.

Proof: Since A is finite, all indices in the following diagram are finite (and, in particular, $Q(A)$ is defined):

$$\begin{array}{ccccc} & & A & & \\ & / & & \backslash & \\ A_\sigma & & & & A_\tau \\ | & & & & | \\ A^\tau & & & & A^\sigma \\ & \backslash & & / & \\ & & (0) & & \end{array}$$

Because $(A : A_\sigma) = (A^\sigma : (0))$ and $(A : A_\tau) = (A^\tau : (0))$ it follows that $(A_\sigma : A^\tau) = (A_\tau : A^\sigma)$—so $Q(A) = 1$. ∎

Suppose that $0 \longrightarrow A \xrightarrow{i} B \xrightarrow{j} C \longrightarrow 0$ is an exact sequence of groups and that σ is an endomorphism of B such that (when i is viewed as inclusion) the subgroup A is stable under σ; then σ induces an endomorphism of C, which we denote also by σ. Since $\sigma \circ i = i \circ \sigma$ and $\sigma \circ j = j \circ \sigma$ we say that σ is an **admissible endomorphism** for the short exact sequence. Clearly, the same terminology applies when there are three endomorphisms $\sigma_A, \sigma_B, \sigma_C$ of $A, B,$ and C, respectively such that $\sigma_B \circ i = i \circ \sigma_A$ and $\sigma_C \circ j = j \circ \sigma_B$.

1-2-3. Proposition. Suppose that σ and τ are admissible endomorphisms of the short exact sequence

$$0 \longrightarrow A \xrightarrow{i} B \xrightarrow{j} C \longrightarrow 0$$

of abelian groups and that $\sigma \circ \tau = \tau \circ \sigma = 0$ (on B and, hence, on A and C too). If any two of $Q(A), Q(B), Q(C)$ are defined, then so is the third, and then

$$Q(B) = Q(A)\,Q(C)$$

Proof: From A, B, C, σ, τ we may construct the differential graded groups $(\bar{A}, d, 1)$, $(\bar{B}, d, 1)$, $(\bar{C}, d, 1)$. The maps i and j determine, in the obvious fashion, mappings (which we also denote by i and j) such that

$$0 \longrightarrow \bar{A} \xrightarrow{i} \bar{B} \xrightarrow{j} \bar{C} \longrightarrow 0$$

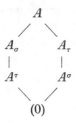

is an exact sequence of differential graded groups with admissible maps. By (1-1-4) and periodicity, we have the closed exact sequence of derived groups

$$\to H_0(\bar{A}) \xrightarrow{i_*} H_0(\bar{B}) \xrightarrow{j_*} H_0(\bar{C}) \xrightarrow{d_*} H_1(\bar{A}) \xrightarrow{i_*} H_1(\bar{B}) \xrightarrow{j_*} H_1(\bar{C}) \xrightarrow{d_*} H_0(\bar{A})$$

If two of $Q(A)$, $Q(B)$, $Q(C)$ are defined, then four of these derived groups are finite; it then follows easily that the two remaining derived groups are finite, so that the third Herbrand quotient is defined.

Now, when all the derived groups are finite, let us put

$$a_0 = \#[d_*(H_1(\bar{C}))] \qquad b_0 = \#[i_*(H_0(\bar{A}))] \qquad c_0 = \#[j_*(H_0(\bar{B}))]$$

$$a_1 = \#[d_*(H_0(\bar{C}))] \qquad b_1 = \#[i_*(H_1(\bar{A}))] \qquad c_1 = \#[j_*(H_1(\bar{B}))]$$

Therefore, exactness yields :

$$\#(H_0(\bar{A})) = a_0 b_0\,, \qquad \#(H_0(\bar{B})) = b_0 c_0\,, \qquad \#(H_0(\bar{C})) = c_0 a_1\,,$$

$$\#(H_1(\bar{A})) = a_1 b_1\,, \qquad \#(H_1(\bar{B})) = b_1 c_1\,, \qquad \#(H_1(\bar{C})) = c_1 a_0\,,$$

so that

$$Q(A)\,Q(C) = \left(\frac{a_0 b_0}{a_1 b_1}\right)\left(\frac{c_0 a_1}{c_1 a_0}\right) = \frac{b_0 c_0}{b_1 c_1} = Q(B)$$

This completes the proof. ▮

In particular, if A is a subgroup of B with $(B : A) < \infty$, then, according to (1-2-2) and (1-2-3), $Q(A)$ is defined if and only if $Q(B)$ is defined, and then $Q(A) = Q(B)$—which is simply another way to express the assertion of (1-2-1).

1-2-4. Exercise. (i) A differential graded group

$$\left(A = \sum A_n\,,\, d = \sum d_n\,,\, +1\right)$$

is said to be **periodic** of period $m > 0$ when $A_n = A_{n+m}$ and $d_n = d_{n+m}$ for all integers n. Let

$$h_n = h_n(A) = \#(H_n(A)) = (\ker d_n : \operatorname{im} d_{n-1})$$

denote the order of the nth derived group. If all A_n are finite and we write $a_n = \#(A_n)$, and if m is even, then

$$\prod_0^{m-1} (h_n)^{(-1)^n} = \prod_0^{m-1} (a_n)^{(-1)^n}$$

(ii) The Herbrand quotient may be generalized as follows. Let $(A, d, +1)$ be a differential graded group of even period m such that all $h_n(A) < \infty$. We call

$$Q(A) = \prod_0^{m-1} \{h_n(A)\}^{(-1)^n}$$

the **Herbrand characteristic** of A. Suppose that

$$0 \longrightarrow A \xrightarrow{i} B \xrightarrow{j} C \longrightarrow 0$$

is an exact sequence of differential graded groups of even period m and that the maps i and j are admissible; if two of $Q(A), Q(B), Q(C)$ are defined, then so is the third and $Q(B) = Q(A)\,Q(C)$.

1-3. G-COMPLEXES (POSITIVE PART)

Let $G = \{1, \sigma, \tau, \ldots\}$ be a finite multiplicative group, and let $\Gamma = \mathbf{Z}[G]$ denote the integer group ring of G. Thus, the elements of the ring Γ are formal sums $\sum_{\sigma \in G} n_\sigma \sigma$, $n_\sigma \in \mathbf{Z}$; addition in Γ is componentwise, while multiplication is determined in the natural way by the multiplication in G—in more detail,

$$\sum_{\sigma \in G} m_\sigma \sigma + \sum_{\sigma \in G} n_\sigma \sigma = \sum_{\sigma \in G} (m_\sigma + n_\sigma)\sigma,$$

and

$$\left(\sum_{\sigma\in G} m_\sigma\sigma\right)\left(\sum_{\sigma\in G} n_\sigma\sigma\right) = \sum_{\sigma\in G}\left(\sum_{\tau\rho=\sigma} m_\tau n_\rho\right)\sigma = \sum_{\sigma\in G}\left(\sum_{\tau\in G} m_\tau n_{\tau^{-1}\sigma}\right)\sigma.$$

We may view G as imbedded in Γ under the identification of $\sigma \in G$ with $1 \cdot \sigma \in \Gamma$. In particular, 1 denotes the identity of both G and Γ. It will be customary for us to rename objects associated with Γ in terms of G. For example, we say that an additive abelian group $A = \{a, b, c,...\}$ is a **G-module** when it is a left unitary Γ-module (**unitary** means that $1 \in \Gamma$ is the identity operator on A). If A is a G-module, then clearly

$$\begin{aligned} \sigma(a+b) &= \sigma a + \sigma b \\ \sigma(\tau a) &= (\sigma\tau)\, a \qquad\qquad (*) \\ 1(a) &= a \end{aligned}$$

Conversely, if G acts on the additive group A according to the rules $(*)$, then A becomes a G-module when the operation of Γ on A is defined by $(\sum_{\sigma\in G} n_\sigma\sigma)(a) = \sum_{\sigma\in G} [n_\sigma(\sigma a)]$.

If the group A is multiplicative, then the action of G is often denoted by a^σ; note that then $(a^\sigma)^\tau = a^{\tau\sigma}$ and $a^{\Sigma_\sigma n_\sigma \sigma} = \prod_\sigma (a^{n_\sigma\sigma})$.

For every G we define the action of G on $\mathbf{Z}$ to be trivial—that is, $\sigma n = n$ for all $\sigma \in G$ and $n \in \mathbf{Z}$—so that $\mathbf{Z}$ becomes a G-module. In the same way, $\mathbf{Q}$ and $\mathbf{Q}/\mathbf{Z}$ (more precisely, their additive groups) will always be viewed as G-modules with trivial action. Of course, the multiplication in Γ always permits us to view Γ as a G-module.

Suppose that $(X = \sum \oplus X_n, \partial = \sum \oplus \partial_n, -1)$ is a differential graded group of homology type, then $(X, \partial, -1)$ is said to a **chain complex over Γ** (or over G) when each X_n is a Γ-module and ∂ (and hence each ∂_n) is a Γ-homomorphism. The chain complex $(X, \partial, -1)$ is said to be : **free** if each X_n is a Γ-free module, **finite free** if each X_n is Γ-free with a finite basis over Γ, **acyclic** if its derived groups are (0) (that is, if im ∂_n = ker ∂_{n-1} for all n), and **augmented** if there exist Γ-homomorphisms $\varepsilon : X_0 \twoheadrightarrow \mathbf{Z}$ (meaning that ε is onto) and $\mu : \mathbf{Z} \rightarrowtail X_{-1}$ (meaning that μ is 1–1) such that $\partial_0 = \mu \circ \varepsilon$. Any finite free, acyclic, and augmented chain complex over the finite group G will be referred to as a **G-complex** and denoted by $(X, \partial, \varepsilon, \mu)$ or simply by X. The standard diagram for a G-complex is the following :

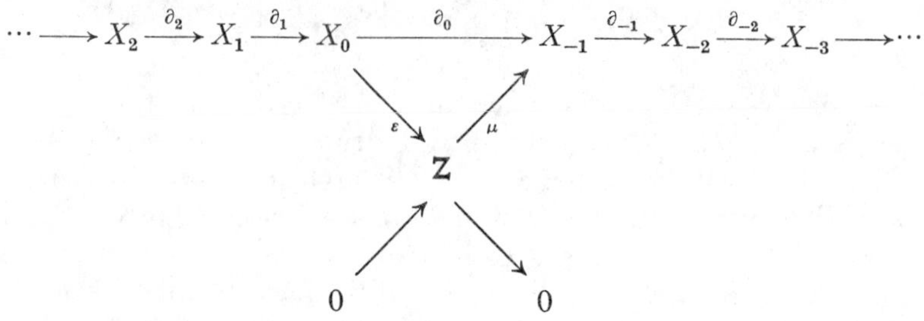

Since ε is onto and μ is 1–1, it follows from $\partial_0 = \mu \circ \varepsilon$ that im ∂_0 = im μ and ker ∂_0 = ker ε, and therefore, a G-complex can be broken up into two exact G-sequences (by a **G-sequence** we mean a sequence of G-modules and G-homomorphisms), namely :

$$\cdots \longrightarrow X_2 \xrightarrow{\partial_2} X_1 \xrightarrow{\partial_1} X_0 \xrightarrow{\varepsilon} \mathbf{Z} \longrightarrow 0$$

which is called the **positive part** of the G-complex, and

$$0 \longrightarrow \mathbf{Z} \xrightarrow{\mu} X_{-1} \xrightarrow{\partial_{-1}} X_{-2} \xrightarrow{\partial_{-2}} X_{-3} \longrightarrow \cdots$$

which is called the **negative part**. Note that, conversely, if we are given a positive part and a negative part, then they can be combined (by putting $\partial_0 = \mu \circ \varepsilon$) to form a G-complex.

1-3-1. Theorem. Given a finite group G, there exists a positive part for a G-complex.

Proof: For $n = 0$, define a symbol $[\cdot]$, call it the **empty cell**, and put

$$X_0 = \Gamma[\cdot]$$

so that X_0 is a finite free Γ-module with a basis consisting of a single element. For $n > 0$, define symbols $[\sigma_1, \ldots, \sigma_n]$ consisting of ordered n-tuples of elements $\sigma_i \in G$, call them **n-cells**, and put

$$X_n = \sum_{\sigma_1, \ldots, \sigma_n \in G} \oplus\, \Gamma[\sigma_1, \ldots, \sigma_n]$$

so that X_n is a finite free Γ-module with the set of all n-cells serving as a basis.

Since ε and the boundary operators ∂_n are to be G-homomorphisms, it suffices to define them on a G-basis—that is, on the cells—and extend linearly. Now define

$$\begin{aligned}
\varepsilon([\cdot]) &= 1 \\
\partial_1([\sigma]) &= \sigma[\cdot] - [\cdot] \\
\partial_2([\sigma_1, \sigma_2]) &= \sigma_1[\sigma_2] - [\sigma_1\sigma_2] + [\sigma_1] \\
&\vdots \\
\partial_n([\sigma_1, \sigma_2, \ldots, \sigma_n]) &= \sigma_1[\sigma_2, \ldots, \sigma_n] \\
&\quad + \sum_{i=1}^{n-1} (-1)^i [\sigma_1, \ldots, \sigma_{i-1}, \sigma_i\sigma_{i+1}, \sigma_{i+2}, \ldots, \sigma_n] \\
&\quad + (-1)^n [\sigma_1, \ldots, \sigma_{n-1}] \qquad n \geqslant 1
\end{aligned}$$

Note that the formulas for ∂_2 and ∂_1 are indeed special cases of the formula for ∂_n. The definition of the boundary operators ∂_n may seem somewhat artificial, but we shall see in (1-5-13) that it arises in a rather natural way.

It remains to show that the sequence

$$\cdots \longrightarrow X_2 \xrightarrow{\partial_2} X_1 \xrightarrow{\partial_1} X_0 \xrightarrow{\varepsilon} \mathbf{Z} \longrightarrow 0 \qquad (*)$$

is exact, and one way to accomplish this is via the following preparations.

Suppose that $f, g : (A, \partial, -1) \longrightarrow (B, \partial, -1)$ are admissible maps of differential graded groups, then f and g are said to be **homotopic** if there exists a homomorphism $D : A \longrightarrow B$ such that $D : A_n \longrightarrow B_{n+1}$ and $f - g = D\partial + \partial D$. In other words, if we put $D_n = D \mid A_n$, we have the diagram

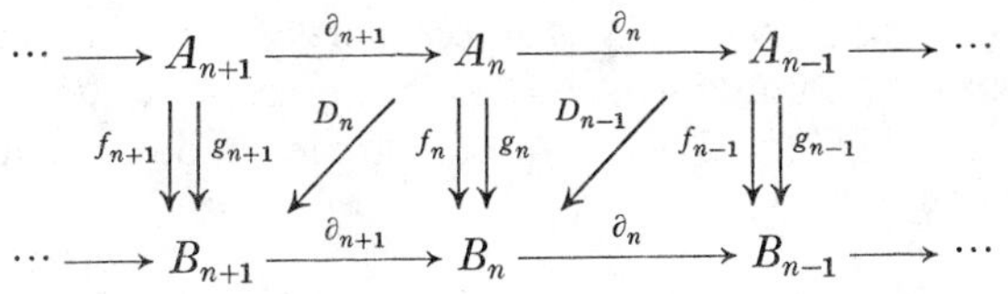

and for each n

$$f_n - g_n = D_{n-1} \circ \partial_n + \partial_{n+1} \circ D_n$$

1-3-2. Proposition. If f and g are homotopic maps of differential graded groups, then $f_* = g_*$.

Proof: Let $a \in A$ be a cycle. Since $\partial a = 0$ we have

$$f(a) - g(a) = (f - g)\, a = \partial Da + D\, \partial a = \partial Da,$$

a boundary. Thus $f(a)$ and $g(a)$ represent the same homology class. ∎

1-3-3. Corollary. Consider the admissible maps $f = 1$ and $g = 0$ of the differential graded group $(A, \partial, -1)$ into itself. If there exists a homotopy D between f and g (such a D is called a **contracting homotopy**), then $H(A) = (0)$.

Proof: From above, $0_* = 1_*$; but 0_* maps $H(A)$ to (0), while 1_* is the identity map of $H(A)$. ∎

This result is applied to the proof of (1-3-1) by showing that the sequence $(*)$ (with 0's adjoined on the right to make it infinite in both directions) is a differential graded group for which there exists a contracting homotopy. For this, we need to define maps $E : \mathbf{Z} \longrightarrow X_0$ and $D_n : X_n \longrightarrow X_{n+1}$, $n \geqslant 0$, such that

$$\begin{array}{llll} (1) & \varepsilon E = 1 & \text{on } \mathbf{Z} & \\ (2) & \partial_1 D_0 + E\varepsilon = 1 & \text{on } X_0 & \\ (3) & \partial_{n+1} D_n + D_{n-1}\partial_n = 1 & \text{on } X_n \quad n \geqslant 1 & \end{array} \qquad (\#)$$

Now, a contracting homotopy is an ordinary group homomorphism (rather than a G-homomorphism) so it suffices to define E and D_n and to verify (1), (2), (3) on $\mathbf{Z}$-bases of the modules in question. Thus we may put

$$E(1) = [\cdot]$$

$$D_0(\sigma[\cdot]) = [\sigma]$$

$$D_n(\sigma[\sigma_1, \ldots, \sigma_n]) = [\sigma, \sigma_1, \ldots, \sigma_n] \qquad n \geqslant 1$$

and extend linearly from these $\mathbf{Z}$-bases. Then $\varepsilon E(1) = \varepsilon([\cdot]) = 1$, which proves (1), and

$$\begin{aligned}(\partial_1 D_0 + E\varepsilon)(\sigma[\cdot]) &= \partial_1([\sigma]) + E\sigma\varepsilon([\cdot]) \\ &= \sigma[\cdot] - [\cdot] + E\sigma(1) \\ &= \sigma[\cdot] - [\cdot] + [\cdot] \\ &= \sigma[\cdot]\end{aligned}$$

which proves (2). As for (3), we check that

$$\begin{aligned}D_{n-1}\partial_n(\sigma[\sigma_1, \dots, \sigma_n]) &= [\sigma\sigma_1, \sigma_2, \dots, \sigma_n] \\ &+ \sum_{i=1}^{n-1} (-1)^i [\sigma, \sigma_1, \dots, \sigma_i\sigma_{i+1}, \dots, \sigma_n] \\ &+ (-1)^n [\sigma, \sigma_1, \dots, \sigma_{n-1}]\end{aligned}$$

and that

$$\begin{aligned}\partial_{n+1}D_n(\sigma[\sigma_1, \dots, \sigma_n]) &= \sigma[\sigma_1, \dots, \sigma_n] - [\sigma\sigma_1, \sigma_2, \dots, \sigma_n] \\ &+ \sum_{i=1}^{n-1} (-1)^{i-1}[\sigma, \sigma_1, \dots, \sigma_i\sigma_{i+1}, \dots, \sigma_n] \\ &+ (-1)^{n+1} [\sigma, \sigma_1, \dots, \sigma_{n-1}]\end{aligned}$$

Therefore, D is a contracting homotopy, provided we are dealing with a differential graded group; hence, in order to complete the proof, it remains to show that the square of the boundary is 0. For this we observe first that from (1) and (2) it follows that $\varepsilon E\varepsilon = \varepsilon = \varepsilon\partial_1 D_0 + \varepsilon E\varepsilon$; therefore, $\varepsilon\partial_1 D_0 = 0$, and since the image of D_0 contains the set of 1-cells, we have $\varepsilon\partial_1 = 0$. From the case $n = 1$ of (3) and from (2) we have

$$\partial_1 = \partial_1\partial_2 D_1 + \partial_1 D_0 \partial_1 = \partial_1\partial_2 D_1 + \partial_1 - E\varepsilon\partial_1 ;$$

therefore, $\partial_1\partial_2 D_1 = 0$, and since all 2-cells are in the image of D_1, $\partial_1\partial_2 = 0$. Finally, suppose inductively that $\partial_{r-1}\partial_r = 0$, $r \geqslant 2$; then multiplying the case $r - 1$ of (3) on the right by ∂_r and the case r of (3) on the left by ∂_r we have

$$\partial_r D_{r-1}\partial_r + D_{r-2}\partial_{r-1}\partial_r = \partial_r = \partial_r\partial_{r+1}D_r + \partial_r D_{r-1}\partial_r$$

Consequently, $\partial_r \partial_{r+1} D_r = 0$, and then $\partial_r \partial_{r+1} = 0$. This completes the proof. ∎

1-3-4. Exercise. By a straightforward computation, show that in the positive part of a G-complex as constructed above, $\varepsilon \circ \partial_1 = 0$ and $\partial_{n-1} \circ \partial_n = 0$ for $n \geqslant 2$.

1-4. G-COMPLEXES (NEGATIVE PART)

1-4-1. Theorem. Any positive part for a G-complex can be completed to a full G-complex. More precisely, if there is given an exact G-sequence

$$\cdots \longrightarrow X_2 \xrightarrow{\partial_2} X_1 \xrightarrow{\partial_1} X_0 \xrightarrow{\varepsilon} \mathbf{Z} \longrightarrow 0$$

in which each X_n is G-free with finite basis, then there exists an exact G-sequence

$$0 \longrightarrow \mathbf{Z} \xrightarrow{\mu} X_{-1} \xrightarrow{\partial_{-1}} X_{-2} \xrightarrow{\partial_{-2}} X_{-3} \longrightarrow \cdots$$

such that each X_n is G-free with finite basis.

Proof: We require some preliminaries :

1-4-2. Proposition. Suppose that A, B, and C are G-modules. Let Hom (A, B) denote the set of all homomorphisms of A into B (as additive groups), and let $\mathrm{Hom}_G\,(A, B)$ denote the set of all G-homomorphisms (that is, Γ-homomorphisms of Γ-modules) of A into B. Then

(1) Hom (A, B) is an additive group which becomes a G-module when the action of G is defined by

$$f^\sigma = \sigma \circ f \circ \sigma^{-1} \qquad f \in \mathrm{Hom}\,(A, B), \quad \sigma \in G$$

(2) $\mathrm{Hom}_G\,(A, B)$ is a subgroup of Hom (A, B); in fact,

$$\mathrm{Hom}_G\,(A, B) = \{f \in \mathrm{Hom}\,(A, B) \mid f^\sigma = f \quad \forall \sigma \in G\}$$

(3) If $f \in \text{Hom}(A, B)$ and $g \in \text{Hom}(B, C)$, then

$$g \circ f \in \text{Hom}(A, C) \qquad \text{and} \qquad (g \circ f)^\sigma = g^\sigma \circ f^\sigma.$$

Proof: Straightforward. Note that to prove that $\text{Hom}(A, B)$ is a G-module one must check that $(f_1 + f_2)^\sigma = f_1^\sigma + f_2^\sigma$, $(f^\sigma)^\tau = f^{\tau\sigma}$, and $f^1 = f$. The action of Γ on $\text{Hom}(A, B)$ is given in our exponential notation by

$$f^{(\Sigma_\sigma n_\sigma \sigma)} = \sum_\sigma n_\sigma f^\sigma$$ ∎

1-4-3. Proposition. Let A, A_1, A_2, B, B_1, B_2 be G-modules, then :

(1) If $\varphi \in \text{Hom}(A_1, A)$ and $\psi \in \text{Hom}(B, B_1)$, then we may define a homomorphism (of additive groups)

$$(\varphi, \psi) : \text{Hom}(A, B) \longrightarrow \text{Hom}(A_1, B_1)$$

by putting for $f \in \text{Hom}(A, B)$

$$(\varphi, \psi) f = \psi \circ f \circ \varphi$$

(2) If in addition, $\varphi_1 \in \text{Hom}(A_2, A_1)$ and $\psi_1 \in \text{Hom}(B_1, B_2)$, then

$$(\varphi_1, \psi_1) \circ (\varphi, \psi) = (\varphi \circ \varphi_1, \psi_1 \circ \psi)$$

(3) (φ, ψ) is additive in each variable; $(\varphi, 0)$ and $(0, \psi)$ are 0-maps; $(1, 1)$ is the identity map.

(4) If φ and ψ are both G-homomorphisms, then (φ, ψ) is a G-homomorphism; symbolically,

$$(\varphi, \psi) \in \text{Hom}_G(\text{Hom}(A, B), \text{Hom}(A_1, B_1))$$

(5) If φ and ψ are both G-homomorphisms, then (φ, ψ) maps $\text{Hom}_G(A, B) \longrightarrow \text{Hom}_G(A_1, B_1)$; symbolically,

$$(\varphi, \psi) \in \text{Hom}(\text{Hom}_G(A, B), \text{Hom}_G(A_1, B_1))$$

Proof: Straightforward verification. ∎

1-4-4. Proposition. Let A, B, C, and $\mathbf{Z}$ be H-modules, where, as usual, the action of G on $\mathbf{Z}$ is trivial, and define $\hat{A} = \operatorname{Hom}(A, \mathbf{Z})$; then

(1) $\hat{A}$ is a G-module.

(2) If A is G-free with finite basis (over Γ), then so is $\hat{A}$.

(3) If $f \in \operatorname{Hom}(A, B)$ and we write $\hat{f} = (f, 1)$, then $\hat{f} \in \operatorname{Hom}(\hat{B}, \hat{A})$. If, moreover, $f \in \operatorname{Hom}_G(A, B)$, then $\hat{f} \in \operatorname{Hom}_G(\hat{B}, \hat{A})$.

(4) If $f, f_1, f_2 \in \operatorname{Hom}(A, B)$ and $g \in \operatorname{Hom}(B, C)$, then

$$\widehat{f_1 + f_2} = \hat{f}_1 + \hat{f}_2, \qquad \widehat{g \circ f} = \hat{f} \circ \hat{g}, \qquad \hat{1} = 1, \qquad \hat{0} = 0.$$

(5) If $f \in \operatorname{Hom}(A, B)$ is an epimorphism, then $\hat{f}$ is a monomorphism.

Proof: (1) follows from (1-4-2), and (3) follows from (1-4-3). As for (4), it too follows from the properties of the symbol (φ, ψ)—for example,

$$\widehat{g \circ f} = (gf, 1) = (f, 1) \circ (g, 1) = \hat{f} \circ \hat{g}.$$

To prove (5), suppose that $\hat{f}(t) = 0 \in \hat{A}$ for some $t \in \hat{B}$. Thus $(f, 1)t = t \circ f = 0$, and since $f(A) = B$ this implies $t(B) = 0$—hence $t = 0$.

It remains to prove (2). By hypothesis, A can be written in the form $A = \sum_{i=1}^{n} \oplus \Gamma a_i$; therefore, $A = \sum_{i,\sigma} \oplus \mathbf{Z}(\sigma a_i)$, so that A is $\mathbf{Z}$-free with basis $\{\sigma a_i \mid i = 1,..., n, \sigma \in G\}$. For $i = 1,..., n$ define $f_i \in \hat{A}$ by putting

$$f_i(\sigma a_j) = \begin{cases} 1 & \sigma = 1, \quad i = j \\ 0 & \text{otherwise} \end{cases}$$

and extend linearly from the $\mathbf{Z}$-basis to all of A. Now,

$$\{f_i^\tau \mid \tau \in G, i = 1,..., n\}$$

is a $\mathbf{Z}$-basis for $\hat{A}$ because

$$f_i^\tau(\sigma a_j) = \tau f_i \tau^{-1} \sigma a_j = \begin{cases} 1 & \sigma = \tau, \quad i = j \\ 0 & \text{otherwise} \end{cases}$$

Therefore, $\{f_i \mid i = 1,\dots, n\}$ is a Γ-basis for $\hat{A}$, since

$$\hat{A} = \sum_{i,\sigma} \oplus \mathbf{Z} f_i^\sigma = \sum_{i,\sigma} \oplus f_i^{\mathbf{Z}\sigma} = \sum_i \oplus f_i^\Gamma \qquad \blacksquare$$

1-4-5. Proposition. For any G-module A we have the following isomorphisms, which shall often be taken as identifications :

(1) $\mathrm{Hom}_G(\Gamma, A) \approx A$ (as additive groups)

(2) $\mathrm{Hom}(\mathbf{Z}, A) \underset{G}{\approx} A$ (as G-modules)

(3) $\hat{\mathbf{Z}} \underset{G}{\approx} \mathbf{Z}$ (as G-modules)

Proof: Any $f \in \mathrm{Hom}_G(\Gamma, A)$ is determined by its action on $1 \in \Gamma$, and the map $f \longrightarrow f(1)$ is an isomorphism of $\mathrm{Hom}_G(\Gamma, A)$ onto A—so (1) is proved. Similarly, the map $f \longrightarrow f(1)$ provides an isomorphism of $\mathrm{Hom}(\mathbf{Z}, A)$ onto A as additive groups. It is a G-isomorphism, since $f^\sigma \longrightarrow f^\sigma(1) = \sigma f \sigma^{-1}(1) = \sigma(f(1))$— which proves (2). Since the action of G on $\hat{\mathbf{Z}} = \mathrm{Hom}(\mathbf{Z}, \mathbf{Z})$ is trivial, (3) follows immediately. ∎

Now, let us return to the proof of (1-4-1). From the exact G-sequence

$$\cdots \longrightarrow X_2 \xrightarrow{\partial_2} X_1 \xrightarrow{\partial_1} X_0 \xrightarrow{\varepsilon} \mathbf{Z} \longrightarrow 0 \qquad (*)$$

we get by dualization the G-sequence

$$0 \longrightarrow \hat{\mathbf{Z}} \xrightarrow{\hat{\varepsilon}} \hat{X}_0 \xrightarrow{\hat{\partial}_0} \hat{X}_1 \xrightarrow{\hat{\partial}_1} \hat{X}_2 \longrightarrow \cdots \qquad (**)$$

By changing notation and using part (3) of (1-4-5) we may write this as

$$0 \longrightarrow \mathbf{Z} \xrightarrow{\mu} X_{-1} \xrightarrow{\partial_{-1}} X_{-2} \xrightarrow{\partial_{-2}} X_{-3} \longrightarrow \cdots \qquad (***)$$

According to (1-4-4), each X_{-n} ($n \geqslant 1$) is G-free with finite basis, and μ is a monomorphism since ε in onto. Again from (1-4-4), we have $\partial_{-1} \circ \mu = 0$, $\partial_{-(n+1)} \circ \partial_{-n} = 0$. Thus (***) (with an infinite string of 0's added on the left) is a differential graded group. To complete the proof of (1-4-1) it suffices, therefore, to show that (∗∗∗) has a contracting homotopy. For this we need :

1-4-6. Proposition. Suppose that $(A, \partial, -1)$ is a differential graded group with $H(A) = \sum \oplus H_n(A) = (0)$. If each A_n is a free abelian group (that is, a free module over $\mathbf{Z}$), then there exists a contracting homotopy.

Proof: For each n, $\partial A_n = \partial_n A_n$ is a submodule of the free $\mathbf{Z}$-module A_{n-1}, so it too is $\mathbf{Z}$-free (see, for example, [42, p. 44]). Hence, there exists a $\mathbf{Z}$-homomorphism $T_{n-1} : \partial A_n \longrightarrow A_n$ which makes the following diagram commute :

$$\begin{array}{ccccc} & & \partial A_n & & \\ & \swarrow^{T_{n-1}} & \downarrow 1 & & \\ A_n & \xrightarrow{\partial_n} & \partial A_n & \longrightarrow & 0 \end{array}$$

It follows that on A_n, $\partial_n \circ (1 - T_{n-1} \circ \partial_n) = \partial_n - \partial_n = 0$, so $\operatorname{im}(1 - T_{n-1} \circ \partial_n) \subset \ker \partial_n = \operatorname{im} \partial_{n+1}$. Now define $D_n : A_n \longrightarrow A_{n+1}$ by putting

$$D_n = T_n \circ (1 - T_{n-1} \circ \partial_n)$$

and check that on A_n we have $D_{n-1} \circ \partial_n + \partial_{n+1} \circ D_n = 1$. Therefore, $D = \sum \oplus D_n$ is a contracting homotopy. ∎

This result applies to the differential graded group determined by (∗). More precisely (using the notation of the proof of (1-3-1)), there exist $\mathbf{Z}$-homomorphisms $E : \mathbf{Z} \longrightarrow X_0$ and $D_n : X_n \longrightarrow X_{n+1}$ which provide a contracting homotopy for (∗) (that is, the relations (#) of the proof of (1-3-1) hold for any positive part of a G-complex). By dualization, we have $\mathbf{Z}$-homomorphisms

$$\hat{E} : \hat{X}_0 \longrightarrow \hat{\mathbf{Z}} \approx \mathbf{Z} \qquad \text{and} \qquad \hat{D}_n : \hat{X}_{n+1} \longrightarrow \hat{X}_n$$

such that

$$\begin{array}{lll} (1) & 1 = \hat{1} = \widehat{\varepsilon E} = \hat{E} \circ \hat{\varepsilon} & \text{on } \hat{\mathbf{Z}} \approx \mathbf{Z} \\ (2) & 1 = \hat{1} = \hat{D}_0 \circ \hat{\partial}_1 + \hat{\varepsilon} \circ \hat{E} & \text{on } \hat{X}_0 \\ (3) & 1 = \hat{1} = \hat{D}_n \circ \hat{\partial}_{n+1} + \hat{\partial}_n \circ \hat{D}_{n-1} & \text{on } \hat{X}_n, \quad n \geqslant 1 \end{array}$$

In other words, the differential graded group determined by $(**)$ has a contracting homotopy. Therefore, the sequence $(***)$ is exact, and the proof of (1-4-1) is complete. ∎

A diagram which illustrates the proof of (1-4-1) and the resulting G-complex is as follows :

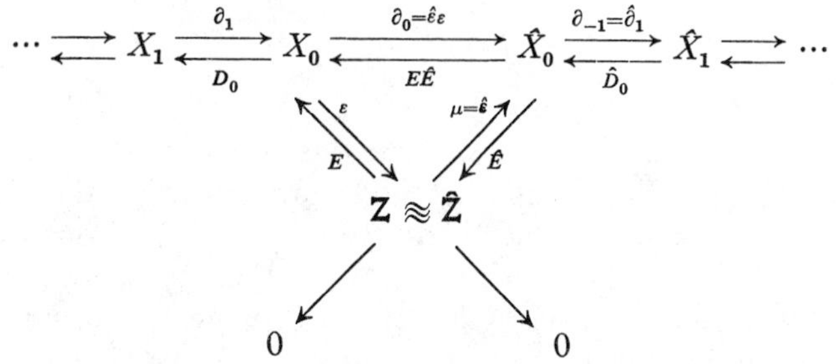

In the case where the positive part is the one constructed in (1-3-1), the negative part just constructed may also be given explicitly. (It may be noted in passing that in this situation, (1-4-6) is not needed for the proof of (1-4-1).) The full G-complex derived from them is known as the **standard G-complex**. Since X_0 has a G-basis consisting of the empty cell $[\cdot]$, it follows from the proof of part (2) of (1-4-4) that $X_{-1} = \hat{X}_0$ has a canonical G-basis consisting of a single element which is denoted by $\langle\cdot\rangle$ and is called **the (−1)-cell**. For any $\tau \in G$, the action of the element $\langle\cdot\rangle^\tau$ of $\hat{X}_0$ on X_0 is given by its action on the $\mathbf{Z}$-basis $\{\sigma[\cdot] \mid \sigma \in G\}$; namely,

$$\langle\cdot\rangle^\tau(\sigma[\cdot]) = \begin{cases} 1 & \sigma = \tau \\ 0 & \text{otherwise} \end{cases}$$

For $n \geqslant 1$, $\{[\sigma_1, \ldots, \sigma_n] \mid \sigma_i \in G\}$ is a G-basis for X_n ; hence, a G-basis for $X_{-(n+1)} = \hat{X}_n$ is the set of all **$-(n+1)$-cells,** $\langle\sigma_1, \ldots, \sigma_n\rangle$. (Note that a $(-n)$-cell is given by an ordered $(n-1)$-tuple of elements of G.) The action of $\hat{X}_n = X_{-(n+1)}$ on X_n is given (see part (2) of (1-4-4)) by

$$\langle\tau_1, \ldots, \tau_n\rangle^\tau(\sigma[\sigma_1, \ldots, \sigma_n]) = \begin{cases} 1 & \sigma = \tau, \quad \sigma_i = \tau_i, \quad i = 1, \ldots, n \\ 0 & \text{otherwise} \end{cases}$$

As for the action of ε, μ, and ∂_0, note first that, by construction, $\varepsilon : X_0 \longrightarrow \mathbf{Z}$ is the G-homomorphism that maps $[\cdot] \longrightarrow 1 \in \mathbf{Z}$, in particular, ε is an element of $\operatorname{Hom}(X_0, \mathbf{Z}) = \hat{X}_0 = X_{-1}$, and for $\sigma \in G$, $\varepsilon : \sigma[\cdot] \longrightarrow 1$. On the other hand, the element $\sum_{\tau \in G} \langle\cdot\rangle^\tau$ of $\hat{X}_0$ also maps $\sigma[\cdot] \longrightarrow 1$. If we denote the element $\sum_{\tau \in G} \tau$ of $\mathbf{Z}[G]$ by S (and call it the **trace element**), then

$$\varepsilon = \sum_{\tau \in G} \langle\cdot\rangle^\tau = \langle\cdot\rangle^S,$$

and by an abuse of notation we write this also as $S\langle\cdot\rangle$. To compute $\mu : \mathbf{Z} \approx \hat{\mathbf{Z}} \longrightarrow \hat{X}_0$, recall that under the identification $\mathbf{Z} \approx \hat{\mathbf{Z}}$, $1 \in \mathbf{Z}$ corresponds to the identity map $1 \in \hat{\mathbf{Z}}$. Consequently, $\mu(1) = \hat{\varepsilon}(1) = (\varepsilon, 1) \circ 1 = 1 \circ \varepsilon = S\langle\cdot\rangle$. Finally, ∂_0 is determined by $\partial_0[\cdot] = \mu\varepsilon[\cdot] = \mu(1) = S\langle\cdot\rangle$. The explicit formulas for negative dimensional boundaries are given in (1-4-7), which summarizes the results of Sections 1-3 and 1-4.

1-4-7. Theorem. A G-complex exists for any finite group G. The **standard G-complex** (sometimes denoted by X^s) is

$$\begin{array}{ccccccccccc}
\cdots \longrightarrow & X_2 & \xrightarrow{\partial_2} & X_1 & \xrightarrow{\partial_1} & X_0 & \xrightarrow{\quad\partial_0\quad} & X_{-1} & \xrightarrow{\partial_{-1}} & X_{-2} & \longrightarrow \cdots \\
& & & & & & \searrow^{\varepsilon} \quad \nearrow_{\mu} & & & & \\
& & & & & & \mathbf{Z} & & & & \\
& & & & & & \nearrow \quad \searrow & & & & \\
& & & & & & 0 \qquad\quad 0 & & & &
\end{array}$$

where

$$X_0 = \Gamma[\cdot] \qquad X_{-n} = \hat{X}_{n-1} \quad n \geqslant 1 \qquad X_{-1} = \Gamma\langle\cdot\rangle$$

$$X_n = \sum_{\sigma_1, \ldots, \sigma_n \in G} \oplus\, \Gamma[\sigma_1, \ldots, \sigma_n] \qquad n > 0$$

$$X_{-n} = \sum_{\sigma_1, \ldots, \sigma_{n-1} \in G} \oplus\, \Gamma\langle\sigma_1, \ldots, \sigma_{n-1}\rangle \qquad n > 1$$

$$\partial_{-n} = \hat{\partial}_n \quad n \geqslant 1 \qquad \varepsilon = S\langle\cdot\rangle = \mu(1) = \partial_0[\cdot]$$

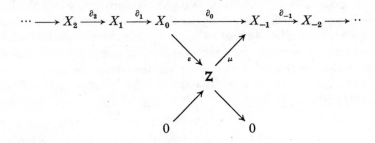

and the boundaries are given by

$$\partial_1([\sigma]) = \sigma[\cdot] - [\cdot]$$

$$\begin{aligned}\partial_n([\sigma_1, \ldots, \sigma_n]) &= \sigma_1[\sigma_2, \ldots, \sigma_n] \\ &+ \sum_{i=1}^{n-1} (-1)^i [\sigma_1, \ldots, \sigma_{i-1}, \sigma_i\sigma_{i+1}, \sigma_{i+2}, \ldots, \sigma_n] \\ &+ (-1)^n [\sigma_1, \ldots, \sigma_{n-1}] \qquad n \geqslant 2\end{aligned}$$

$$\partial_{-1}\langle\cdot\rangle = \sum_{\tau\in G} \langle\tau\rangle^{\tau^{-1}} - \sum_{\tau\in G} \langle\tau\rangle$$

$$\begin{aligned}\partial_{-n}(\langle\tau_1, \ldots, \tau_{n-1}\rangle) &= \sum_{\tau\in G} \langle\tau, \tau_1, \ldots, \tau_{n-1}\rangle^{\tau^{-1}} \\ &+ \sum_{i=1}^{n-1} (-1)^i \sum_{\tau\in G} \langle\tau_1, \ldots, \tau_{i-1}, \tau_i\tau^{-1}, \tau, \tau_{i+1}, \ldots, \tau_{n-1}\rangle \\ &+ (-1)^n \sum_{\tau\in G} \langle\tau_1, \ldots, \tau_{n-1}, \tau\rangle \qquad n \geqslant 2\end{aligned}$$

Proof: Only the negative dimensional boundary formulas remain to be proved, and the formula for ∂_{-1} is a special case of that for ∂_{-n}. (Note that in the expression for X_{-n} we abuse notation and write $\Gamma\langle\sigma_1, \ldots, \sigma_{n-1}\rangle$ for $\langle\sigma_1, \ldots, \sigma_{n-1}\rangle^{\Gamma}$.) It suffices to verify that both sides of the formula for the G-homomorphism ∂_{-n} have the same action on a $\mathbf{Z}$-basis of X_n. Thus

$$\begin{aligned}&\partial_{-n}\langle\tau_1, \ldots, \tau_{n-1}\rangle(\sigma[\sigma_1, \ldots, \sigma_n]) \\ &= \hat{\partial}_n\langle\tau_1, \ldots, \tau_{n-1}\rangle(\sigma[\sigma_1, \ldots, \sigma_n]) \\ &= \langle\tau_1, \ldots, \tau_{n-1}\rangle\, \partial_n(\sigma[\sigma_1, \ldots, \sigma_n]) \\ &= \langle\tau_1, \ldots, \tau_{n-1}\rangle(\sigma\sigma_1[\sigma_2, \ldots, \sigma_n]) \\ &+ \sum_{i=1}^{n-1} (-1)^i \langle\tau_1, \ldots, \tau_{n-1}\rangle\, (\sigma[\sigma_1, \ldots, \sigma_i\sigma_{i+1}, \ldots, \sigma_n]) \\ &+ (-1)^n \langle\tau_1, \ldots, \tau_{n-1}\rangle(\sigma[\sigma_1, \ldots, \sigma_{n-1}])\end{aligned}$$

and when the right-hand side of the formula is evaluated at $\sigma[\sigma_1, \ldots, \sigma_n]$, it too gives an expression of form

$$A + \sum_{i=1}^{n-1} (-1)^i B_i + C.$$

In each case, it is easy to check that

$$A = \begin{cases} 1 & \sigma\sigma_1 = 1, \quad \tau_1 = \sigma_2, \ldots, \tau_{n-1} = \sigma_n \\ 0 & \text{otherwise} \end{cases}$$

$$B = \begin{cases} 1 & \sigma = 1, \quad \tau_1 = \sigma_1, \ldots, \tau_{i-1} = \sigma_{i-1}, \quad \tau_i = \sigma_i\sigma_{i+1}, \\ & \tau_{i+1} = \sigma_{i+2}, \ldots, \tau_{n-1} = \sigma_n \\ 0 & \text{otherwise} \end{cases}$$

$$C = \begin{cases} 1 & \sigma = 1, \quad \tau_1 = \sigma_1, \ldots, \tau_{n-1} = \sigma_{n-1} \\ 0 & \text{otherwise} \end{cases}$$ ∎

1-4-8. Exercise. How do the results of Sections 1-3 and 1-4 carry over for an infinite group G? (Of course, the same question is appropriate for future results, too.) In particular, the positive part of a G-complex exists (in another terminology, this says that a free resolution of $\mathbf{Z}$ over Γ exists) with the understanding that the free Γ-modules need not have finite bases. What about the negative part of a G-complex?

1-5. COHOMOLOGY GROUPS OF G IN A

Suppose that G is a finite group and that A is a G-module; let $(X, \partial, \varepsilon, \mu)$ be any G-complex and consider the additive group $\mathrm{Hom}_G(X, A) = \sum \oplus \mathrm{Hom}_G(X_n, A)$. Since $\partial : X \longrightarrow X$ and $1 : A \longrightarrow A$ are both G-homomorphisms, it follows from (1-4-3) that $\delta = (\partial, 1)$ is an endomorphism of $\mathrm{Hom}_G(X, A)$ and that $\delta^2 = (\partial, 1) \circ (\partial, 1) = (\partial^2, 1) = (0, 1) = 0$. There also exist homomorphisms $\delta_n = (\partial_n, 1) : \mathrm{Hom}_G(X_{n-1}, A) \longrightarrow \mathrm{Hom}_G(X_n, A)$ with

$$\delta_{n+1} \circ \delta_n = 0 \qquad \text{and} \qquad \delta = \sum \oplus \delta_n .$$

In other words,

$$(\mathrm{Hom}_G(X, A) = \sum \oplus \mathrm{Hom}_G(X_n, A), \qquad \delta = \sum \oplus \delta_n, \qquad +1)$$

is a differential graded group of cohomology type which may also be written as the sequence :

$$\cdots \longrightarrow \mathrm{Hom}_G(X_{n-1}, A) \xrightarrow{\delta_n} \mathrm{Hom}_G(X_n, A) \xrightarrow{\delta_{n+1}} \mathrm{Hom}_G(X_{n+1}, A) \longrightarrow \cdots$$

We call $\mathscr{C}^n = \mathscr{C}^n(G, A) = \mathrm{Hom}_G(X_n, A)$ the group of **n-cochains** of G in A, $\mathscr{Z}^n = \mathscr{Z}^n(G, A) = \{f \in \mathscr{C}^n \mid \delta f = 0\}$ the group of **n-cocycles** of G in A, and $\mathscr{B}^n = \mathscr{B}^n(G, A) = \delta\mathscr{C}^{n-1}$ the group of **n-coboundaries** of G in A. The nth derived group $H_n(\mathrm{Hom}_G(X, A)) = \mathscr{Z}^n/\mathscr{B}^n$ is known as the **nth cohomology group of G in A**; it is denoted by $H^n(G, A)$. Two n-cocycles f and g are said to be **cohomologous** (denoted $f \sim g$) when they differ by a coboundary—that is, when they determine the same element in the cohomology group $H^n(G, A)$.

Strictly speaking, the cohomology groups (and the other objects too) should be denoted in such a way as to indicate their dependence on the G-complex; thus, we shall write $H^n(G, A, X)$ or $H^n_X(G, A)$ for $H^n(G, A)$ when necessary. It will be shown, in Section 2-1, that the cohomology groups are independent (up to isomorphism) of the choice of G-complex, so that in the current discussion, where the objective is to describe the lower-dimensional cohomology groups, it is sufficient to make use solely of the standard G-complex. The following notations will be used :

$$A^G = \{a \in A \mid \sigma a = a \quad \forall \sigma \in G\}$$

$$SA = \{Sa \mid a \in A\} \qquad \text{where} \quad S = \sum_{\sigma \in G} \sigma \in \Gamma$$

$$A_S = \{a \in A \mid Sa = 0\}$$

$$I = \left\{\sum_{\sigma \in G} n_\sigma \sigma \in \Gamma \mid \sum_{\sigma \in G} n_\sigma = 0\right\} = \left\{\sum_{\sigma \in G} \oplus \mathbf{Z}(\sigma - 1)\right\}$$

1-5-1. Proposition.

$$H^2(G, A) \approx \frac{\{\text{factor sets}\}}{\{\text{splitting factor sets}\}}.$$

Proof: An n-cochain (with respect to the standard complex $X = X^s$) is completely determined by its values on the n-cells. For an $f \in \mathscr{C}^2 = \mathrm{Hom}_G(X_2, A)$, its coboundary, $\delta f \in \mathscr{C}^3$, is a function on the 3-cells, which, in virtue of (1-4-7), is given by

$$\begin{aligned}\delta f[\sigma, \tau, \rho] &= f\partial[\sigma, \tau, \rho] \\ &= \sigma f[\tau, \rho] - f[\sigma\tau, \rho] + f[\sigma, \tau\rho] - f[\sigma, \tau] \qquad \forall \sigma, \tau, \rho \in G\end{aligned}$$

Thus, f is a cocycle if and only if it satisfies

$$\sigma f[\tau, \rho] + f[\sigma, \tau\rho] = f[\sigma\tau, \rho] + f[\sigma, \tau] \qquad \forall \sigma, \tau, \rho \in G$$

Furthermore, the 2-cochain f is a coboundary if and only if it is of the form $f = \delta g$ for some $g \in \mathscr{C}^1$—thus, it must satisfy

$$f[\sigma, \tau] = \sigma g[\tau] - g[\sigma\tau] + g[\sigma] \qquad \forall \sigma, \tau \in G$$

Functions of two variables satisfying the cocycle identity were known classically as **factor sets**, while those satisfying the coboundary identity were known as **splitting factor sets**. They arose in the study of both group extensions (which we will consider in Chapter V) and simple algebras (see, for example, [*12*], [*4*, Chapter 8]). The group A was usually multiplicative, $f[\sigma, \tau]$ was written as $a_{\sigma,\tau}$, and the action of G on A was written exponentially; thus, in this notation, the function of two variables $\{a_{\sigma,\tau}\}$ is a factor set (i.e. cocycle) when it satisfies

$$a_{\tau,\rho}^{\sigma} a_{\sigma,\tau\rho} = a_{\sigma\tau,\rho} a_{\sigma,\tau} \qquad \forall \sigma, \tau, \rho \in G$$

and $\{a_{\sigma,\tau}\}$ is a splitting factor set (i.e. coboundary) when there is a function $\{b_\sigma\}$ of one variable such that

$$a_{\sigma,\tau} = \frac{b_\tau^\sigma b_\sigma}{b_{\sigma\tau}} \qquad \forall \sigma, \tau \in G \quad \blacksquare$$

1-5-2. Proposition.

$$H^1(G, A) \approx \frac{\{\text{crossed homomorphisms}\}}{\{\text{principal crossed homomorphisms}\}}$$

Proof: A standard 1-cochain f is a cocycle $\Longleftrightarrow$

$$\delta f[\sigma, \tau] = \sigma f[\tau] - f[\sigma\tau] + f[\sigma] = 0 \text{ for all } \sigma, \tau \in G$$

—or what is then same, if and only if

$$f[\sigma\tau] = \sigma f[\tau] + f[\sigma] \qquad \forall \sigma, \tau \in G$$

Such functions $f: G \longrightarrow A$ are (for obvious reasons) called **crossed homomorphisms.**

Furthermore, the 1-cochain f is a coboundary $\Longleftrightarrow$ there exists $f \in \mathscr{C}^0$ such that $f = \delta g \Longleftrightarrow$

$$f[\sigma] = \delta g[\sigma] = g\partial[\sigma] = g(\sigma[\cdot] - [\cdot]) = \sigma(g[\cdot]) - g[\cdot] \quad \text{for all} \quad \sigma \in G.$$

Thus, if we set up the natural 1–1 correspondence $\mathscr{C}^0 \longleftrightarrow A$ given by $g \longleftrightarrow a = h[\cdot]$, then f is a coboundary if and only if there exists $a \in A$ such that

$$f[\sigma] = (\sigma - 1)\, a \qquad \forall \sigma \in G$$

Such functions $f: G \longrightarrow A$ are called **principal crossed homomorphisms.** ▮

We shall write $\hat{G}$ for the character group of G (since G is finite, we have $\hat{G} = \operatorname{Hom}(G, \mathbf{Q}/\mathbf{Z})$ where $\mathbf{Q}/\mathbf{Z}$ is additive) and G^c for the commutator subgroup of G. It then follows from above that :

1-5-3. Corollary. If G acts trivially on A, then

$$H^1(G, A) \approx \operatorname{Hom}(G, A).$$

In particular,

$$H^1(G, \mathbf{Z}) = (0)$$

$$H^1\left(G, \frac{\mathbf{Q}}{\mathbf{Z}}\right) \approx \hat{G} \approx \widehat{\left(\frac{G}{G^c}\right)} \approx \frac{G}{G^c}$$

1-5-4. Theorem. (Noether's equations). If K/F is a finite Galois extension with Galois group G, then

$$H^1(G, K^*) = (1)$$

Proof: Let f be a standard 1-cocycle of G in the multiplicative group of K; so $f : G \longrightarrow K^*$ satisfies $f(\sigma\tau) = (\sigma f(\tau))(f(\sigma))$ for all $\sigma, \tau \in G$. We recall (see, for example, [3, p. 35]) that distinct automorphisms $\sigma_1, \ldots, \sigma_n$ of a field K are independent—which means that if $\alpha_i \in K$ are such that $\sum_1^n \alpha_i \sigma_i(\beta) = 0$ for all $\beta \in K$, then $\alpha_i = 0$ for $i = 1, \ldots, n$. Consequently, there exists $\beta \in K^*$ such that

$$\gamma = \sum_{\tau \in G} f(\tau)\, \tau(\beta) \neq 0$$

Then, for any $\sigma \in G$

$$\sigma\gamma = \sum_{\tau \in G} (\sigma f(\tau))(\sigma\tau(\beta)) = \sum_{\tau \in G} \frac{f(\sigma\tau)}{f(\sigma)}\, \sigma\tau(\beta)$$

so that

$$(\sigma\gamma) f(\sigma) = \sum_{\tau \in G} f(\sigma\tau)(\sigma\tau(\beta)) = \gamma$$

This means that $\gamma^{\sigma-1} = (\sigma\gamma)/\gamma = 1/(f(\sigma))$ so that the function $\sigma \longrightarrow 1/(f(\sigma))$ is a boundary; hence, f is a coboundary. ∎

1-5-5. Corollary. (Hilbert's theorem 90). Suppose that K/F is a cyclic extension of degree n, and let σ be a generator of $G = \mathcal{G}(K/F)$. If $\alpha \in K$ has $N_{K \to F}\alpha = 1$, then there exists $\beta \in K^*$ such that $\alpha = \beta^{1-\sigma}$.

Proof: Define a 1-cocycle $f : G \longrightarrow K^*$ by putting

$$f(1) = 1,\ f(\sigma) = \alpha,\ f(\sigma^2) = \alpha(\sigma\alpha), \ldots,\ f(\sigma^{n-1}) = \alpha(\sigma\alpha) \cdots (\sigma^{n-2}\alpha)$$ ∎

1-5-6. Proposition. $H^0(G, A) \approx A^G/(SA)$.

Proof: The mapping $f \longrightarrow f[\cdot]$ is an isomorphism of $\mathscr{C}^0$ onto A. Given $f \in \mathscr{C}^0$ we have

$$\delta f[\sigma] = \sigma f[\cdot] - f[\cdot]$$

If we put $f[\cdot] = a$, then

$$f \text{ is a 0-cocycle} \iff (\sigma - 1)\,a = 0 \quad \forall \sigma \in G$$

and our mapping induces an isomorphism of $\mathscr{Z}^0$ onto A^G.

Furthermore, the mapping $g \longrightarrow g\langle\cdot\rangle$ is an isomorphism of $\mathscr{C}^{-1}$ onto A. Given $g \in \mathscr{C}^{-1}$ we have

$$\delta g[\cdot] = g(S\langle\cdot\rangle) = S(g\langle\cdot\rangle)$$

Thus, for $f \in \mathscr{C}^0$,

$$\begin{aligned} f \text{ is a 0-coboundary} &\iff \text{there exists } g \in \mathscr{C}^{-1} \text{ with } a = f[\cdot] = S(g\langle\cdot\rangle) \\ &\iff a \in SA \end{aligned}$$

and the map $f \longrightarrow f[\cdot]$ also induces an isomorphism of $\mathscr{B}^0$ onto SA. ▮

1-5-7. Corollary. If the group G of order n acts trivially on A, then $H^0(G, A) \approx A/(nA)$. In particular,

$$H^0(G, \mathbf{Z}) \approx \frac{\mathbf{Z}}{n\mathbf{Z}} \qquad H^0\left(G, \frac{\mathbf{Q}}{\mathbf{Z}}\right) = (0)$$

1-5-8. Corollary. If K/F is a Galois extension with Galois group G, then

$$H^0(G, K^*) \approx \frac{F^*}{N_{K\to F}K^*}$$

Proof: Since F is the fixed field of G we have $K^{*G} = F^*$, and because the module is multiplicative, the action of S is that of the norm from K to F. ▮

1-5-9. Proposition. $H^{-1}(G, A) \approx A_S/(IA)$.

Proof: Consider (as in (1-5-6)) the isomorphism $f \longrightarrow f\langle\cdot\rangle$ of $\mathscr{C}^{-1}$ onto A. Then for $f \in \mathscr{C}^{-1}$,

$$f \text{ is a } -1 \text{ cocycle} \iff S(f\langle\cdot\rangle) = 0$$

and we get an induced isomorphism of $\mathscr{Z}^{-1}$ onto A_S.

Furthermore, for $g \in \mathscr{C}^{-2}$ we have (see (1-4-7))

$$\delta g\langle\cdot\rangle = \sum_{\tau \in G} \tau^{-1} g\langle\tau\rangle - \sum_{\tau \in G} g\langle\tau\rangle = \sum_{\tau \in G} (\tau^{-1} - 1)(g\langle\tau\rangle)$$

and therefore, for $f \in \mathscr{C}^{-1}$,

$$\begin{aligned} f \text{ is a } -1 \text{ coboundary} &\Longleftrightarrow \text{there exists } g \in \mathscr{C}^{-2} \\ &\qquad \text{with } f\langle\cdot\rangle = \sum_{\tau \in G} (\tau^{-1} - 1)(g\langle\tau\rangle) \\ &\Longleftrightarrow f\langle\cdot\rangle \in I \cdot A \end{aligned}$$

Thus, we have an induced isomorphism of $\mathscr{B}^{-1}$ onto IA. ∎

1-5-10. Corollary. If G has order n, then

$$H^{-1}(G, \mathbf{Z}) = (0) \qquad H^{-1}(G, \mathbf{Q}/\mathbf{Z}) \approx \left(\frac{1}{n}\mathbf{Z}\right)\Big/\mathbf{Z}$$

1-5-11. Corollary. If K/F is a Galois extension with Galois group G, then

$$H^{-1}(G, K^*) \approx \frac{\{\alpha \in K^* \mid N_{K \to F}\, \alpha = 1\}}{\{\prod_{\sigma \in G} \alpha_\sigma^{\sigma - 1} \mid \alpha_\sigma \in K^*\}}$$

In particular, if K/F is cyclic, then

$$H^{-1}(G, K^*) = (0)$$

Proof: The first part is a translation of (1-5-9) to multiplicative notation. In the cyclic case, apply (1-5-5), after observing that if σ is a generator of G and A is any G-module, then $(\sigma^i - 1)A \subset (\sigma - 1)A$ for all $i \geqslant 1$. ∎

1-5-12. Remark. In proving the preceding results, we have made use of the coboundary formulas in low dimensions. For convenience and future reference, we state them for all dimensions; of course, they are immediate consequences of the boundary formulas given in (1-4-7).

Let $f \in \mathscr{C}^n$ be an n-cochain of G in A with respect to the standard G-complex, then for all $\sigma_1, \sigma_2, \ldots, \sigma_n \in G$

$$(\delta f)[\sigma_1] = \sigma_1(f[\cdot]) - f[\cdot] \qquad n = 0$$

$$(\delta f)[\sigma_1, \sigma_2] = \sigma_1 f[\sigma_2] - f[\sigma_1\sigma_2] + f[\sigma_1] \qquad n = 1$$

$$\begin{aligned}(\delta f)[\sigma_1, \sigma_2, \sigma_3] &= \sigma_1 f[\sigma_1, \sigma_3] - f[\sigma_1\sigma_2, \sigma_3] \\ &\quad + f[\sigma_1, \sigma_2\sigma_3] - f[\sigma_1, \sigma_2] \qquad n = 2\end{aligned}$$

$$(\delta f)([\cdot]) = S(f\langle\cdot\rangle) \qquad n = -1$$

$$(\delta f)(\langle\cdot\rangle) = \sum_{\sigma\in G} (\sigma^{-1} - 1)(f\langle\sigma\rangle) \qquad n = -2$$

The cases $n = 0, 1, 2$ may be viewed as special cases of the general formula :

$$\begin{aligned}(\delta f)[\sigma_1, \ldots, \sigma_{n+1}] &= \sigma_1 f[\sigma_2, \ldots, \sigma_{n+1}] \\ &\quad + \sum_{i=1}^{n} (-1)^i f[\sigma_1, \ldots, \sigma_{i-1}, \sigma_i\sigma_{i+1}, \sigma_{i+2}, \ldots, \sigma_{n+1}] \\ &\quad + (-1)^{n+1} f[\sigma_1, \ldots, \sigma_n] \qquad n \geqslant 0\end{aligned}$$

The case $n = -2$ may be viewed as a special case of

$$\begin{aligned}&(\delta f)\langle\sigma_1, \sigma_2, \ldots, \sigma_{|n|-2}\rangle \\ &\quad = \sum_{\sigma\in G} \sigma^{-1} f\langle\sigma, \sigma_1, \ldots, \sigma_{|n|-2}\rangle \\ &\qquad + \sum_{i=1}^{|n|-2} (-1)^i \sum_{\sigma\in G} f\langle\sigma_1, \ldots, \sigma_{i-1}, \sigma_i\sigma^{-1}, \sigma, \sigma_{i+1}, \ldots, \sigma_{|n|-2}\rangle \\ &\qquad + (-1)^{|n|-1} \sum_{\sigma\in G} f\langle\sigma_1, \ldots, \sigma_{|n|-2}, \sigma\rangle \qquad n \leqslant -2\end{aligned}$$

1-5-13. Remark. The cohomology groups of G in A were introduced by Eilenberg and MacLane [22] (inspired in part by the work of Hochschild [32] on cohomology of associative algebras) according to the following procedure.

For each $n \geqslant 0$ consider functions Φ which map ordered $(n+1)$-tuples of elements of G into A—that is,

$$\Phi : \underbrace{G \times \cdots \times G}_{n+1} \longrightarrow A.$$

Those which are homogeneous in the sense that

$$\Phi(\sigma\sigma_0, \sigma\sigma_1, \ldots, \sigma\sigma_n) = \sigma\Phi(\sigma_0, \sigma_1, \ldots, \sigma_n) \qquad \forall \sigma, \sigma_0, \ldots, \sigma_n \in G$$

are called n-dimensional homogeneous cochains; they form a group denoted by $\mathfrak{C}^n = \mathfrak{C}^n(G, A)$. For $\Phi \in \mathfrak{C}^n$ define $\delta\Phi \in \mathfrak{C}^{n+1}$ by

$$(\delta\Phi)(\sigma_0, \ldots, \sigma_{n+1}) = \sum_{i=0}^{n+1} (-1)^i \, \Phi(\sigma_0, \ldots, \not{\sigma}_i, \ldots, \sigma_{n+1})$$

where $\not{\sigma}_i$ indicates that this term is omitted. δ (which is called the coboundary) is clearly a homomorphism of $\mathfrak{C}^n \longrightarrow \mathfrak{C}^{n+1}$, and it satisfies $\delta \circ \delta = 0$. Denote the kernel of δ by $\mathfrak{Z}^n$ (called the group of n-cocycles) and the image of δ by $\mathfrak{B}^{n+1}$ (called the group of $(n+1)$-coboundaries). Put $\mathfrak{B}^0 = (0)$. Now $\mathfrak{B}^n \subset \mathfrak{Z}^n$, and the group $\mathfrak{H}^n(G, A) = \mathfrak{Z}^n/\mathfrak{B}^n$ is called the nth cohomology group of G in A.

Consider further the group of nonhomogeneous n-cochains, $n \geqslant 1$; these are the functions ϕ which map ordered n-tuples of elements of G into A, with no restrictions on ϕ. Define a nonhomogeneous 0-cochain to be simply, an element of A; so A is the group of nonhomogeneous 0-cochains. For each $n \geqslant 0$ this group may be identified with $\mathfrak{C}^n$ because it is isomorphic to it under the correspondence $\Phi \longleftrightarrow \phi$ given by

$$\Phi(\sigma_0, \sigma_1, \ldots, \sigma_n) = \sigma_0\phi(\sigma_0^{-1}\sigma_1, \sigma_1^{-1}\sigma_2, \ldots, \sigma_{n-1}^{-1}\sigma_n)$$

$$\phi(\sigma_1, \ldots, \sigma_n) = \Phi(1, \sigma_1, \sigma_1\sigma_2, \ldots, \sigma_1\sigma_2 \cdots \sigma_n)$$

Note that for $n = 0$, Φ corresponds to the element $\phi = \Phi(1)$ of A. When the coboundary δ is transferred to the nonhomogeneous situation—that is, so that $\delta\Phi \longleftrightarrow \delta\phi$—it turns out that

$$\begin{aligned}(\delta\phi)(\sigma_1, \ldots, \sigma_{n+1}) = {} & \sigma_1\phi(\sigma_2, \ldots, \sigma_{n+1}) \\ & + \sum_{i=1}^{n} (-1)^i \, \phi(\sigma_1, \ldots, \sigma_{i-1}, \sigma_i\sigma_{i+1}, \sigma_{i+2}, \ldots, \sigma_{n+1}) \\ & + (-1)^{n+1} \phi(\sigma_1, \ldots, \sigma_n)\end{aligned}$$

For $n = 0$, this specializes to $(\delta\phi)(\sigma_1) = \sigma_1\phi - \phi$ where $\phi \in A$. Then, under identification, $\mathfrak{Z}^n$ becomes the group of nonhomogeneous cocycles, $\mathfrak{B}^n$ becomes the group of nonhomogeneous coboundaries, and $\mathfrak{H}^n(G, A) = \mathfrak{Z}^n/\mathfrak{B}^n$.

For $n \geqslant 1$, these cohomology groups are clearly identical with those treated earlier in terms of the standard complex. In fact, this explains the origins of the positive part of the standard complex and its boundary operator. For $n = 0$, the cohomology group defined in terms of the nonhomogeneous cochains differs from the standard complex cohomology group—it is A^G, since there are no 0-dimensional coboundaries.

It should be noted that the foregoing discussion (except for the comparison of H^0 and $\mathfrak{H}^0$) applies when G is infinite.

1-6. PROBLEMS AND SUPPLEMENTS

Let R denote a ring with 1; in particular, if G is a multiplicative group, finite or infinite, then R may be taken as the integer group ring $\mathbf{Z}[G]$ (of course, for infinite G, $\mathbf{Z}[G]$ consists of the elements $\sum_{\sigma\in G} n_\sigma \sigma$, $n_\sigma \in \mathbf{Z}$, $n_\sigma = 0$ for almost all σ, which are added and multiplied in the obvious fashion) or the group algebra

$$k[G] = \left\{\sum_{\sigma\in G} a_\sigma \sigma \mid a_\sigma \in k\right\}$$

of G over the field k. By any R-module, we mean a left unitary R-module. In this section, **unless otherwise specified**, we are concerned with R-modules and R-homomorphisms.

1-6-1. (i) Let $f: A \longrightarrow B$ be a homomorphism, and write $\operatorname{coker} f = B/\operatorname{im} f$, $\operatorname{coim} f = A/\ker f$. Then $\operatorname{coim} f \approx \operatorname{im} f$ and we have exact sequences

$$0 \longrightarrow \ker f \longrightarrow A \longrightarrow \operatorname{coim} f \longrightarrow 0$$

$$0 \longrightarrow \operatorname{im} f \longrightarrow B \longrightarrow \operatorname{coker} f \longrightarrow 0$$

$$0 \longrightarrow \ker f \longrightarrow A \xrightarrow{f} B \longrightarrow \operatorname{coker} f \longrightarrow 0$$

(ii) **Five Lemma.** Consider the commutative diagram

$$\begin{array}{ccccccccc} A_2 & \xrightarrow{f_2} & A_1 & \xrightarrow{f_1} & A_0 & \xrightarrow{f_0} & A_{-1} & \xrightarrow{f_{-1}} & A_{-2} \\ {\scriptstyle\varphi_2}\downarrow & & {\scriptstyle\varphi_1}\downarrow & & {\scriptstyle\varphi_0}\downarrow & & {\scriptstyle\varphi_{-1}}\downarrow & & {\scriptstyle\varphi_{-2}}\downarrow \\ B_2 & \xrightarrow{g_2} & B_1 & \xrightarrow{g_1} & B_0 & \xrightarrow{g_0} & B_{-1} & \xrightarrow{g_{-1}} & B_{-2} \end{array}$$

with exact rows.

(a) If φ_1 and φ_{-1} are monomorphisms, and φ_2 is an epimorphism, then φ_0 is a monomorphism.

(b) If φ_1 and φ_{-1} are epimorphisms, and φ_{-2} is a monomorphism, then φ_0 is an epimorphism.

(iii) Consider the commutative diagram with exact rows

$$\begin{array}{ccccccccc} 0 & \longrightarrow & A & \longrightarrow & B & \longrightarrow & C & \longrightarrow & 0 \\ & & {\scriptstyle f}\downarrow & & {\scriptstyle g}\downarrow & & {\scriptstyle h}\downarrow & & \\ 0 & \longrightarrow & A' & \longrightarrow & B' & \longrightarrow & C' & \longrightarrow & 0 \end{array}$$

(a) If f and h are epimorphisms, then so is g.

(b) If f and h are monomorphisms, then so is g.

(c) If f and h are isomorphisms onto, then so is g.

(iv) **Four Lemma.** Consider the commutative diagram with exact rows

$$\begin{array}{ccccccc} A_2 & \xrightarrow{f_2} & A_1 & \xrightarrow{f_1} & A_0 & \xrightarrow{f_0} & A_{-1} \\ {\scriptstyle\varphi_2}\downarrow & & {\scriptstyle\varphi_1}\downarrow & & {\scriptstyle\varphi_0}\downarrow & & {\scriptstyle\varphi_{-1}}\downarrow \\ B_2 & \xrightarrow{g_2} & B_1 & \xrightarrow{g_1} & B_0 & \xrightarrow{g_0} & B_{-1} \end{array}$$

(a) If φ_2 is an epimorphism and φ_{-1} is a monomorphism, then $\ker \varphi_0 = f_1(\ker \varphi_1)$ and $\operatorname{im} \varphi_1 = g_1^{-1}(\operatorname{im} \varphi_0)$. In particular, if φ_1 is a monomorphism, then so is φ_0, and if φ_0 is an epimorphism, then so is φ_1.

(b) Use part (a) to prove the five lemma.

(v) Consider the commutative diagram with exact rows

$$\begin{array}{ccccc} A_1 & \xrightarrow{f_1} & A_0 & \xrightarrow{f_0} & A_{-1} \\ {\scriptstyle\varphi_1}\downarrow & & {\scriptstyle\varphi_0}\downarrow & & {\scriptstyle\varphi_{-1}}\downarrow \\ B_1 & \xrightarrow{g_1} & B_0 & \xrightarrow{g_0} & B_{-1} \end{array}$$

(a) If f_0 and φ_1 are epimorphisms and φ_0 is a monomorphism, then φ_{-1} is a monomorphism.

(b) If g_1 and φ_{-1} are monomorphisms and φ_0 is an epimorphism, then φ_1 is an epimorphism.

(c) If g_1 is a monomorphism, then we have an exact sequence

$$\ker \varphi_1 \longrightarrow \ker \varphi_0 \longrightarrow \ker \varphi_{-1}$$

(d) If f_0 is an epimorphism then we have an exact sequence

$$\operatorname{coker} \varphi_1 \longrightarrow \operatorname{coker} \varphi_0 \longrightarrow \operatorname{coker} \varphi_{-1}$$

(e) If g_1 is a monomorphism and f_0 is an epimorphism, then we have an exact sequence

$$\ker \varphi_1 \longrightarrow \ker \varphi_0 \longrightarrow \ker \varphi_{-1} \longrightarrow \operatorname{coker} \varphi_1 \longrightarrow \operatorname{coker} \varphi_0 \longrightarrow \operatorname{coker} \varphi_{-1}$$

1-6-2. Consider the commutative diagram of modules

$$\begin{array}{ccccccccc} & & 0 & & 0 & & 0 & & \\ & & \downarrow & & \downarrow & & \downarrow & & \\ 0 & \longrightarrow & A' & \longrightarrow & A & \longrightarrow & A'' & \longrightarrow & 0 \\ & & \downarrow & & \downarrow & & \downarrow & & \\ 0 & \longrightarrow & B' & \longrightarrow & B & \longrightarrow & B'' & \longrightarrow & 0 \\ & & \downarrow & & \downarrow & & \downarrow & & \\ 0 & \longrightarrow & C' & \longrightarrow & C & \longrightarrow & C'' & \longrightarrow & 0 \\ & & \downarrow & & \downarrow & & \downarrow & & \\ & & 0 & & 0 & & 0 & & \end{array}$$

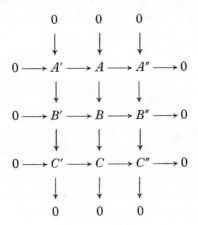

in which the columns are exact. If the first two rows are exact, or if the last two rows are exact, then so is the remaining row. If the first and third rows are exact, find a condition that will guarantee the exactness of the middle row.

1-6-3. (i) Consider a collection of modules M_α where α runs over some arbitrary indexing set $\mathcal{A}$. The **(external) direct product** $\prod_\alpha M_\alpha$ is a module consisting of all maps $f : \mathcal{A} \longrightarrow \bigcup_\alpha M_\alpha$ such that $f(\alpha) = m_\alpha \in M_\alpha$, and with the natural definitions of addition and scalar multiplication; it is often convenient to denote the generic element f of $\prod M_\alpha$ by $\prod m_\alpha$ or $\{m_\alpha\}$, so the module operations are then componentwise. There is a submodule of $\prod_\alpha M_\alpha$ consisting of those $\{m_\alpha\}$ for which almost all $m_\alpha = 0$; it is called the **(external) direct sum** and denoted by $\sum_\alpha \oplus M_\alpha$ or $\coprod_\alpha M_\alpha$.

For each $\beta \in \mathcal{A}$ there is a map $i_\beta : M_\beta \longrightarrow \prod M_\alpha$ which takes $m_\beta \in M_\beta$ to the element with m_β as the β-component and 0 elsewhere, and a map $\pi_\beta : \prod M_\alpha \longrightarrow M_\beta$, the projection on the β-coordinate. It is clear that $\pi_\alpha \circ i_\beta = \delta_{\alpha\beta}$ where by $\delta_{\alpha\beta} : M_\beta \longrightarrow M_\alpha$ we mean the 0-map when $\alpha \neq \beta$ and the identity map when $\alpha = \beta$. It is also clear that $i_\beta : M_\beta \longrightarrow \sum \oplus M_\alpha$ and $\pi_\beta : \sum \oplus M_\alpha \longrightarrow M_\beta$, and that $\sum_\alpha i_\alpha \circ \pi_\alpha$ is the identity map of $\sum \oplus M_\alpha$.

(ii) Suppose that we have modules M_α, $\alpha \in \mathcal{A}$ and M. If for each α we have homomorphisms $\varphi_\alpha : M_\alpha \longrightarrow M$ and $\psi_\alpha : M \longrightarrow M_\alpha$, and such that $\psi_\beta \circ \varphi_\alpha = \delta_{\alpha\beta}$, then we say that $\{M_\alpha \xrightarrow{\varphi_\alpha} M \xrightarrow{\psi_\alpha} M_\alpha\}$ is a **direct family**. Given such a direct family we have homomorphisms $\varphi : \sum \oplus M_\alpha \longrightarrow M$ and $\psi : M \longrightarrow \prod M_\alpha$ given by

$$\varphi\left(\sum_\alpha m_\alpha\right) = \sum_\alpha \varphi_\alpha(m_\alpha) \qquad \text{and} \qquad \psi(m) = \{\psi_\alpha(m)\}.$$

The map φ is a monomorphism; if φ is an isomorphism onto, we say that $\{M_\alpha \xrightarrow{\varphi_\alpha} M \xrightarrow{\psi_\alpha} M_\alpha\}$ is a **complete representation of** $\boldsymbol{M}$ **as a direct sum**. The direct family $\{M_\alpha \xrightarrow{\varphi_\alpha} M \xrightarrow{\psi_\alpha} M_\alpha\}$ is a complete representation of M as a direct sum $\Longleftrightarrow \sum_\alpha \varphi_\alpha \circ \psi_\alpha$ if the identity; moreover, the ψ_α's can then be recaptured from the φ_α's.

If ψ is an isomorphism onto, we say that $\{M_\alpha \xrightarrow{\varphi_\alpha} M \xrightarrow{\psi_\alpha} M_\alpha\}$ is a **complete representation of *M* as a direct product.** In such a situation, the φ_α's can be recaptured from the ψ_α's.

When the indexing set $\mathcal{A}$ is finite, then direct sums and products are identical; also the direct family $\{M_\alpha \xrightarrow{\varphi_\alpha} M \xrightarrow{\psi_\alpha} M_\alpha\}$ is a complete representation of M as a direct sum $\Longleftrightarrow$ it is a complete representation of M as a direct product.

(iii) Let $\{M_\alpha \xrightarrow{\varphi_\alpha} M \xrightarrow{\psi_\alpha} M_\alpha\}$, $\alpha \in \mathcal{A}$, be a complete representation of M as a direct sum, and let $\{N_\beta \xrightarrow{\lambda_\beta} N \xrightarrow{\mu_\beta} N_\beta\}$, $\beta \in \mathcal{B}$, be a complete representation of N as a direct product. If $\text{Hom}_R(M, N)$ denotes the additive group of all R-homomorphisms of M into N, then

$$\{\text{Hom}_R(M_\alpha, N_\beta) \xrightarrow{(\psi_\alpha, \lambda_\beta)} \text{Hom}_R(M, N) \xrightarrow{(\varphi_\alpha, \mu_\beta)} \text{Hom}_R(M_\alpha, N_\beta)\}$$

with $\alpha \in \mathcal{A}$, $\beta \in \mathcal{B}$ is a complete representation of $\text{Hom}_R(M, N)$ as a direct product (of abelian groups). Symbolically, one may write

$$\text{Hom}_R\Big(\sum_\alpha \oplus M_\alpha, \prod_\beta N_\beta\Big) \approx \prod_{\alpha,\beta} \text{Hom}_R(M_\alpha, N_\beta)$$

What is the situation for

$$\text{Hom}_R\Big(\prod M_\alpha, \prod N_\beta\Big), \qquad \text{Hom}_R\Big(\prod M_\alpha, \sum \oplus N_\beta\Big)$$

and

$$\text{Hom}\Big(\sum \oplus M_\alpha, \sum \oplus N_\beta\Big)?$$

1-6-4. (i) The monomorphism $i : A \longrightarrow B$ is said to be **direct** when iA is a direct summand of B—that is, when there exists a module C such that $B = iA \oplus C$. Another way to express this is to say that the exact sequence $0 \longrightarrow A \xrightarrow{i} B$ **splits;** and this is true $\Longleftrightarrow$ there exists a homomorphism $g : B \longrightarrow A$ such that $g \circ i$ is the identity map on A; furthermore, in this situation, $B = \text{im}\, i \oplus \ker g$.

(ii) The epimorphism $j : B \longrightarrow C$ is said to be **direct** when $\ker j$ is a direct summand of B. Another way to express this is to say that the exact sequence $B \xrightarrow{j} C \longrightarrow 0$ **splits;** and this is true $\Leftrightarrow$ there exists a homomorphism $f : C \longrightarrow B$ such that $j \circ f$ is the identity map of C; furthermore, in this situation, $B = \operatorname{im} f \oplus \ker j$.

(iii) Consider the short exact sequence

$$0 \longrightarrow A \longrightarrow B \longrightarrow C \longrightarrow 0,$$

then $0 \longrightarrow A \longrightarrow B$ splits $\Leftrightarrow$ $B \longrightarrow C \longrightarrow 0$ splits—and, in this situation, we say that the short exact sequence **splits**.

(iv) Consider the sequence $0 \longrightarrow M_1 \xrightarrow{f_1} M \xrightarrow{g_2} M_2 \longrightarrow 0$; it is exact and splits $\Leftrightarrow$ there exist maps $f_2 : M_2 \longrightarrow M$ and $g_1 : M \longrightarrow M_1$ such that $\{M_i \xrightarrow{f_i} M \xrightarrow{g_i} M_i\}$, $i = 1, 2$, is a complete representation of M as a direct sum.

1-6-5. (i) A module P is said to be **projective** if for any diagram

$$\begin{array}{ccccc} & & P & & \\ & \swarrow & \downarrow f & & \\ B & \xrightarrow{g} & C & \longrightarrow & 0 \end{array}$$

with the row exact there exists a homomorphism $h : P \longrightarrow B$ such that $g \circ h = f$—that is, such that the triangle commutes. This condition may be replaced by the requirement that for any diagram

$$\begin{array}{ccc} & & P \\ & & \downarrow f \\ B & \xrightarrow{g} & C \end{array}$$

in which $f(P) \subset g(B)$ there exists a homomorphism $h : P \longrightarrow B$ such that $g \circ h = f$.

(ii) If $0 \longrightarrow A \xrightarrow{i} B \xrightarrow{j} C$ is any exact sequence and P is any module then

$$0 \longrightarrow \operatorname{Hom}_R(P, A) \xrightarrow{(1,i)} \operatorname{Hom}_R(P, B) \xrightarrow{(1,j)} \operatorname{Hom}_R(P, C)$$

is an exact sequence of abelian groups. Moreover, P is projective

$\Longleftrightarrow$ for any short exact sequence $0 \longrightarrow A \xrightarrow{i} B \xrightarrow{j} C \longrightarrow 0$ the sequence

$$0 \longrightarrow \operatorname{Hom}_R(P, A) \xrightarrow{(1,i)} \operatorname{Hom}_R(P, B) \xrightarrow{(1,j)} \operatorname{Hom}_R(P, C) \longrightarrow 0$$

is exact.

(iii) A direct sum $P = \sum_\alpha \oplus P_\alpha$ is projective $\Longleftrightarrow$ each direct summand P_α is projective.

(iv) A module P is projective $\Longleftrightarrow$ P is a direct summand of a free module.

(v) A module P is projective $\Longleftrightarrow$ every exact sequence $0 \longrightarrow A \longrightarrow B \longrightarrow P \longrightarrow 0$ splits.

(vi) Given exact sequences $0 \longrightarrow A' \longrightarrow A \longrightarrow A'' \longrightarrow 0$, $0 \longrightarrow M' \longrightarrow P' \longrightarrow A' \longrightarrow 0$, and $0 \longrightarrow M'' \longrightarrow P'' \longrightarrow A'' \longrightarrow 0$ with P' and P'' projective, then these may be imbedded in a commutative diagram

$$\begin{array}{ccccccccc}
 & & 0 & & 0 & & 0 & & \\
 & & \downarrow & & \downarrow & & \downarrow & & \\
0 & \longrightarrow & M' & \longrightarrow & M & \longrightarrow & M'' & \longrightarrow & 0 \\
 & & \downarrow & & \downarrow & & \downarrow & & \\
0 & \longrightarrow & P' & \longrightarrow & P & \longrightarrow & P'' & \longrightarrow & 0 \\
 & & \downarrow & & \downarrow & & \downarrow & & \\
0 & \longrightarrow & A' & \longrightarrow & A & \longrightarrow & A'' & \longrightarrow & 0 \\
 & & \downarrow & & \downarrow & & \downarrow & & \\
 & & 0 & & 0 & & 0 & &
\end{array}$$

in which the rows and columns are exact, P is projective, and the middle row splits.

1-6-6. (i) A free module is projective. Every module A is a quotient of a projective module—that is, A can be imbedded in an exact sequence $0 \longrightarrow B \longrightarrow P \longrightarrow A \longrightarrow 0$ with P projective.

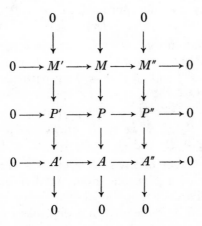

(ii) For $n > 1$ consider the ring $\mathbf{Z}_n = \mathbf{Z}/(n\mathbf{Z})$. For each divisor r of n, put $r' = n/r$, and consider the ideals $r\mathbf{Z}_n$ and $r'\mathbf{Z}_n$ in $\mathbf{Z}_n$ along with the exact sequence of $\mathbf{Z}_n$ modules

$$0 \longrightarrow r'\mathbf{Z}_n \longrightarrow \mathbf{Z}_n \longrightarrow r\mathbf{Z}_n \longrightarrow 0$$

This sequence splits $\Longleftrightarrow$ $(r, r') = 1 \Longleftrightarrow$ the $\mathbf{Z}_n$-module $r\mathbf{Z}_n$ is projective. Give examples of projective modules which are not free.

(iii) If $R = \mathbf{Z}$ (so that R-modules are the same as abelian groups), then a module is projective $\Longleftrightarrow$ it is free.

1-6-7. (i) An additive abelian group D is said to be **divisible** when for each $x \in D$ and every integer $n \neq 0$ there exists an element $y \in D$ such that $ny = x$—in other words, the map $n : D \longrightarrow D$ which takes $z \longrightarrow nz$ is an epimorphism. Suppose that B is a subgroup of the abelian group A and that D is a divisible group, then any homomorphism $f_0 : B \longrightarrow D$ can be extended to a homomorphism $f : A \longrightarrow D$.

(ii) If, in addition, $n : D \longrightarrow D$ is not a monomorphism for any $n > 1$, then given any $a \in A$ with $a \notin B$ and f_0 as above, it is possible to choose f so that $f(a) \neq 0$. In particular, this statement holds for $D = \mathbf{Q}/\mathbf{Z}$.

(iii) Let $B^\perp = \{f \in \operatorname{Hom}(A, D) \mid f(B) = 0\}$. Then for subgroups B_1 and B_2 of A we have, under the hypotheses of (ii), $B_1 < B_2 \Longrightarrow B_2^\perp < B_1^\perp$ and $B_1^\perp = B_2^\perp \Longrightarrow B_1 = B_2$.

(iv) A direct sum of divisible groups is divisible. A homomorphic image of a divisible group is divisible. The additive group of $\mathbf{Q}$ is divisible; thus the additive group of $\mathbf{Z}$ and any free abelian group can be imbedded in a divisible group. Finally, any abelian group can be imbedded in a divisible group.

1-6-8. (i) A module Q is said to be **injective** if for any diagram

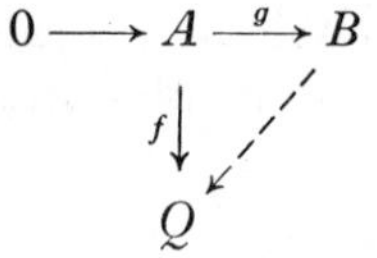

with the row exact there exists a homomorphism $h : B \longrightarrow Q$ such that the triangle commutes. This condition may be replaced by the requirement that for any diagram

$$\begin{array}{ccc} A & \xrightarrow{g} & B \\ {\scriptstyle f}\downarrow & & \\ Q & & \end{array}$$

in which $\ker g \subset \ker f$ there exists a homomorphism $h : B \longrightarrow Q$ such that $h \circ g = f$.

(ii) If $A \xrightarrow{i} B \xrightarrow{j} C \longrightarrow 0$ is any exact sequence and Q is any module, then

$$0 \longrightarrow \text{Hom}_R(C, Q) \xrightarrow{(j,1)} \text{Hom}_R(B, Q) \xrightarrow{(i,1)} \text{Hom}_R(A, Q)$$

is an exact sequence of abelian groups. Moreover, Q is injective $\Longleftrightarrow$ for any short exact sequence $0 \longrightarrow A \xrightarrow{i} B \xrightarrow{j} C \longrightarrow 0$ the sequence

$$0 \longrightarrow \text{Hom}_R(C, Q) \xrightarrow{(j,1)} \text{Hom}_R(B, Q) \xrightarrow{(i,1)} \text{Hom}_R(A, Q) \longrightarrow 0$$

is exact.

(iii) A direct product of modules is injective $\Longleftrightarrow$ each factor is injective.

(iv) If $R = \mathbf{Z}$, then our modules are simply abelian groups, and a module Q is injective $\Longleftrightarrow$ the abelian group Q is divisible. In particular, (see 1-6-7) the $\mathbf{Z}$-module $\mathbf{Q}/\mathbf{Z}$ is injective.

1-6-9. (i) An R-module Q is injective $\Longleftrightarrow$ for each left ideal I of R (with I viewed as a left R-module) there exists, for each $f \in \text{Hom}_R(I, Q)$ an element $q \in Q$ such that $f(a) = aq$ for all $a \in I$ $\Longleftrightarrow$ for each left ideal I of R, every $f \in \text{Hom}_R(I, Q)$ can be extended to an element of $\text{Hom}_R(R, Q)$ (where R is viewed as a left R-module).

(ii) If X is a left R-module and Y is an abelian group (i.e. a $\mathbf{Z}$-module; right or left is irrelevant), then $\text{Hom}(X, Y)$ may be made into a right R-module by

$$(fr)(x) = f(rx) \qquad f \in \text{Hom}(X, Y) \quad r \in R$$

If X is a right R-module, then Hom (X, Y) may be made into a left R-module by

$$(rf)(x) = f(xr) \qquad f \in \text{Hom}\,(X, Y) \quad r \in R$$

(In this connection, see (3-7-8)).

(iii) If the $\mathbf{Z}$-module Y is injective, then the R-module Hom (R, Y) is injective (here R may be viewed as either a left or a right R-module). Every R-module can be imbedded in an injective module; in fact, the injective R-module may be taken of form Hom (R, Y) where Y is an injective $\mathbf{Z}$-module.

(iv) The module Q is injective $\Leftrightarrow$ every exact sequence $0 \longrightarrow Q \longrightarrow B \longrightarrow C \longrightarrow 0$ splits.

(v) Given exact sequences $0 \longrightarrow A' \longrightarrow A \longrightarrow A'' \longrightarrow 0$, $0 \longrightarrow A' \longrightarrow Q' \longrightarrow N' \longrightarrow 0$ and $0 \longrightarrow A'' \longrightarrow Q'' \longrightarrow N'' \longrightarrow 0$ with Q' and Q'' injective, then these may be imbedded in a commutative diagram

$$\begin{array}{ccccccccc}
 & & 0 & & 0 & & 0 & & \\
 & & \downarrow & & \downarrow & & \downarrow & & \\
0 & \longrightarrow & A' & \longrightarrow & A & \longrightarrow & A'' & \longrightarrow & 0 \\
 & & \downarrow & & \downarrow & & \downarrow & & \\
0 & \longrightarrow & Q' & \longrightarrow & Q & \longrightarrow & Q'' & \longrightarrow & 0 \\
 & & \downarrow & & \downarrow & & \downarrow & & \\
0 & \longrightarrow & N' & \longrightarrow & N & \longrightarrow & N'' & \longrightarrow & 0 \\
 & & \downarrow & & \downarrow & & \downarrow & & \\
 & & 0 & & 0 & & 0 & &
\end{array}$$

in which the rows and columns are exact, Q is injective and the middle row splits.

1-6-10. Compute the contracting homotopies D_n in negative dimensions for the standard G-complex.

1-6-11. Suppose that G is a group, finite or infinite, and that A is a free G-module (with finite or infinite basis). Is $\hat{A}$ a free G-module? How does this affect the notion of a G-complex, and the definition of cohomology groups?

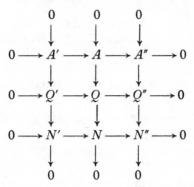

CHAPTER

II

Mappings of Cohomology Groups

Having defined the cohomology groups of G in A in Chapter I we turn in this chapter to an investigation of how changes in the group or in the module (or both) induce changes in the cohomology groups. For this, the notion of homomorphism of pairs is extremely useful. It leads, by specialization, to three important mappings of cohomology groups—namely, restriction, inflation, and conjugation. It also serves as a convenient tool for proving that cohomology groups are independent of the complex.

Another mapping of cohomology groups is the corestriction, or transfer. Although it cannot be defined via homomorphism of pairs, it has many properties in common with the other mappings.

With an eye to future applications, and for historical reasons too, we conclude this chapter with a discussion of explicit formulas for the various mappings of cohomology groups—all in terms of standard complexes.

2-1. HOMOMORPHISM OF PAIRS

2-1-1. Proposition. Let G and G' be finite groups, and let A and B be G-modules. Suppose that $\lambda : G' \longrightarrow G$ is a homomorphism; then

(1) A may be made into a G'-module by putting

$$\sigma' a = (\lambda\sigma')\, a \qquad \sigma' \in G', \quad a \in A$$

(2) Any G-homomorphism of A into B is also a G'-homomorphism; that is,

$$\mathrm{Hom}_G (A, B) \subset \mathrm{Hom}_{G'} (A, B)$$

Proof: Straightforward. ▮

Let us introduce the symbol (G, A) (and call it a **pair**) to signify that A is a G-module. Suppose that (G', A') is another pair. If we have a homomorphism $\lambda : G' \longrightarrow G$ (so that A becomes a G'-module) and a G'-homomorphism $f : A \longrightarrow A'$ then the composite object (λ, f) is called a **homomorphism of pairs**; symbolically, we write

$$(\lambda, f) : (G, A) \longrightarrow (G', A')$$

2-1-2. Proposition. Suppose that $(\lambda, f) : (G, A) \longrightarrow (G', A')$ and $(\lambda', f') : (G', A') \longrightarrow (G'', A'')$ are homomorphisms of pairs, then $(\lambda \circ \lambda', f' \circ f) : (G, A) \longrightarrow (G'', A'')$ is a homomorphism of pairs; symbolically, we write

$$(\lambda', f') \circ (\lambda, f) = (\lambda \circ \lambda', f' \circ f)$$

Proof: Straightforward. ▮

We shall show that if $(\lambda, f) : (G, A) \longrightarrow (G', A')$ is a homomorphism of pairs, $(X, \partial, -1)$ is any G-complex, and $(X', \partial', -1)$ is any G'-complex then (for each n) there is determined in a canonical way, a homomorphism of cohomology groups

$$(\lambda, f)_{X,X'} : H^n(G, A, X) \longrightarrow H^n(G', A', X')$$

The basic step in this process consists of the contruction of maps $\Lambda_n : X'_n \longrightarrow X_n$ associated with λ. For this it is convenient to make some preliminary remarks.

2-1-3. Proposition. Let A and B be G-modules, then the trace of a homomorphism of A into B is a G-homomorphism—that is

$$S : \operatorname{Hom}(A, B) \longrightarrow \operatorname{Hom}_G(A, B) = (\operatorname{Hom}(A, B))^G$$

where $S(f) = \sum_{\sigma \in G} f^\sigma$.

Proof: For every $\sigma \in G$, $\sigma S = S = S\sigma \in \mathbf{Z}[G]$, so that for any G-module A we have $S : A \longrightarrow A^G$ ($Sa = \sum_{\sigma \in G} \sigma a$, $a \in A$). Since $\operatorname{Hom}(A, B)$ is a G-module, S maps $\operatorname{Hom}(A, B)$ into $(\operatorname{Hom}(A, B))^G$. Now, if $f \in \operatorname{Hom}(A, B)$ and $\sigma \in G$, then

$$f^\sigma = f \Longleftrightarrow \sigma f \sigma^{-1} = f \Longleftrightarrow \sigma f = f \sigma,$$

so that $(\operatorname{Hom}(A, B))^G = \operatorname{Hom}_G(A, B)$. ▌

A G-module A is said to be **G-regular** if there exists a $\mathbf{Z}$-submodule B of A such that

$$A = \sum_{\sigma \in G} \oplus \sigma B$$

2-1-4. Proposition. (1) If A is G-free, then it is G-regular.

(2) If $A = \sum_{\sigma \in G} \oplus \sigma B$ is G-regular and B is $\mathbf{Z}$-free, then A is G-free.

(3) If A is G-regular, then the identity map 1_A of A is the trace of an endomorphism of A; in more detail, there exists $\pi \in \operatorname{Hom}(A, A)$ such that

$$\pi^\sigma \circ \pi^\tau = \delta_{\sigma,\tau} \pi^\tau \qquad \text{and} \qquad 1_A = \sum_{\sigma \in G} \pi^\sigma = S(\pi)$$

Proof: (1) and (2) are straightforward. As for (3), write $A = \sum_{\sigma \in G} \oplus \sigma B$ and let π be the projection of A on B. Thus, if $a \in A$ is expressed uniquely in the form $a = \sum_{\sigma \in G} \sigma b_\sigma$ with $b_\sigma \in B$ then $\pi(a) = b_1$. For $\tau \in G$, $\pi^\tau(a) = \tau \pi \tau^{-1}(\sum_\sigma \sigma b_\sigma) = \tau b_\tau$, so that π^τ is the projection on τB, and π satisfies the requirements. ▌

In dealing with the mappings A_n we shall refer to the following diagram (which is based on any G'-complex X' and any G-complex X) :

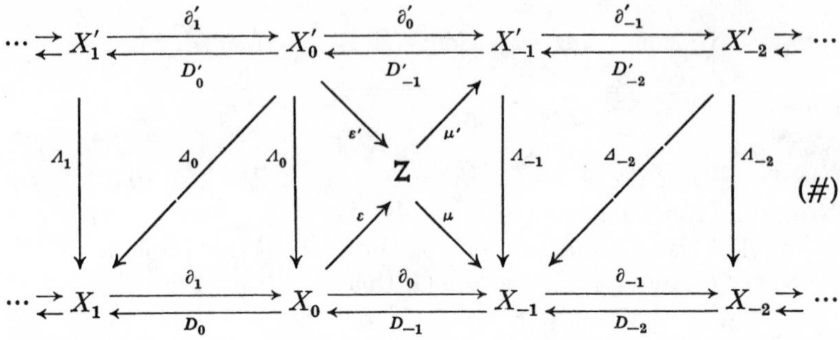

(#)

Here $\partial', \varepsilon', \mu'$ are G'-homomorphisms; $\partial, \varepsilon, \mu$ are G-homomorphisms, and, hence (when the X_n are viewed as G'-modules) they are also G'-homomorphisms. The mappings D and D' are $\mathbf{Z}$-homomorphisms; they are contracting homotopies (which exist, by (1-4-6)). The mappings Λ_n and Δ_n will be introduced in the course of the discussion. We shall write $S = \sum_{\sigma \in G} \sigma$ and $S' = \sum_{\sigma' \in G'} \sigma'$ for the G and G'-traces, respectively.

2-1-5. Lemma. For any homomorphism $\lambda : G' \longrightarrow G$ there exist G'-homomorphisms Λ_n for $n \geqslant 0$ such that $\varepsilon \circ \Lambda_0 = \varepsilon'$ and $\partial_{n+1} \circ \Lambda_{n+1} = \Lambda_n \circ \partial'_{n+1}$.

Proof: Since X'_0 is G'-free and ε is onto, there exists a G'-homomorphism $\Lambda_0 : X'_0 \longrightarrow X_0$ such that $\varepsilon \circ \Lambda_0 = \varepsilon'$. Note that $\partial_0 \Lambda_0 \partial'_1 = \mu\varepsilon\Lambda_0\partial'_1 = \mu\varepsilon'\partial'_1 = 0$. Now, suppose inductively that Λ_n is defined and that $\partial_n \Lambda_n \partial'_{n+1} = 0$. Since X'_{n+1} is G'-free it is G'-regular, so by (2-1-4) there exists $\pi'_{n+1} \in \operatorname{Hom}(X'_{n+1}, X'_{n+1})$ such that $S'(\pi'_{n+1}) = 1$ (the identity map on X'_{n+1}). Put

$$\Lambda_{n+1} = S'(D_n \Lambda_n \partial'_{n+1} \pi'_{n+1})$$

Thus, Λ_{n+1} is a G'-homomorphism, and using the fact that G'-homomorphisms slip through the trace S', we have

$$\begin{aligned}
\partial_{n+1}\Lambda_{n+1} &= S'(\partial_{n+1} D_n \Lambda_n \partial'_{n+1} \pi'_{n+1}) \\
&= S'(\Lambda_n \partial'_{n+1} \pi'_{n+1}) - S'(D_{n-1}\partial_n \Lambda_n \partial'_{n+1} \pi'_{n+1}) \\
&\qquad \text{(since } \partial_{n+1} D_n + D_{n-1}\partial_n = 1) \\
&= \Lambda_n \partial'_{n+1} S'(\pi'_{n+1}) = \Lambda_n \partial'_{n+1}
\end{aligned}$$

Finally, $\partial_{n+1}\Lambda_{n+1}\partial'_{n+2} = \Lambda_n\partial'_{n+1}\partial'_{n+2} = 0$, so the induction hypothesis is satisfied. ▮

Under ordinary circumstances it is impossible to continue such a process and define Λ_n for $n < 0$; however, we have :

2-1-6. Lemma. If $\lambda : G' \longrightarrow G$ is a monomorphism, then there exist G'-homomorphisms Λ_n such that $\varepsilon \circ \Lambda_0 = \varepsilon'$ and $\partial_{n+1} \circ \Lambda_{n+1} = \Lambda_n \circ \partial'_{n+1}$ for all n.

Proof: We have Λ_0 such that $\varepsilon \circ \Lambda_0 = \varepsilon'$ and $\partial_0\Lambda_0\partial'_1 = 0$. Suppose, inductively, that for $n \leqslant 0$ we have Λ_n with $\partial_n\Lambda_n\partial'_{n+1} = 0$. Since G' may be viewed as a subgroup of G, it follows that each X_n is G'-free, and hence, G'-regular; thus for each n there exists $\pi_n \in \text{Hom}\,(X_n\,,\, X_n)$ such that $S'(\pi_n) = 1$. It is now easy to verify that

$$\Lambda_{n-1} = S'(\pi_{n-1}\partial_n\Lambda_n D'_{n-1})$$

satisfies the requirements. It may be noted in passing that it follows easily that $\mu = \Lambda_{-1} \circ \mu'$. ▮

Let us extend the meaning of the term G-complex to include either our customary full G-complex X or the positive part of a full G-complex—which we shall, for present purposes, call a **half G-complex** and write as

$$\begin{array}{l} \cdots \longrightarrow X_2 \xrightarrow{\partial_2} X_1 \xrightarrow{\partial_1} X_0 \xrightarrow{0} (0) \longrightarrow (0) \longrightarrow \cdots \\ \qquad\qquad\qquad\qquad\qquad\quad \searrow^{\varepsilon} \\ \qquad\qquad\qquad\qquad\qquad\qquad\quad \mathbf{Z} \\ \qquad\qquad\qquad\qquad\qquad\qquad\qquad \searrow \\ \qquad\qquad\qquad\qquad\qquad\qquad\qquad\quad 0 \end{array}$$

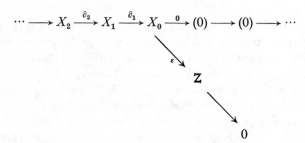

(Note that the row is exact except at X_0.) Then the differential graded group $(\text{Hom}_G\,(X, A), \delta, +1)$ is the usual one when the full G-complex is used; but when the half G-complex is used, then for $n < 0$, $\text{Hom}_G\,(X_n\,, A) = (0)$ and $\delta_n = 0$. Consequently, in the half G-complex case, the derived groups are the customary cohomology groups $H^n(G, A, X)$ for $n > 0$—while, for $n = 0$ the derived group is $H^0(G, A, X) = \mathscr{Z}^0(G, A, X)$ (the group of 0-cocycles), and for $n < 0$, $H^n(G, A, X) = (0)$. For all practical purposes, these groups $H^n(G, A, X)$ for $n < 0$ may be ignored—and in all statements which apply to them there will be no harm in taking the view that they are not defined.

At this point, the notation $H^n(G, A, X)$ has several possible meanings; eventually (that is, beginning in Section 2-3) the notation $H^n(G, A, X)$ will be reserved exclusively for the groups computed via the full complex.

Consider a homomorphism $\lambda : G' \longrightarrow G$. If λ is not a monomorphism then, given full complexes X' and X, the mappings Λ_n exist, in general, only for $n \geqslant 0$; therefore, in such a situation we can deal only with half-complexes, and the diagram (#) is replaced by

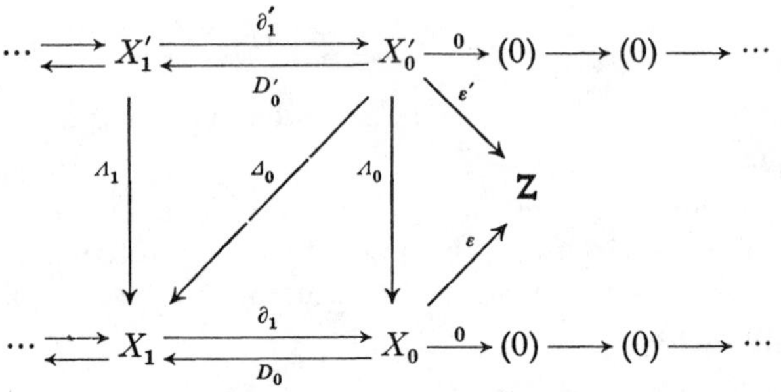

(##)

On the other hand, if λ is a monomorphism, then the mappings Λ_n exist for all n; thus, in this situation, we have the choice of dealing with full complexes X', X, and the associated diagram (#) or with half complexes and the associated diagram (##).

In the forthcoming discussion, we shall deal with these **three possibilities simultaneously**, and indicate this by saying that we are dealing with complexes in the "extended sense." In the same vein, a statement about $H^n(G, A, X)$ for all n refers to all $n \in \mathbf{Z}$ for the case of a full complex and to all $n \geqslant 0$ in the case of a half complex.

2-1-7. Lemma. If we put $\Lambda = \sum \oplus \Lambda_n$ then any two Λ's are G'-homotopic.

Proof: Let Λ and Λ' be two such Λ's and put $\Xi = \Lambda - \Lambda'$. In all cases, Ξ is an admissible map of differential graded groups (with operator group G'); in particular, $\partial_{n+1}\Xi_{n+1} = \Xi_n\partial'_{n+1}$. We shall construct G'-homomorphisms $\Delta_n : X'_n \longrightarrow X_{n+1}$ such that

$$\Xi_n = \partial_{n+1}\Delta_n + \Delta_{n-1}\partial'_n \qquad (*)$$

First, let us put, in all cases,

$$\Delta_0 = S'(D_0\Xi_0\pi'_0) \qquad \text{and} \qquad \Delta_{-1} = 0$$

where the notation is that of (2-1-5) and (2-1-6). With ∂_0 and D_{-1} coming from the full complex in all cases, we have

$$\begin{aligned}\partial_1\Delta_0 = S'(\partial_1 D_0\Xi_0\pi'_0) &= S'(\Xi_0\pi'_0) - S'(D_{-1}\partial_0\Xi_0\pi'_0)\\ &= \Xi_0 S'(\pi'_0) - S'(D_{-1}\mu\varepsilon\Xi_0\pi'_0)\\ &= \Xi_0 \qquad\qquad (\text{since } \varepsilon\Xi_0 = 0)\end{aligned}$$

This takes care of $n = 0$ in $(*)$. For $n > 0$, we proceed by induction. Thus, assume that $(*)$ holds for n, and put

$$\Delta_{n+1} = S'\{D_{n+1}(\Xi_{n+1} - \Delta_n\partial'_{n+1})\,\pi'_{n+1}\} \qquad n > 0$$

Then using the fact that

$$\begin{aligned}\partial_{n+1}(\Xi_{n+1} - \Delta_n\partial'_{n+1}) &= \partial_{n+1}\Xi_{n+1} - \partial_{n+1}\Delta_n\partial'_{n+1}\\ &= \Xi_n\partial'_{n+1} - \partial_{n+1}\Delta_n\partial'_{n+1}\\ &= \partial_{n+1}\Delta_n\partial'_{n+1} + \Delta_{n-1}\partial'_n\partial'_{n+1} - \partial_{n+1}\Delta_n\partial'_{n+1}\\ &= 0\end{aligned}$$

it follows that

$$\begin{aligned}\partial_{n+2}\Delta_{n+1} &= S'\{(\Xi_{n+1} - \Delta_n\partial'_{n+1})\,\pi'_{n+1}\} - S'\{D_n\partial_{n+1}(\Xi_{n+1} - \Delta_n\partial'_{n+1})\,\pi'_{n+1})\\ &= \Xi_{n+1} - \Delta_n\partial'_{n+1}\end{aligned}$$

so that $(*)$ holds for $n + 1$.

If we are dealing with either of the half-complex cases, put

$\Delta_n = 0$ for $n < -1$, and we are finished. If $\lambda : G' \longrightarrow G$ is a monomorphism and we are dealing with the full complex case, put

$$\Delta_{n-1} = S'\{\pi_n(\Xi_n - \partial_{n+1}\Delta_n)\, D'_{n-1}\} \qquad n < 0$$

It is then easy to verify by induction that $(*)$ holds for all $n < 0$. ▮

2-1-8. Theorem. Suppose that $(\lambda, f) : (G, A) \longrightarrow (G', A')$ is a homomorphism of pairs; let X be any G-complex, and let X' be any G'-complex (both in the extended sense). Then there is determined a homomorphism

$$(\lambda, f)_{X,X'} : H^n(G, A, X) \longrightarrow H^n(G', A', X')$$

for all n. Moreover, the following properties hold:

(1) $(1, 1)_{X,X}$ is the identity map.

(2) If, in addition, $(\lambda', f') : (G', A') \longrightarrow (G'', A'')$ is a homomorphism of pairs, and X'' is any G''-complex, then

$$(\lambda\lambda', f'f)_{X,X''} = (\lambda', f')_{X',X''} \circ (\lambda, f)_{X,X'}$$

(3) If, in addition, $(\lambda, g) : (G, A) \longrightarrow (G', A')$ is a homomorphism of pairs, then

$$(\lambda, f + g)_{X,X'} = (\lambda, f)_{X,X'} + (\lambda, g)_{X,X'}$$

Proof: Let $\Lambda : X' \longrightarrow X$ be a G'-homomorphism, which commutes with boundaries, associated with λ. We have therefore, with the notation of (1-4-3), a homomorphism of $\mathrm{Hom}_{G'}(X, A)$ into $\mathrm{Hom}_{G'}(X', A')$. Since $\mathrm{Hom}_G(X, A) \subset \mathrm{Hom}_{G'}(X, A)$ and $(\Lambda, f) \circ \delta = (\Lambda, f) \circ (\partial, 1) = (\partial\Lambda, f) = (\Lambda\partial', f) = (\partial', 1) \circ (\Lambda, f) = \delta' \circ (\Lambda, f)$, the mapping

$$(\Lambda, f) : \mathrm{Hom}_G(X, A) \longrightarrow \mathrm{Hom}_{G'}(X', A')$$

is an admissible map of differential graded groups. In more detail, we have the commutative diagram

$$
\begin{array}{ccccc}
\cdots \longrightarrow & \mathrm{Hom}_G(X_{n-1}, A) & \xrightarrow{\delta_n = (\partial_n, 1)} & \mathrm{Hom}_G(X_n, A) & \longrightarrow \\
& \Big\downarrow{\scriptstyle (\Lambda_{n-1}, f)} & & \Big\downarrow{\scriptstyle (\Lambda_n, f)} & \\
\cdots \longrightarrow & \mathrm{Hom}_{G'}(X'_{n-1}, A') & \xrightarrow{\delta'_n = (\partial'_n, 1)} & \mathrm{Hom}_{G'}(X'_n, A') & \longrightarrow
\end{array}
$$

$$
\begin{array}{ccc}
\xrightarrow{\delta_{n+1}} & \mathrm{Hom}_G(X_{n+1}, A) & \longrightarrow \cdots \\
& \Big\downarrow{\scriptstyle (\Lambda_{n+1}, 1)} & \\
\xrightarrow{\delta'_{n+1}} & \mathrm{Hom}\text{-}\ (X'_{n+1}, A') & \longrightarrow \cdots
\end{array}
$$

The homomorphism of cohomology groups induced by (Λ, f) (that is, $(\Lambda, f)_*$ in the notation of (1-1-1)) is denoted by $(\lambda, f)_{X,X'}$.

Furthermore, $(\lambda, f)_{X,X'}$ does not depend on the choice of Λ. In fact, if $\Lambda' : X' \longrightarrow X$ is another G'-homomorphism associated with λ, then Λ and Λ' are G'-homotopic—that is, $\Lambda - \Lambda' = \partial\Delta + \Delta\partial'$ with $\Delta : X'_n \longrightarrow X_{n+1}$. It follows that (Λ, f) and (Λ', f) are homotopic. (Homotopies were treated in Section 1-3 for homology only, but all statements may be transferred to cohomology.) In more detail, for the map

$$(\Delta, f) : \mathrm{Hom}_G(X_{n+1}, A) \longrightarrow \mathrm{Hom}_{G'}(X'_n, A)$$

we have

$$(\Lambda, f) - (\Lambda', f) = (\Delta, f)\,\delta + \delta'(\Delta, f)$$

so that $(\Lambda, f)_* = (\Lambda', f)_*$.

If $G' = G$, $A' = A$, $\lambda = 1$, $f = 1$, and $X' = X$, then we may take $\Lambda = 1$, so that (Λ, f) is the identity, and then so is $(1, 1)_{X,X}$. This proves (1).

As for (2), suppose that $\Lambda' : X'' \longrightarrow X'$ corresponds to λ'. Then $\Lambda\Lambda' : X'' \longrightarrow X$ corresponds to $\lambda\lambda'$—that is, it is a G''-homomorphism which maps $X''_n \longrightarrow X_n$, and such that $\Lambda\Lambda'\partial'' = \partial\Lambda\Lambda'$, $\varepsilon\Lambda_0\Lambda'_0 = \varepsilon''$. From $(\Lambda\Lambda', f'f) = (\Lambda', f') \circ (\Lambda, f)$ and (1-1-1) it follows that $(\lambda\lambda', f'f)_{X,X''} = (\lambda', f')_{X',X''} \circ (\lambda, f)_{X,X'}$.

Finally, (3) is immediate from $(\Lambda, f+g) = (\Lambda, f) + (\Lambda, g)$. This completes the proof. ▮

2-1-9. Theorem. Let $\lambda : G' \longrightarrow G$ be a homomorphism and consider the diagram

$$\begin{array}{ccccccccc}
0 & \longrightarrow & A & \xrightarrow{i} & B & \xrightarrow{j} & C & \longrightarrow & 0 \\
 & & {\scriptstyle f}\big\downarrow & & {\scriptstyle g}\big\downarrow & & {\scriptstyle h}\big\downarrow & & \\
0 & \longrightarrow & A' & \xrightarrow{i'} & B' & \xrightarrow{j'} & C' & \longrightarrow & 0
\end{array}$$

Suppose that the top row is an exact G-sequence, that the bottom row is an exact G'-sequence, that

$$(\lambda, f) : (G, A) \longrightarrow (G', A') \qquad (\lambda, g) : (G, B) \longrightarrow (G', B')$$

$$(\lambda, h) : (G, C) \longrightarrow (G', C')$$

are homomorphisms of pairs, and that the diagram is commutative. Let X and X' be any G-complex and any G'-complex (in the extended sense), respectively. Then the following diagram is commutative and has exact rows :

$$\begin{array}{ccccccc}
\cdots \xrightarrow{\delta_*} & H_X^n(G, A) & \xrightarrow{(1,i)_{X,X}} & H_X^n(G, B) & \xrightarrow{(1,j)_{X,X}} & H_X^n(G, C) & \longrightarrow \\
 & {\scriptstyle (\lambda,f)_{X,X'}}\big\downarrow & & {\scriptstyle (\lambda,g)_{X,X'}}\big\downarrow & & {\scriptstyle (\lambda,h)_{X,X'}}\big\downarrow & \\
\cdots \xrightarrow{\delta'_*} & H_{X'}^n(G', A') & \xrightarrow{(1,i')_{X',X'}} & H_X^n(G', B') & \xrightarrow{(1,j')_{X',X'}} & H_{X'}^n(G', C') & \longrightarrow
\end{array}$$

$$\begin{array}{ccc}
\xrightarrow{\delta_*} & H_X^{n+1}(G, A) & \longrightarrow \cdots \\
 & {\scriptstyle (\lambda,f)_{X,X'}}\big\downarrow & \\
\xrightarrow{\delta'_*} & H_{X'}^{n+1}(G', A') & \longrightarrow \cdots
\end{array}$$

Furthermore, if $\bar{X}$ is another G-complex and $\bar{X}'$ is another G'-complex, then the diagram of cohomology groups which they determine may be connected to the one above via the maps $(1, 1)_{X,\bar{X}}$ and $(1, 1)_{X',\bar{X}'}$, and the resulting three dimensional diagram is commutative.

Proof: Consider the diagram

$$\begin{array}{ccccccccc} 0 \longrightarrow & \mathrm{Hom}_G(X, A) & \xrightarrow{(1,i)} & \mathrm{Hom}_G(X, B) & \xrightarrow{(1,j)} & \mathrm{Hom}_G(X, C) & \longrightarrow 0 \\ & \downarrow{\scriptstyle (A,f)} & & \downarrow{\scriptstyle (A,g)} & & \downarrow{\scriptstyle (A,h)} & \\ 0 \longrightarrow & \mathrm{Hom}_{G'}(X', A') & \xrightarrow{(1,i')} & \mathrm{Hom}_{G'}(X', B') & \xrightarrow{(1,j')} & \mathrm{Hom}_{G'}(X', C') & \longrightarrow 0 \end{array}$$

All the groups are differential graded groups of cohomology type, all the maps are admissible, and the diagram is commutative. Moreover, the top row is exact at $\mathrm{Hom}_G(X, C)$ in virtue of the fact that X is G-free. The top row is also exact at $\mathrm{Hom}_G(X, A)$ and $\mathrm{Hom}_G(X, B)$ (for this, no hypotheses on X are needed). In the same way, the bottom row is exact. An application of (1-1-4), with obvious changes of notation, yields the desired result.

As for the remaining part, it is immediate from the above and part (2) of (2-1-8). This completes the proof. ▮

In connection with (2-1-9), it is important to understand clearly (the details are obvious and are left to the reader) how the coboundary maps $\delta_* : H^n(G, C) \longrightarrow H^{n+1}(G, A)$ arise from the exact sequence

$$0 \longrightarrow \mathrm{Hom}_G(X, A) \xrightarrow{(1,i)} \mathrm{Hom}_G(X, B) \xrightarrow{(1,j)} \mathrm{Hom}_G(X, C) \longrightarrow 0$$

2-2. INDEPENDENCE OF THE COMPLEX

As a first application of the results on homomorphisms of pairs, we have, with our conventions about the meaning of G-complex and of $H^n(G, A, X)$ still in force :

2-2-1. Proposition. The cohomology groups of G in A are **independent of the complex**; this means that if X and $\bar{X}$ are any two G-complexes, then, for each n, the mapping

$$1_{X,\bar{X}} = (1, 1)_{X,\bar{X}} : H^n_X(G, A) \longrightarrow H^n_{\bar{X}}(G, A)$$

is an isomorphism, and its inverse is $1_{\bar{X},X}$. Such an isomorphism, $1_{X,\bar{X}}$, will be called **canonical**.

Proof: Immediate from (2-1-8). ∎

One may also define independence of the complex by imposing additional requirements on the behavior of the canonical maps $1_{X,\bar{X}}$ with respect to certain other mappings; however, we prefer to state these properties separately as they arise.

It may be noted at this point that the augmentation in the definition of G-complex serves, along with other hypotheses, to guarantee that the cohomology groups are independent of the complex.

2-2-2. Remark. In virtue of this result, we shall view $H^n(G, A)$, for each n, as an "abstract" group which has many concrete realizations—namely, the groups $H_X^n(G, A)$ for each choice of G-complex X—all of which are isomorphic to each other in a canonical way. Given this point of view, we shall also consider "abstract" homomorphisms (with concrete realizations) between abstract cohomology groups. Thus, by a homomorphism

$$\Psi : H^n(G_1, A_1) \longrightarrow H^n(G_2, A_2)$$

we mean that for any G_1-complex X_1 and any G_2-complex X_2 there exists a homomorphism

$$\Psi_{X_1,X_2} : H_{X_1}^n(G_1, A_1) \longrightarrow H_{X_2}^n(G_2, A_2)$$

(which is a concrete realization of Ψ) such that if $\bar{X}_1$ and $\bar{X}_2$ are any G_1 and G_2 complexes, respectively, then the following diagram commutes :

$$\begin{array}{ccc} H_{X_1}^n(G_1, A_1) & \xrightarrow{\Psi_{X_1,X_2}} & H_{X_2}^n(G_2, A_2) \\ {\scriptstyle 1_{X_1,\bar{X}_1}}\downarrow & & \downarrow{\scriptstyle 1_{X_2,\bar{X}_2}} \\ H_{\bar{X}_1}^n(G_1, A_1) & \xrightarrow{\Psi_{\bar{X}_1,\bar{X}_2}} & H_{\bar{X}_2}^n(G_2, A_2) \end{array}$$

This commutative diagram leads us to say that Ψ is **independent of the complex**—since Ψ is independent of the choice of complex in the same sense that this is true for the cohomology groups themselves. An obvious illustration of a way in which such maps Ψ can arise is given by the following :

2-2-3. Proposition. If $(\lambda, f) : (G, A) \longrightarrow (G', A')$ is a homomorphism of pairs, then, for each n, there exists a homomorphism

$$(\lambda, f)_* : H^n(G, A) \longrightarrow H^n(G', A')$$

We shall also write

$$f_* = (1, f)_* : H^n(G, A) \longrightarrow H^n(G', A')$$

Proof: This means that if X and $\bar{X}$ are any G-complexes, and X' and $\bar{X}'$ are any G'-complexes then the following diagram commutes :

$$\begin{array}{ccc} H^n_X(G, A) & \xrightarrow{(\lambda,f)_{X,X'}} & H^n_{X'}(G', A') \\ {\scriptstyle 1_{X,\bar{X}}}\downarrow & & {\scriptstyle 1_{X',\bar{X}'}}\downarrow \\ H^n_{\bar{X}}(G, A) & \xrightarrow{(\lambda,f)_{\bar{X},\bar{X}'}} & H^n_{\bar{X}'}(G', A') \end{array}$$

and this is immediate from (2-1-8).

If $\lambda = 1$ (so, in particular, $G = G'$), then the same G-complex X may be used, and in this situation the map associated with $\lambda = 1$ may be taken as $\Lambda = 1$; thus $f_* = (1, f)_*$ is induced by the cochain map

$$(1, f) : \mathrm{Hom}_G (X, A) \longrightarrow \mathrm{Hom}_G (X, A') \qquad \blacksquare$$

2-2-4. Proposition. If $0 \longrightarrow A \xrightarrow{i} B \longrightarrow C \xrightarrow{j} 0$ is an exact G-sequence, then we have an exact sequence

$$\cdots \longrightarrow H^n(G, A) \xrightarrow{i_*} H^n(G, B) \xrightarrow{j_*} H^n(G, C) \xrightarrow{\delta_*} H^{n+1}(G, A) \longrightarrow \cdots$$

which is independent of the complex.

Proof: By this we mean that if X and $\bar{X}$ are any G-complexes, then the following diagram is commutative and has exact rows :

$$\begin{array}{ccccccccc}
\cdots \longrightarrow & H^n_X(G, A) & \xrightarrow{(1,i)_{X.X}} & H^n_X(G, B) & \xrightarrow{(1,j)_{X.X}} & H^n_X(G, C) & \xrightarrow{\delta_*} & H^{n+1}_X(G, A) & \longrightarrow \cdots \\
& \downarrow{\scriptstyle 1_{X.\mathfrak{X}}} & & \downarrow{\scriptstyle 1_{X.\mathfrak{X}}} & & \downarrow{\scriptstyle 1_{X.\mathfrak{X}}} & & \downarrow{\scriptstyle 1_{X.\mathfrak{X}}} & \\
\cdots \longrightarrow & H^n_{\mathfrak{X}}(G, A) & \xrightarrow{(1,i)_{X.\mathfrak{X}}} & H^n_{\mathfrak{X}}(G, B) & \xrightarrow{(1,j)_{\mathfrak{X}.\mathfrak{X}}} & H^n_{\mathfrak{X}}(G, C) & \xrightarrow{\delta_*} & H^{n+1}_{\mathfrak{X}}(G, A) & \longrightarrow \cdots
\end{array}$$

This is immediate from (2-1-9). Because (2-2-3) takes care of the commutativities in which δ_* does not appear, this is sometimes stated as : the canonical maps commute with the coboundaries arising from exact sequences. ∎

If cohomology is done with respect to the full complex, then this result combined with (1-5-6) and (1-5-9) leads to an exact sequence

$$\cdots \longrightarrow H^{-2}(G, B) \longrightarrow H^{-2}(G, C) \longrightarrow \frac{A_S}{IA} \longrightarrow \frac{B_S}{IB} \longrightarrow \frac{C_S}{IC} \longrightarrow \frac{A^G}{SA} \longrightarrow \frac{B^G}{SB} \longrightarrow \frac{C^G}{SC} \longrightarrow H^1(G, A) \longrightarrow \cdots \tag{\#}$$

where the explicit meaning of the maps in dimensions -1 and 0 remains to be discussed. On the other hand, if cohomology is done with respect to the half complex, the resulting exact sequence is

$$0 \longrightarrow A^G \longrightarrow B^G \longrightarrow C^G \longrightarrow H^1(G, A) \longrightarrow H^1(G, B) \longrightarrow \cdots$$

2-2-5. Proposition. Let the hypotheses be as in (2-1-9); then the following diagram is commutative and has exact rows :

$$\begin{array}{ccccccccc}
\cdots \to & H^n(G, A) & \xrightarrow{i_*} & H^n(G, B) & \xrightarrow{j_*} & H^n(G, C) & \xrightarrow{\delta_*} & H^{n+1}(G, A) & \to \cdots \\
& \downarrow{\scriptstyle (\lambda,f)_*} & & \downarrow{\scriptstyle (\lambda,g)_*} & & \downarrow{\scriptstyle (\lambda,h)_*} & & \downarrow{\scriptstyle (\lambda,f)_*} & \\
\cdots \to & H^n(G', A') & \xrightarrow{i'_*} & H^n(G', B') & \xrightarrow{j'_*} & H^n(G', C') & \xrightarrow{\delta'_*} & H^{n+1}(G', A') & \to \cdots
\end{array}$$

Proof: Apply (2-1-9). This result may be stated loosely as: homomorphisms of pairs commute with the coboundaries arising from exact sequences. ▮

2-2-6. Remark. Now, let us consider the cohomology groups $H^0(G, A)$ and $H^{-1}(G, A)$. It is understood, of course, that here we deal with full G-complexes. In (1-5-6) and (1-5-9) we have encountered isomorphisms $f \longrightarrow f[\cdot]$ and $g \longrightarrow g\langle\cdot\rangle$ of $\mathscr{C}^0(G, A, X^s)$ onto A and of $\mathscr{C}^{-1}(G, A, X^s)$ onto A, respectively. Let us denote the inverse isomorphisms by $\dot{\kappa}$ and $\dot{\eta}$, respectively. (When it becomes necessary to emphasize that the standard complex X^s is being used, we shall write $\dot{\kappa}_s$ and $\dot{\eta}_s$.) Then, as seen in the proofs of (1-5-6) and (1-5-9), $\dot{\kappa}$ and $\dot{\eta}$ provide isomorphisms:

$$\dot{\kappa} : \begin{cases} A \rightarrowtail\!\!\!\twoheadrightarrow \mathscr{C}^0_{X^s}(G, A) \\ A^G \rightarrowtail\!\!\!\twoheadrightarrow \mathscr{Z}^0_{X^s}(G, A) \\ SA \rightarrowtail\!\!\!\twoheadrightarrow \mathscr{B}^0_{X^s}(G, A) \end{cases}$$

$$\dot{\eta} : \begin{cases} A \rightarrowtail\!\!\!\twoheadrightarrow \mathscr{C}^{-1}_{X^s}(G, A) \\ A_S \rightarrowtail\!\!\!\twoheadrightarrow \mathscr{Z}^{-1}_{X^s}(G, A) \\ IA \rightarrowtail\!\!\!\twoheadrightarrow \mathscr{B}^{-1}_{X^s}(G, A) \end{cases}$$

To be explicit—if $a \in A$, then $\dot{\kappa}a \in \mathscr{C}^0_{X^s}$ and $\dot{\eta}a \in \mathscr{C}^{-1}_{X^s}$ are given by

$$(\dot{\kappa}a)[\cdot] = a \qquad (\dot{\eta}a)\langle\cdot\rangle = a$$

From $\dot{\kappa}$ and $\dot{\eta}$ there now arise, in the natural way, epimorphisms

$$\kappa : A^G \twoheadrightarrow H^0_{X^s}(G, A) \qquad \eta : A_S \twoheadrightarrow H^{-1}_{X^s}(G, A)$$

with kernels SA and IA, respectively, and isomorphisms

$$\bar{\kappa} : \frac{A_G}{SA} \rightarrowtail\!\!\!\twoheadrightarrow H^0_{X^s}(G, A) \qquad \bar{\eta} : \frac{A_S}{IA} \rightarrowtail\!\!\!\twoheadrightarrow H^{-1}_{X^s}(G, A)$$

(Of course, strictly speaking, these maps should be denoted by

κ_s, η_s, $\bar{\kappa}_s$, $\bar{\eta}_s$.) Note that for standard cocycles $f \in \mathscr{Z}^0$ and $g \in \mathscr{Z}^{-1}$ their cohomology classes are $\kappa(f[\cdot])$ and $\eta(g\langle\cdot\rangle)$, respectively.

For an arbitrary full G-complex X we may define

$$\kappa_X : A^G \twoheadrightarrow H_X^0(G, A) \qquad \eta_X : A_S \twoheadrightarrow H_X^{-1}(G, A)$$

by putting

$$\kappa_X = 1_{X^s, X} \circ \kappa_s \qquad \eta_X = 1_{X^s, X} \circ \eta_s$$

so that κ_X is an epimorphism with kernel SA and η_X is an epimorphism with kernel IA. It is clear that

$$\kappa_s = 1_{X, X^s} \circ \kappa_X \qquad \eta_s = 1_{X, X^s} \circ \eta_X$$

and that if X' is any other G-complex, then

$$\kappa_{X'} = 1_{X, X'} \circ \kappa_X \qquad \eta_{X'} = 1_{X, X'} \circ \eta_X$$

In view of this, we have abstract homomorphisms (the notation for which should cause no serious confusion)

$$\kappa : A^G \twoheadrightarrow H^0(G, A) \qquad \eta : A_S \twoheadrightarrow H^{-1}(G, A)$$

which are independent of the complex and whose kernels are SA and IA, respectively. It then follows that there exist isomorphisms

$$\bar{\kappa} : \frac{A^G}{SA} \rightarrowtail\!\!\!\twoheadrightarrow H^0(G, A) \qquad \bar{\eta} : \frac{A_S}{IA} \rightarrowtail\!\!\!\twoheadrightarrow H^{-1}(G, A)$$

As a rule, we shall make statements only about κ and η and leave it to the reader to carry them over to $\bar{\kappa}$ and $\bar{\eta}$. Note that $\dot{\kappa} = \dot{\kappa}_s$ and $\dot{\eta} = \dot{\eta}_s$ are not carried over for an arbitrary complex X.

2-2-7. Proposition. Let $f : A \longrightarrow B$ be a G-homomorphism of G-modules; then the following diagrams commute :

$$\begin{array}{ccc} A^G & \xrightarrow{f} & B^G \\ \kappa\downarrow & & \kappa\downarrow \\ H^0(G, A) & \xrightarrow{f_*} & H^0(G, B) \end{array} \qquad \begin{array}{ccc} A_S & \xrightarrow{f} & B_S \\ \eta\downarrow & & \eta\downarrow \\ H^{-1}(G, A) & \xrightarrow{f_*} & H^{-1}(G, B) \end{array}$$

Proof: Note first that f maps A^G into B^G, SA into SB, A_S into B_S, and IA into IB. We consider only the dimension 0 case (since the proof in dimension -1 goes the same way) and must show that for any G-complexes X and X' the following diagram commutes:

$$\begin{array}{ccc} A^G & \xrightarrow{\quad f \quad} & B^G \\ {\scriptstyle \kappa_X}\downarrow & & \downarrow{\scriptstyle \kappa_{X'}} \\ H^0_X(G, A) & \xrightarrow{(1,f)_{X,X'}} & H^0_{X'}(G, A) \end{array}$$

For this consider the diagram

$$\begin{array}{ccccccc} & & A^G & \xrightarrow{\qquad f \qquad} & B^G & & \\ & \swarrow{\scriptstyle \kappa_X} & \downarrow{\scriptstyle \kappa_s} & & \downarrow{\scriptstyle \kappa_s} & \searrow{\scriptstyle \kappa_{X'}} & \\ H^0_X(G, A) & \xrightarrow{1_{X,X^s}} & H^0_{X^s}(G, A) & \xrightarrow{(1,f)_{X^s,X^s}} & H^0_{X^s}(G, B) & \xrightarrow{1_{X^s,X'}} & H^0_{X'}(G, B) \end{array}$$

Since the triangles commute, it follows that it suffices to prove that the square commutes. In order to compute $(1, f)_{X^s,X^s}$ we may use $\Lambda = 1 : X^s \longrightarrow X^s$ as the map associated with $1 : G \longrightarrow G$. Now the desired commutativity is a consequence of the fact that for $a \in A^G$

$$((1, f) \circ \dot{\kappa} a)[\cdot] = f(a) = (\dot{\kappa}(fa))[\cdot] \qquad \blacksquare$$

2-2-8. Proposition. Suppose that

$$0 \longrightarrow A \xrightarrow{i} B \xrightarrow{j} C \longrightarrow 0$$

is an exact G-sequence; then the following diagram commutes:

$$\begin{array}{ccccccccccc} A_S & \xrightarrow{i} & B_S & \xrightarrow{j} & C_S & & A^G & \xrightarrow{i} & B^G & \xrightarrow{j} & C^G \\ {\scriptstyle \eta}\downarrow & & {\scriptstyle \eta}\downarrow & & {\scriptstyle \eta}\downarrow & & {\scriptstyle \kappa}\downarrow & & {\scriptstyle \kappa}\downarrow & & {\scriptstyle \kappa}\downarrow \\ H^{-1}(G) & \xrightarrow{i_*} & H^{-1}(B) & \xrightarrow{j_*} & H^{-1}(C) & \xrightarrow{\delta_*} & H^0(G) & \xrightarrow{i_*} & H^0(B) & \xrightarrow{j_*} & H^0(C) \end{array}$$

Moreover, to compute δ_*, we have for $c \in C_S$

$$\delta_*(\eta c) = \kappa a \qquad \text{where} \quad c = jb, \quad ia = Sb$$

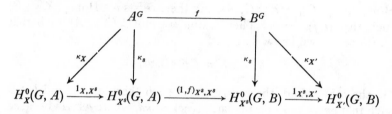

Proof: Only the formula for δ_* requires proof, and for this it suffices to use the standard complex and examine

$$C_S \xrightarrow{\eta_s} H^{-1}_{X^s}(G, C) \xrightarrow{\delta_*} H^0_{X^s}(G, A) \xleftarrow{\kappa_s} A^G$$

Let us in this proof (and, henceforth, whenever possible) omit explicit reference in the notation to the standard complex. Now, recall that δ_* arises from the exact sequence

$$0 \longrightarrow \mathrm{Hom}_G (X, A) \xrightarrow{(1,i)} \mathrm{Hom}_G (X, B) \xrightarrow{(1,j)} \mathrm{Hom}_G (X, C) \longrightarrow 0$$

In particular, $\dot{\eta}c \in \mathrm{Hom}_G (X_{-1}, C) \subset \mathrm{Hom}_G (X, C)$ (since $\mathrm{Hom}_G (X_{-1}, C) = \mathscr{C}^{-1}(G, A)$) and there exists $u \in \mathrm{Hom}_G (X_{-1}, B)$ such that $(i, j)u = \dot{\eta}c$. Then taking $v \in \mathrm{Hom}_G (X_0, A)$ such that $(1, i)v = \delta u$, we know that

$$\delta_*(\eta c) = v + \mathscr{B}^0(G, A)$$

On the other hand, there exist $a \in A$ such that $\dot{\kappa}a = v$ and $b \in B$ such that $\dot{\eta}b = u$. In particular, $\delta_*(\eta c) = \kappa a$, where $a \in A^G$ because v is a cocycle. Since the following diagrams commute :

$$\begin{array}{ccc} B & \xrightarrow{\quad j \quad} & C \\ \dot{\eta}\downarrow & & \downarrow\dot{\eta} \\ \mathrm{Hom}_G(X_{-1}, B) & \xrightarrow{(1,j)} & \mathrm{Hom}_G(X_{-1}, C) \end{array} \qquad \begin{array}{ccc} A & \xrightarrow{\quad i \quad} & B \\ \dot{\kappa}\downarrow & & \downarrow\dot{\kappa} \\ \mathrm{Hom}_G(X_0, A) & \xrightarrow{(1,i)} & \mathrm{Hom}_G(X_0, B) \end{array}$$

it follows that $\dot{\eta}c = (1, j)u = (1, j)(\dot{\eta}b) = \dot{\eta}(jb)$, so that $jb = c$, and

$$\begin{aligned} ia &= (i \circ \dot{\kappa}a)[\cdot] = ((1, i) \circ \dot{\kappa}a)[\cdot] = (\delta u)[\cdot] = (u \circ \partial)[\cdot] \\ &= (\dot{\eta}b)(S\langle\cdot\rangle) = S((\eta b)\langle\cdot\rangle) = Sb. \end{aligned}$$

This completes the proof. ∎

2-2-9. Remark. This result permits us to define a homomorphism $\delta_* : C_S/(IC) \longrightarrow A^G/(SA)$ by going around the diagram

$$\begin{array}{ccc} \dfrac{C_S}{IC} & & \dfrac{A^G}{SA} \\ \bar{\eta}\downarrow & & \uparrow\bar{\kappa}^{-1} \\ H^{-1}(G, C) & \xrightarrow{\delta_*} & H^0(G, A) \end{array}$$

Thus, $\delta_* = \bar{\kappa}^{-1} \circ \delta_* \circ \bar{\eta}$ maps $c + IC \longrightarrow a + SA$ where $c \in C_S$, $c = jb$, $Sb = ia$. This mapping δ_* may also be defined directly, and is sometimes called the Tate linking (see, for example [75] and (3-7-7)). By applying these results to (#) we have the exact sequence

$$\begin{aligned}
\cdots &\longrightarrow H^{-2}(A) \xrightarrow{i_*} H^{-2}(B) \xrightarrow{j_*} H^{-2}(C) \xrightarrow{\delta_*} \frac{A_S}{IA} \\
&\longrightarrow \frac{B_S}{IB} \xrightarrow{\bar{j}} \frac{C_S}{IC} \xrightarrow{\delta_*} \frac{A^G}{SA} \xrightarrow{\bar{i}} \frac{B^G}{SB} \\
&\longrightarrow \frac{C^G}{SC} \xrightarrow{\delta_*} H^1(A) \xrightarrow{i_*} H^1(B) \xrightarrow{j_*} H^1(C) \longrightarrow \cdots
\end{aligned}$$

Here, the first δ_* is the composite

$$H^{-2}(C) \xrightarrow{\delta_*} H^{-1}(A) \xrightarrow{\bar{\eta}^{-1}} \frac{A_S}{IA},$$

$\bar{i}$ and $\bar{j}$ are the natural maps induced by i and j, respectively, and the last δ_* is the composite $C^G/(SC) \xrightarrow{\bar{\kappa}} H^0(C) \xrightarrow{\delta_*} H^1(A)$.

2-2-10. Exercise. Let A be a G-module, let $(X^s, \partial^s, \varepsilon^s, \mu^s)$ be the standard G-complex, and let $(X, \partial, \varepsilon, \mu)$ be any full G-complex.

(a) Suppose that $\Lambda : X^s \longrightarrow X$ is suitable for $\lambda = 1 : G \longrightarrow G$ in computing $1_{X,X^s}$. If we write

$$\Lambda_0([\cdot]) = x_0 \in X_0 \qquad \text{and} \qquad \Lambda_{-1}(\langle\cdot\rangle) = x_{-1} \in X_{-1},$$

then $\varepsilon x_0 = 1$ and $Sx_{-1} = \mu(1)$. Conversely, given $x_0 \in X_0$, $x_{-1} \in X_{-1}$ such that $\varepsilon x_0 = 1$, $Sx_{-1} = \mu(1)$, there exists $\Lambda : X^s \longrightarrow X$ such that $\Lambda_0([\cdot]) = x_0$, $\Lambda_{-1}(\langle\cdot\rangle) = x_{-1}$.

(b) Let $x_0 \in X_0$, $x_{-1} \in X_{-1}$ be such that $\varepsilon x_0 = 1$, $Sx_{-1} = \mu(1)$, then for any $u \in \mathscr{Z}_X^0(G, A)$ and $v \in \mathscr{Z}_X^{-1}(G, A)$ we have

$$\kappa_X(ux_0) = u + \mathscr{B}_X^0(G, A) \qquad \eta_X(vx_{-1}) = v + \mathscr{B}_X^{-1}(G, A)$$

Thus κ_X and η_X can be described without reference to the standard complex.

(c) Consider the G-isomorphism $\omega : A \rightarrowtail\!\!\twoheadrightarrow \operatorname{Hom}(\mathbf{Z}, A)$ such that for $a \in A$, $\omega a = \omega_a$ where $\omega_a(1) = a$. The restriction of ω to A^G is an isomorphism $\omega : A^G \rightarrowtail\!\!\twoheadrightarrow (\operatorname{Hom}(\mathbf{Z}, A))^G = \operatorname{Hom}_G(\mathbf{Z}, A)$, and for $a \in A^G$

$$\kappa_X(a) = \omega_a \varepsilon + \mathscr{B}_X^0(G, A)$$

(d) Let $(\lambda, f) : (G, A) \longrightarrow (G', A')$ be a homomorphism of pairs with λ a monomorphism; then the following diagram commutes

$$\begin{array}{ccc} A^G & \xrightarrow{\ f\ } & A'^{G'} \\ {\scriptstyle\kappa}\downarrow & & \downarrow{\scriptstyle\kappa} \\ H^0(G, A) & \xrightarrow{(\lambda, f)_*} & H^0(G', A') \end{array}$$

2-3. CONJUGATION, RESTRICTION, AND INFLATION

In this section we introduce certain mappings of cohomology groups that arise from homomorphisms of pairs. Henceforth, by a G-complex we shall mean, once again, a full G-complex and the notation $H^n(G, A)$ shall be reserved for the cohomology groups computed with respect to a full complex. If a half-complex is to be considered, this hypothesis will be stated explicitly; in this situation we may still use the notation $H^n(G, A)$ for $n \geqslant 1$.

Suppose that A is a G-module. We shall use exponential notation for the action of the inner automorphisms of the multiplicative group G. In other words, for $\sigma, \tau \in G$ we write

$$\tau^\sigma = \sigma\tau\sigma^{-1}$$

so that $(\tau_1\tau_2)^\sigma = \tau_1^\sigma\tau_2^\sigma$ and $(\tau^{\sigma_1})^{\sigma_2} = \tau^{\sigma_2\sigma_1}$. Suppose that H is a subgroup of G, then every $\sigma \in G$ determines a conjugate subgroup

$$H^\sigma = \sigma H \sigma^{-1}$$

of H. Of course, A is an H-module and also an H^σ-module. Suppose further that B is an H-submodule of A, then clearly

$$\sigma B = \{\sigma b \mid b \in B\}$$

is an H^σ-submodule of A. Now, conjugation by σ^{-1} provides an isomorphism of H^σ onto H—$\sigma^{-1} : H^\sigma \rightarrowtail\!\!\twoheadrightarrow H$—so that by using this map, B becomes, in the canonical way, an H^σ-module for which the action of H^σ is given by

$$(\rho^\sigma)\, b = (\rho^\sigma)^{\sigma^{-1}}\, b = \rho b \qquad \rho \in H, \quad b \in B$$

Moreover, the map $\sigma : B \longrightarrow \sigma B$ is an H^σ-homomorphism since

$$\sigma(\rho^\sigma b) = \sigma(\rho b) = \sigma\rho\sigma^{-1}\sigma b = \rho^\sigma(\sigma b) \qquad \rho \in H, \quad b \in B$$

This means that we have a homomorphism of pairs

$$(\sigma^{-1}, \sigma) : (H, B) \longrightarrow (H^\sigma, \sigma B) \tag{\#}$$

and because σ^{-1} is a monomorphism there arise (by (2-2-3)) mappings of cohomology groups

$$(\sigma^{-1}, \sigma)_* : H^n(H, B) \longrightarrow H^n(H^\sigma, \sigma B) \qquad n \in \mathbf{Z}$$

We shall write $\sigma_* = (\sigma^{-1}, \sigma)_*$ and call it **conjugation** by σ. We have therefore :

2-3-1. Proposition. Let A be a G-module, let H be a subgroup of G, and let B be an H-submodule of A; then every $\sigma \in G$ determines an isomorphism

$$\sigma_* = (\sigma^{-1}, \sigma)_* : H^n(H, B) \longrightarrow H^n(H^\sigma, \sigma B) \qquad n \in \mathbf{Z} \quad (\#\#)$$

such that $(\sigma\tau)_* = \sigma_* \circ \tau_*$ (that is, conjugation is transitive), 1_* is the identity, and $(\sigma^{-1})_*$ is the isomorphism inverse to σ_*.

Moreover, if $\rho \in H$ (so that $H^\rho = H$ and $\rho B = B$), then ρ_* is the identity map of $H^n(H, B)$.

Proof: The first part is immediate from (2-1-8). As for the second part, let X be any G-complex. Then it is clear that X is an H-complex, and an H^σ complex, too—so X may be used for the cohomology of both H and H^σ, and we can be fairly explicit about the map

$$(\sigma^{-1}, \sigma)_{X,X} : H^n_X(H, B) \longrightarrow H^n_X(H^\sigma, \sigma B)$$

More precisely, in order to compute $(\sigma^{-1}, \sigma)_{X,X}$ we need an H^σ-homomorphism $\Lambda : X \longrightarrow X$ associated with $\lambda = \sigma^{-1} : H^\sigma \longrightarrow H$ (where the first X is viewed as an H^σ-complex, and the second X is viewed as an H-complex which becomes an H^σ-module via the map λ). We assert that $\Lambda = \sigma^{-1}$ will do. In fact, $\varepsilon\Lambda_0 = \varepsilon$ since for $x \in X_0$, $\varepsilon\Lambda_0 x = \varepsilon\sigma^{-1}x = \sigma^{-1}\varepsilon x = \varepsilon x$; also $\Lambda \circ \partial = \partial \circ \Lambda$ because ∂ is a G-homomorphism; and finally, Λ is an H^σ-homomorphism since for $x \in X$, $\rho \in H$ we have $\Lambda(\rho^\sigma x) = \sigma^{-1}\sigma\rho\sigma^{-1}x = \rho(\sigma^{-1}x)$ and $\rho^\sigma(\Lambda x) = \rho^\sigma(\sigma^{-1}x) = \rho(\sigma^{-1}x)$ by the way in which the second X becomes an H^σ-module. Thus, the homomorphism of pairs (#) leads to the cochain map

$$(\sigma^{-1}, \sigma) : \mathrm{Hom}_H(X, B) \longrightarrow \mathrm{Hom}_{H^\sigma}(X, \sigma B)$$

which in turn determines the mapping of cohomology groups

$$(\sigma^{-1}\sigma)_{X,X} : H_X^n(H, B) \longrightarrow H_X^n(H^\sigma, \sigma B)$$

This means that if $u \in \mathrm{Hom}_H(X, B)$ is a cocycle, then its cohomology class is mapped by $(\sigma^{-1}, \sigma)_{X,X}$ to the cohomology class represented by the cocycle $(\sigma^{-1}, \sigma)u = \sigma u \sigma^{-1} = u^\sigma \in \mathrm{Hom}_{H^\sigma}(X, \sigma B)$.

Now, if $\sigma \in H$, then $u^\sigma = u$ because (as was noted in the proof of (2-1-3)) $\mathrm{Hom}_H(X, B) = (\mathrm{Hom}(X, B))^H$. It follows that the cochain map (σ^{-1}, σ) is the identity, and then so is $(\sigma^{-1}, \sigma)_{X,X}$ This means that $\sigma_* = 1$, and the proof is complete. ▮

2-3-2. Corollary. For every $\sigma \in G$ the map

$$\sigma_* : H^n(G, A) \longrightarrow H^n(G, A) \qquad n \in \mathbf{Z}$$

is the identity.

2-3-3. Corollary. If H is a normal subgroup of G and B is a G-submodule of A, then the factor group G/H acts on $H^n(H, B)$—that is, $H^n(H, B)$ is a G/H module.

Another mapping of cohomology groups may be defined as follows. Let A be a G-module, and let H be any subgroup of G. If, as usual, we let 1 denote the identity map of A and $i = i_{H \to G}$ denote the inclusion map from H into G then clearly

$$(i, 1) : (G, A) \longrightarrow (H, A)$$

is a homomorphism of pairs. Since i is a monomorphism there arise mappings of cohomology groups

$$(i, 1)_* : H^n(G, A) \longrightarrow H^n(H, A) \qquad n \in \mathbf{Z}$$

We shall write res $= \text{res}_{G \to H} = (i, 1)_*$ and call it the **restriction** from G to H.

2-3-4. Remark. To deal with restriction concretely one takes any G-complex X and any H-complex X' and examines the map

$$\text{res}_{X,X'} = (i, 1)_{X,X'} : H_X^n(G, A) \longrightarrow H_{X'}^n(H, A)$$

As an illustration, let X be any G-complex. Then, as observed in the proof of (2-3-1), X is also an H-complex, and it is not hard to describe $\text{res}_{X,X}$. In fact, in order to compute $\text{res}_{X,X}$ we need an H-homomorphism $\Lambda : X \longrightarrow X$ associated with $\lambda = i : H \longrightarrow G$ (here the first X is an H-complex and the second X is a G-complex which becomes an H-module via the map λ). It is clear that $\Lambda = 1$ will do. Thus, the homomorphism of pairs $(i, 1)$ leads to the cochain map

$$(1, 1) : \text{Hom}_G (X, A) \longrightarrow \text{Hom}_H (X, A)$$

(note that this map is simply inclusion) and this determines the mapping of cohomology groups

$$\text{res}_{X,X} : H_X^n(G, A) \longrightarrow H_X^n(H, A) \qquad n \in \mathbf{Z}$$

This means that if $u \in \text{Hom}_G (X, A)$ is a cocycle, then its cohomology class is mapped by $\text{res}_{X,X}$ to the cohomology class represented by the cocycle $u \in \text{Hom}_H (X, A)$.

2-3-5. Proposition. Let H be a subgroup of G and let A be a G-module, then there exists a map

$$\text{res}_{G \to H} : H^n(G, A) \longrightarrow H^n(H, A) \qquad n \in \mathbf{Z}$$

Moreover, restriction is transitive—in other words, if H' is a subgroup of H, then

$$\text{res}_{G \to H'} = \text{res}_{H \to H'} \circ \text{res}_{G \to H}$$

Proof: Let X, $\bar{X}$ and X' be G, H and H'-complexes, respectively. Then, from (2-1-8) we have

$$(i_{H' \to G}, 1)_{X,X'} = (i_{H' \to H}, 1)_{\bar{X},X'} \circ (i_{H \to G}, 1)_{X,\bar{X}}$$

which proves transitivity. ∎

Still another mapping of cohomology groups may be defined in the following manner. Let A be a G-module and suppose that H is a normal subgroup of G. If $a \in A^H$ and $\sigma \in G$, then $\sigma a \in A^H$—because, given $\rho \in H$ there exists $\rho' \in H$ such that $\rho\sigma = \sigma\rho'$, so that $\rho(\sigma a) = \sigma\rho' a = \sigma a$. Furthermore, $(\sigma\rho)a = \sigma a$, so that each coset σH has a well-defined action on A^H. In other words, A^H may be viewed as a G/H-module. Let $\pi : G \longrightarrow G/H$ be the canonical map (we also write $\pi(\sigma) = \bar{\sigma}$) and let $i : A^H \longrightarrow A$ be the inclusion map. When A^H is made into a G-module via the map π, then i is a G-homomorphism, and consequently

$$(\pi, i) : (G/H, A^H) \longrightarrow (G, A)$$

is a homomorphism of pairs. Since i is not 1–1, there arise mappings of cohomology groups

$$(\pi, i)_* : H^n(G/H, A^H) \longrightarrow H^n(G, A) \qquad n \geqslant 1$$

We shall write $\inf = \inf_{(G/H) \to G} = (\pi, i)_*$ and call it the **inflation** map. The inflation map is not considered for $n = 0$ because of our convention about the meaning of $H^n(G, A)$. In virtue of the transitivity principle of (2-1-8) we have therefore :

2-3-6. Proposition. Let H be a normal subgroup of G and let A be a G-module; then there exists a map

$$\mathrm{inf}_{(G/H)\to G} : H^n(G/H, A^H) \longrightarrow H^n(G, A) \qquad n \geqslant 1$$

Moreover, inflation is transitive; more precisely, if $H_2 \subset H_1 \subset G$ and H_1, H_2 are normal subgroups of G then the following diagram commutes :

$$\begin{array}{ccc} H^n(G/H_1, A^{H_1}) & \xrightarrow{\mathrm{inf}} & H^n(G/H_2, A^{H_2}) \\ & \searrow^{\mathrm{inf}} & \downarrow{\scriptstyle \mathrm{inf}} \\ & & H^n(G, A) \end{array} \qquad n \geqslant 1$$

We may also note that the homomorphisms of pairs (σ^{-1}, σ), $(i, 1)$, and (π, i) commute with each other, so that application of the transitivity principle shows that conjugation, restriction, and inflation commute with each other. In detail, we have :

2-3-7. Proposition. Consider the groups $H_2 \subset H_1 \subset G$ and the G-module A, and let B be an H_1-submodule of A (so that B is also an H_2-submodule of A); then for $\sigma \in G$ the following diagram commutes :

$$\begin{array}{ccc} H^n(H_1, B) & \xrightarrow{\mathrm{res}} & H^n(H_2, B) \\ {\scriptstyle \sigma_*}\downarrow & & {\scriptstyle \sigma_*}\downarrow \\ H^n(H_1^\sigma, \sigma B) & \xrightarrow{\mathrm{res}} & H^n(H_2^\sigma, \sigma B) \end{array} \qquad n \in \mathbf{Z}$$

If, in addition, H_2 is normal in H_1, then the following diagram commutes :

$$\begin{array}{ccc} H^n(H_1/H_2, B^{H_2}) & \xrightarrow{\mathrm{inf}} & H^n(H_1, B) \\ {\scriptstyle \sigma_*}\downarrow & & {\scriptstyle \sigma_*}\downarrow \\ H^n(H_1^\sigma/H_2^\sigma, (\sigma B)^{H_2^\sigma}) & \xrightarrow{\mathrm{inf}} & H^n(H_1^\sigma, \sigma B) \end{array} \qquad n \geqslant 1$$

If, in addition, H_2 is normal in G, then the following diagram commutes :

$$\begin{array}{ccc} H^n(G/H_2, A^{H_2}) & \xrightarrow{\text{inf}} & H^n(G, A) \\ \text{res}\downarrow & & \text{res}\downarrow \\ H^n(H_1/H_2, A^{H_2}) & \xrightarrow{\text{inf}} & H^n(H_1, A) \end{array} \qquad n \geqslant 1$$

2-3-8. Remark. It is clear that (2-1-9) and (2-2-5) carry over to σ_*, res, and inf; it is only necessary to restate the hypotheses carefully. We shall express this by saying that conjugation, restriction, and inflation commute with the coboundaries arising from short exact sequences.

2-4. THE TRANSFER OR CORESTRICTION

In this section we define still another mapping of cohomology groups and discuss some of its elementary properties.

Suppose that H is a subgroup of G, and let $G = \bigcup_{i=1}^{m} \sigma_i H$ be a left coset decomposition—so $(G : H) = m$. For any G-module A, we define a function

$$S_{H\to G}(a) = \sum_{1}^{m} \sigma_i a \qquad a \in A^H$$

This function is called the **trace** from H to G, since it generalizes our previous trace $S = S_{\{1\}\to G}$. Of course, $S_{H\to G}$ may be viewed as multiplication by the element $\sum_1^m \sigma_i$ of $\mathbf{Z}[G]$. A few of the basic properties of the trace are given in :

2-4-1. Proposition. Let H be a subgroup of G, and let A, B, C be G-modules, then :

(1) $S_{H\to G}$ is a well-defined (that is, it is independent of the choice of representatives for the cosets) homomorphism of $A^H \longrightarrow A^G$.

(2) The trace is transitive; that is, if $H' \subset H \subset G$, then $S_{H\to G} \circ S_{H'\to H} = S_{H'\to G}$.

(3) $S = S_{\{1\}\to G} : A \longrightarrow A^G$ is a G-homomorphism.

(4) $S_{H\to G}$ is a homomorphism of $\mathrm{Hom}_H(A, B) \longrightarrow \mathrm{Hom}_G(A, B)$.

(5) If $f \in \mathrm{Hom}_G(A, B)$, then on A^H we have $S_{H\to G} \circ f = f \circ S_{H\to G}$.

(6) If $f \in \mathrm{Hom}_G(A, B)$, then $S_{H\to G} f = (G : H)f$.

(7) If $f \in \mathrm{Hom}_H(A, B)$ and $g \in \mathrm{Hom}_G(B, C)$, then

$$S_{H\to G}(g \circ f) = g \circ S_{H\to G}(f).$$

(8) If $f \in \mathrm{Hom}_G(A, B)$ and $g \in \mathrm{Hom}_H(B, C)$, then

$$S_{H\to G}(g \circ f) = (S_{H\to G} g) \circ f.$$

Proof: Entirely straightforward; for example, (4) is immediate from (1) and the fact that $\mathrm{Hom}(A, B)$ is a G-module with

$\mathrm{Hom}_G(A, B) = (\mathrm{Hom}(A, B))^G$ and $\mathrm{Hom}_H(A, B) = (\mathrm{Hom}(A, B))^H$. ▮

Now, let X be any (full) G-complex. Then X is also an H-complex and we have a homomorphism

$$S_{H\to G} : \mathrm{Hom}_H(X, A) \longrightarrow \mathrm{Hom}_G(X, A)$$

For $u \in \mathrm{Hom}_H(X_n, A)$ we note that

$$\delta \circ S_{H\to G} u = (S_{H\to G} u) \circ \partial = S_{H\to G}(u\partial) = S_{H\to G}(\delta u),$$

so that $\delta \circ S_{H\to G} = S_{H\to G} \circ \delta$. This means that $S_{H\to G}$ is an admissible map of differential graded groups, and determines, therefore, a homomorphism, called the **transfer** or **corestriction**, of cohomology groups

$$(\mathrm{cor}_{H\to G})_{X,X} = \mathrm{cor}_{X,X} : H^n_X(H, A) \longrightarrow H^n_X(G, A) \qquad n \in \mathbf{Z}$$

For an arbitrary H-complex X' we define

$$(\mathrm{cor}_{H\to G})_{X',X} = \mathrm{cor}_{X',X} : H^n_{X'}(H, A) \longrightarrow H^n_X(G, A) \qquad n \in \mathbf{Z}$$

as the composite map

$$H^n_{X'}(H, A) \xrightarrow{1^H_{X',X}} H^n_X(H, A) \xrightarrow{\mathrm{cor}_{X,X}} H^n_X(G, A)$$

In order to define an abstract map $\mathrm{cor}_{H\to G} : H^n(H, A) \longrightarrow H^n(G, A)$ we make use of the following result :

2-4-2. Lemma. Let $(\lambda, f) : (G, A) \longrightarrow (G', A')$ be a homomorphism of pairs, and consider subgroups $H \subset G$ and $H' \subset G'$. Suppose that $\lambda(H') \subset H$ (so that $(\lambda, f) : (H, A) \longrightarrow (H', A')$ is a homomorphism of pairs), that $\lambda(G')H = G$, and that $(G' : H') = (G : H) = m$. Then for any full G, G', H, H' complexes $X, X', \bar{X}, \bar{X}'$, respectively, the following diagram commutes for all $n \in \mathbf{Z}$ when λ is a monomorphism, and for $n \geqslant 1$ otherwise :

$$\begin{array}{ccc} H^n_{\bar{X}}(H, A) & \xrightarrow{(\lambda,f)_{\bar{X},\bar{X}'}} & H^n_{\bar{X}'}(H', A') \\ {\scriptstyle \mathrm{cor}_{\bar{X},X}}\downarrow & & \downarrow{\scriptstyle \mathrm{cor}_{\bar{X}',X'}} \\ H^n_X(G, A) & \xrightarrow{(\lambda,f)_{X,X'}} & H^n_{X'}(G', A') \end{array}$$

Proof: Using X and X' as H and H' complexes, respectively, let us consider the diagram

$$\begin{array}{ccc} H^n_{\bar{X}}(H, A) & \xrightarrow{(\lambda,f)_{\bar{X},\bar{X}'}} & H^n_{\bar{X}'}(H', A') \\ {\scriptstyle 1^H_{\bar{X},X}}\downarrow & & \downarrow{\scriptstyle 1^{H'}_{\bar{X}',X'}} \\ H^n_X(H, A) & \xrightarrow{(\lambda,f)_{X,X'}} & H^n_{X'}(H', A') \\ {\scriptstyle \mathrm{cor}_{X,X}}\downarrow & & \downarrow{\scriptstyle \mathrm{cor}_{X',X'}} \\ H^n_X(G, A) & \xrightarrow{(\lambda,f)_{X,X'}} & H^n_{X'}(G', A') \end{array}$$

The top rectangle commutes by (2-1-8), so it remains to show that the bottom rectangle commutes. For this, let $\Lambda : X' \longrightarrow X$ be any G'-homomorphism suitable for λ in the computation of the lower $(\lambda, f)_{X,X'}$. This same Λ may then be used as the H'-homomorphism of $X' \longrightarrow X$ in the computation of the upper $(\lambda, f)_{X,X'}$. Thus, it suffices to show that we have commutativity in the cochain diagram

$$\begin{array}{ccc} \mathrm{Hom}_H(X, A) & \xrightarrow{(\Lambda,f)} & \mathrm{Hom}_{H'}(X', A') \\ {\scriptstyle S_{H\to G}}\downarrow & & \downarrow{\scriptstyle S_{H'\to G'}} \\ \mathrm{Hom}_G(X, A) & \xrightarrow{(\Lambda,f)} & \mathrm{Hom}_{G'}(X', A') \end{array}$$

Let $G' = \bigcup_1^m \sigma_i' H'$ be a left coset decomposition; so

$$\lambda(G') = \bigcup_1^m \lambda(\sigma_i')\,\lambda(H'),$$

and if we put $\lambda(\sigma_i') = \sigma_i$, then $G = \lambda(G')H = \bigcup_1^m \sigma_i H$. Since $(G : H) = m$, this is a left coset decomposition. Now, for $u \in \mathrm{Hom}_H(X, A)$ we have

$$\begin{aligned}(S_{H'\to G'} \circ (\Lambda, f))\, u &= \sum_{i=1}^m (fu\Lambda)^{\sigma_i'} = \sum_{i=1}^m \sigma_i' fu\Lambda\sigma_i'^{-1} \\ &= \sum_{i=1}^m f\sigma_i u\sigma_i^{-1}\Lambda = f\left(\sum_{i=1}^m \sigma_i u \sigma_i^{-1}\right)\Lambda \\ &= f \circ (S_{H\to G} u) \circ \Lambda = ((\Lambda, f) \circ S_{H\to G})\, u.\end{aligned}$$

This completes the proof. ▮

2-4-3. Proposition. Let H be a subgroup of G, then for any G-module A there exists a homomorphism

$$\mathrm{cor} = \mathrm{cor}_{H\to G} : H^n(H, A) \longrightarrow H^n(G, A) \qquad n \in \mathbf{Z}$$

which is independent of the choice of complexes.

Proof: Immediate from (2-4-2). ▮

2-4-4. Proposition. The transfer map is transitive; that is, if $H' \subset H \subset G$, then

$$\mathrm{cor}_{H'\to G} = \mathrm{cor}_{H\to G} \circ \mathrm{cor}_{H'\to H}$$

Proof: We must show that if $X, \bar{X}, \bar{\bar{X}}$ are complexes for G, H, H', respectively, then

$$(\mathrm{cor}_{H'\to G})_{\bar{\bar{X}},X} = (\mathrm{cor}_{H\to G})_{\bar{X},X} \circ (\mathrm{cor}_{H'\to H})_{\bar{\bar{X}},\bar{X}}$$

Since X may be used as a complex for all three groups, and since $S_{H' \to G} = S_{H \to G} \circ S_{H' \to H}$, it follows that

$$(\mathrm{cor}_{H' \to G})_{X,X} = (\mathrm{cor}_{H \to G})_{X,X} \circ (\mathrm{cor}_{H' \to H})_{X,X}$$

Therefore,

$$\begin{aligned}
(\mathrm{cor}_{H \to G})_{\tilde{X},X} \circ (\mathrm{cor}_{H' \to H})_{\bar{\bar{X}},\tilde{X}} &= (\mathrm{cor}_{H \to G})_{X,X} \circ 1^{H}_{\bar{\bar{X}},X} \circ (\mathrm{cor}_{H' \to H})_{\tilde{X},\tilde{X}} \circ 1^{H'}_{\bar{\bar{X}},\tilde{X}} \\
&= (\mathrm{cor}_{H \to G})_{X,X} \circ (\mathrm{cor}_{H' \to H})_{X,X} \circ 1^{H'}_{\tilde{X},X} \circ 1^{H'}_{\bar{\bar{X}},\tilde{X}} \\
&= (\mathrm{cor}_{H' \to G})_{X,X} \circ 1^{H'}_{\bar{\bar{X}},X} \\
&= (\mathrm{cor}_{H' \to G})_{\bar{\bar{X}},X}
\end{aligned}$$

and the proof is complete. ▌

In virtue of (2-4-2), it is easy to see that, under suitable hypotheses, the transfer map commutes with conjugation, restriction, and inflation; more precisely, we have :

2-4-5. Proposition. Consider the groups $H_2 \subset H_1 \subset G$ and the G-module A. If B is an H_1-submodule of A and $\sigma \in G$, then the following diagram commutes :

$$\begin{array}{ccc}
H^n(H_2, B) & \xrightarrow{\mathrm{cor}} & H^n(H_1, B) \\
{\scriptstyle \sigma_*}\downarrow & & {\scriptstyle \sigma_*}\downarrow \\
H^n(H_2^\sigma, \sigma B) & \xrightarrow{\mathrm{cor}} & H^n(H_1^\sigma, \sigma B)
\end{array} \qquad n \in \mathbf{Z}$$

If, in addition, H_2 is normal in G, then the following diagram commutes :

$$\begin{array}{ccc}
H^n(H_1/H_2, A^{H_2}) & \xrightarrow{\mathrm{cor}} & H^n(G/H_2, A^{H_2}) \\
{\scriptstyle \mathrm{inf}}\downarrow & & \downarrow{\scriptstyle \mathrm{inf}} \\
H^n(H_1, A) & \xrightarrow{\mathrm{cor}} & H^n(G, A)
\end{array} \qquad n \geqslant 1$$

2-4-6. Proposition. Let H_1 and H_2 be subgroups of G such that H_1H_2 is a group; then for any G-module A, the following diagram commutes :

$$\begin{array}{ccc} H^n(H_2, A) & \xrightarrow{\text{cor}} & H^n(H_1H_2, A) \\ {\scriptstyle\text{res}}\downarrow & & \downarrow{\scriptstyle\text{res}} \\ H^n(H_1 \cap H_2, A) & \xrightarrow{\text{cor}} & H^n(H_1, A) \end{array} \qquad n \in \mathbf{Z}.$$

Proof: Consider the homomorphism of pairs

$$(i, 1) : (H_1H_2, A) \longrightarrow (H_2, A)$$

Then

$$i(H_1 \cap H_2) \subset H_2, \qquad i(H_1)\, H_2 = H_1H_2,$$

and

$$(H_1H_2 : H_2) = (H_1 : H_1 \cap H_2),$$

so that (2-4-2) applies. ∎

2-4-7. Proposition. The transfer commutes with the coboundary arising from short exact sequences; that is, if

$$0 \longrightarrow A \xrightarrow{i} B \xrightarrow{j} C \longrightarrow 0$$

is an exact G-sequence and $H \subset G$ then the following diagram commutes :

$$\begin{array}{ccccccccc} \cdots \longrightarrow & H^n(H, A) & \xrightarrow{i_*} & H^n(H, B) & \xrightarrow{j_*} & H^n(H, C) & \xrightarrow{\delta_*} & H^{n+1}(H, A) & \longrightarrow \cdots \\ & {\scriptstyle\text{cor}}\downarrow & & {\scriptstyle\text{cor}}\downarrow & & {\scriptstyle\text{cor}}\downarrow & & {\scriptstyle\text{cor}}\downarrow & \\ \cdots \longrightarrow & H^n(G, A) & \xrightarrow{i_*} & H^n(G, B) & \xrightarrow{j_*} & H^n(G, C) & \xrightarrow{\delta_*} & H^{n+1}(G, A) & \longrightarrow \cdots \end{array}$$

Proof: Let X be any G-complex. From the commutative diagram with exact rows

$$\begin{array}{ccccccccc} 0 \longrightarrow & \mathrm{Hom}_H(X, A) & \xrightarrow{(1,i)} & \mathrm{Hom}_H(X, B) & \xrightarrow{(1,j)} & \mathrm{Hom}_H(X, C) & \longrightarrow 0 \\ & {\scriptstyle S_{H\to G}}\downarrow & & {\scriptstyle S_{H\to G}}\downarrow & & {\scriptstyle S_{H\to G}}\downarrow & \\ 0 \longrightarrow & \mathrm{Hom}_G(X, A) & \xrightarrow{(1,i)} & \mathrm{Hom}_G(X, B) & \xrightarrow{(1,j)} & \mathrm{Hom}_G(X, C) & \longrightarrow 0 \end{array}$$

we derive the desired commutativity when the cohomology groups of both rows are computed with respect to X. If X' is any H-complex, then the desired commutativity (when the cohomology groups of H are with respect to X' and those of G are with respect to X) is gotten by use of $1^H_{X',X}$. ▮

2-4-8. Proposition. Consider $H \subset G$ and the G-modules A and B. If $f \in \mathrm{Hom}_H(A, B)$, then the following diagram commutes :

$$\begin{array}{ccc} H^n(G, A) & \xrightarrow{(S_{H\to G}f)_*} & H^n(G, B) \\ \text{res}\downarrow & & \uparrow\text{cor} \\ H^n(H, A) & \xrightarrow{f_*} & H^n(H, B) \end{array} \qquad n \in \mathbf{Z}$$

Proof: According to (2-4-1), $S_{H\to G} f \in \mathrm{Hom}_G(A, B)$. Let X be any G-complex; it suffices to prove our result when all cohomology groups are computed with respect to X. From (2-3-4), we know that $\Lambda = 1 : X \longrightarrow X$ may be used for the computation of $(\mathrm{res}_{G\to H})_{X,X}$. Then the proof is complete as soon as one checks the commutativity of the cochain diagram

$$\begin{array}{ccc} \mathrm{Hom}_G(X, A) & \xrightarrow{(1, S_{H\to G}f)} & \mathrm{Hom}_G(X, B) \\ (1,1)\downarrow & & \uparrow S_{H\to G} \\ \mathrm{Hom}_H(X, A) & \xrightarrow{(1, f)} & \mathrm{Hom}_H(X, B) \end{array}$$ ▮

2-4-9. Corollary. If $H \subset G$ and A is a G-module, then for any $\alpha \in H^n(G, A)$

$$\mathrm{cor}_{H\to G}\,\mathrm{res}_{G\to H}\,\alpha = (G : H)\,\alpha$$

Proof: In (2-4-8) take $A = B$ and $f = 1$. ▮

2-4-10. Exercise. Let A be a G-module, and let H be a subgroup of G; then the following diagram commutes :

$$\begin{array}{ccc} A^H & \xrightarrow{S_{H\to G}} & A^G \\ \kappa\downarrow & & \downarrow\kappa \\ H^0(H, A) & \xrightarrow{\text{cor}} & H^0(G, A) \end{array}$$

2-5. EXPLICIT FORMULAS

The mappings of cohomology groups defined in the preceding sections have concrete realization when standard complexes are used, and this section is devoted to the derivation of explicit formulas for these maps.

First of all, let us consider a framework within which inflation, restriction, and conjugation are included. Thus, suppose that $(\lambda, f) : (G, A) \longrightarrow (G', A')$ is a homomorphism of pairs. Let X and X' denote the **standard complexes** for G and G', respectively. Then knowledge of an explicit formula for a G'-homomorphism $\Lambda : X' \longrightarrow X$ suitable for λ determines an explicit formula for the map of cochains

$$(\Lambda, f) : \operatorname{Hom}_G (X, A) \longrightarrow \operatorname{Hom}_{G'} (X', A')$$

which, in turn, describes

$$(\lambda, f)_{X,X'} : H^n_X(G, A) \longrightarrow H^n_{X'}(G', A')$$

Putting $\Gamma = \mathbf{Z}[G]$ and $\Gamma' = \mathbf{Z}[G']$ we recall (see Section 2-1) that Λ must make the following diagram commutative :

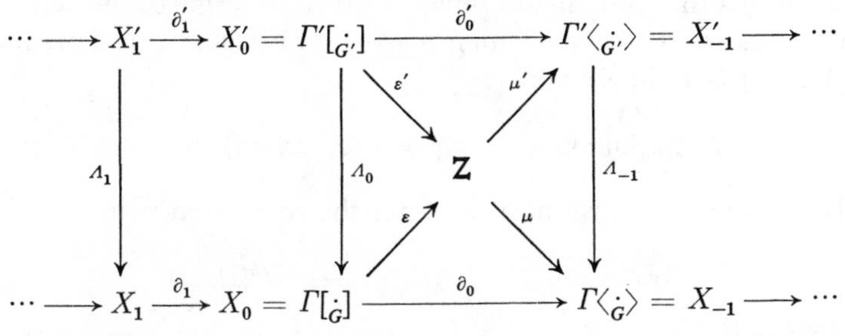

To define Λ it suffices to do so on the G'-cells and then extend G'-linearly. If we put

$$\Lambda_0([\dot{}_{G'}]) = [\dot{}_G]$$

$$\Lambda_n([\sigma'_1, \ldots, \sigma'_n]) = [\lambda\sigma'_1, \ldots, \lambda\sigma'_n] \qquad n \geqslant 1$$

then $\varepsilon \circ \Lambda_0 = \varepsilon'$ and, as in easily checked, $\Lambda_n \circ \partial'_{n+1} = \partial_{n+1} \circ \Lambda_{n+1}$ on the G'-cells; thus we have the positive part of Λ.

For arbitrary λ, the negative dimensional Λ's do not exist, in general. However, if λ is a monomorphism, then (as in (2-1-6)) the negative dimensional Λ's can be found; the details may be left to the reader, because we shall later find such Λ's, in the cases of interest to us, by another method.

Now, consider the inflation map

$$\inf = \inf_{(G/H)\to G} : H^n(G/H, A^H) \longrightarrow H^n(G, A)$$

It arises from the homomorphism of pairs

$$(\pi, i) : (G/H, A^H) \longrightarrow (G, A)$$

where $\pi(\sigma) = \bar{\sigma}$. Let X^G and $X^{G/H}$ be the standard complexes; if the $\Lambda : X^G \longrightarrow X^{G/H}$ suitable for $\lambda = \pi$ constructed according to the foregoing is denoted by $\Lambda^{\inf}$, then

$$\Lambda_0^{\inf}([\dot{_G}]) = [\dot{_{G/H}}]$$

$$\Lambda_n^{\inf}([\sigma_1, ..., \sigma_n]) = [\bar{\sigma}_1, ..., \bar{\sigma}_n] \qquad n \geqslant 1$$

We shall be somewhat careless and write $\inf = (\Lambda^{\inf}, i)$—so that this cochain map inf induces the cohomological map $(\inf_{(G/H)\to G})_{X^{G/H}, X^G}$. This means that the inflation map is given in terms of the standard complexes by the cochain map $u \longrightarrow \inf u$ where u is a standard cochain of G/H in A^H and

$$(\inf u)([\sigma_1, ..., \sigma_n] = u([\bar{\sigma}_1, ..., \bar{\sigma}_n]) \qquad n \geqslant 1$$

In the same way, we may consider the restriction map

$$\operatorname{res} = \operatorname{res}_{G\to H} : H^n(G, A) \longrightarrow H^n(H, A)$$

It arises from the homomorphism of pairs $(i, 1) : (G, A) \longrightarrow (H, A)$, and if we let $\Lambda^{\operatorname{res}} : X^H \longrightarrow X^G$ be the map of standard complexes suitable for $\lambda = i$ described above, then

$$\Lambda_0^{\operatorname{res}}([\dot{_H}]) = [\dot{_G}]$$

$$\Lambda_n^{\operatorname{res}}([\rho_1, ..., \rho_n]) = [\rho_1, ..., \rho_n] \quad n \geqslant 1, \ \rho_1, ..., \rho_n \in H$$

We write res $= (\Lambda^{\text{res}}, 1)$ so that this cochain map res induces $(\text{res}_{G\to H})_{X^G, X^H}$. Since $\Lambda^{\text{res}}_{-1}$ exists (see (2-1-6)), although we have not found it explicitly, we see that the restriction map in dimensions $\geqslant 0$ is given in terms of the standard complexes by the cochain map $u \longrightarrow \text{res}\, u$ where u is a standard cochain of G in A and

$$(\text{res}\, u)([\dot{_H}]) = u[\dot{_G}]$$

$$(\text{res}\, u)([\rho_1, ..., \rho_n]) = u([\rho_1, ..., \rho_n]) \qquad n \geqslant 1, \quad \rho_1, ..., \rho_n \in H$$

Finally, in order to describe conjugation by σ, we consider the homomorphism of pairs $(\sigma^{-1}, \sigma) : (H, B) \longrightarrow (H^\sigma, \sigma B)$, let $\Lambda^\sigma : X^{H^\sigma} \longrightarrow X^H$ denote the map of standard complexes suitable for $\lambda = \sigma^{-1}$ described earlier, write $\text{con}_\sigma = (\Lambda^\sigma, \sigma)$ for the cochain map, and leave the details to the reader. To summarize these results, we have :

2-5-1. Proposition. For positive dimensions, inflation, restriction, and conjugation are determined by standard complex cochain maps of form :

$$(\inf u)([\sigma_1, ..., \sigma_n]) = u([\bar{\sigma}_1, ..., \bar{\sigma}_n]) \qquad n \geqslant 1, \quad \sigma_1, ..., \sigma_n \in G$$

$$(\text{res}\, u)([\dot{_H}]) = u([\dot{_G}])$$

$$(\text{res}\, u)([\rho_1, ..., \rho_n]) = u([\rho_1, ..., \rho_n]) \qquad n \geqslant 1, \quad \rho_1, ..., \rho_n \in H$$

$$(\text{con}_\sigma\, u)([\dot{_{H^\sigma}}]) = \sigma u([\dot{_H}])$$

$$(\text{con}_\sigma\, u)([\rho_1, ..., \rho_n]) = \sigma u([\sigma^{-1}\rho_1\sigma, ..., \sigma^{-1}\rho_n\sigma]) \qquad n \geqslant 1, \quad \rho_1, ..., \rho_n \in H^\sigma$$

The transfer map does not come from a homomorphism of pairs, so that determination of a standard complex cochain map which induces it involves a return to first principles. We recall that if X^H and X^G are the standard complexes, then the cohomological map we seek to make explicit is

$$(\text{cor}_{H\to G})_{X^H, X^G} = (\text{cor}_{H\to G})_{X^G, X^G} \circ 1^H_{X^H, X^G}$$

Consequently, the desired standard complex cochain map, which we denote by cor, is the composite

$$\text{Hom}_H(X^H, A) \xrightarrow{(\Lambda^{\text{cor}}, 1)} \text{Hom}_H(X^G, A) \xrightarrow{S_{H\to G}} \text{Hom}_G(X^G, A)$$

where $\Lambda^{\text{cor}} : X^G \longrightarrow X^H$ is an H-homomorphism suitable for $\lambda = 1$ in the computation of $(1, 1)_{X^H, X^G}$. In other words, for a standard n-cochain u of H in A, we have

$$\text{cor}\, u = S_{H\to G}(u \circ \Lambda_n^{\text{cor}}) \qquad n \in \mathbf{Z} \quad (\#)$$

Now, let us turn to the definition of $\Lambda_n^{\text{cor}} : X_n^G \longrightarrow X_n^H$ for $n \geqslant 0$. Let $C = \{c\}$ denote the set of right cosets of H in G, and for each coset c choose, once and for all, a representative $\bar{c} \in c$. There is no harm in requiring that for $c = H$, $\bar{c} = 1$. We may write

$$G = \bigcup_{c \in C} c = \bigcup_{c} H\bar{c}$$

and every element σ of G has a unique expression $\sigma = \rho\bar{c}$ with $\rho \in H$. If we write $\Gamma_G = \mathbf{Z}[G]$ and $\Gamma_H = \mathbf{Z}[H]$, then

$$X_0^G = \Gamma_G([\dot{_G}]) = \sum_{c \in C} \oplus\, \Gamma_H(\bar{c}[\dot{_G}])$$

so that the set $\{\bar{c}[\dot{_G}] \mid c \in C\}$ constitutes an H-basis for X_0^G. In the same way, $\{\bar{c}[\sigma_1, \ldots, \sigma_n] \mid c \in C,\ \sigma_1, \ldots, \sigma_n \in G\}$ is an H-basis for X_n^G, $n \geqslant 1$. Because Λ_0^{cor} must satisfy $\varepsilon^H \circ \Lambda_0^{\text{cor}} = \varepsilon^G$, and because it suffices to define Λ_0^{cor} on an H-basis, it follows from

$$\varepsilon^H\{\Lambda_0^{\text{cor}}(\bar{c}[\dot{_G}])\} = \varepsilon^G(\bar{c}[\dot{_G}]) = \bar{c}\varepsilon^G([\dot{_G}]) = \bar{c} \cdot 1 = 1 = \varepsilon^H([\dot{_H}])$$

that Λ_0^{cor} should be defined by

$$\Lambda_0^{\text{cor}}(\bar{c}[\dot{_G}]) = [\dot{_H}]$$

From the requirement that $\partial_1^H \circ \Lambda_1^{\text{cor}} = \Lambda_0^{\text{cor}} \circ \partial_1^G$ and the fact that for $c \in C$, $\sigma \in G$ the elements $\bar{c}\sigma$ and $\overline{c\sigma}$ belong to the same coset $c\sigma$, (so $\bar{c}\sigma\overline{c\sigma}^{-1} \in H$) we have

$$\begin{aligned}
\partial_1^H\{\Lambda_1^{\text{cor}}(\bar{c}[\sigma_1])\} &= \Lambda_0^{\text{cor}}\partial_1^G(\bar{c}[\sigma_1]) = \Lambda_0^{\text{cor}}\bar{c}\partial_1^G([\sigma_1]) \\
&= \Lambda_0^{\text{cor}}\{\bar{c}\sigma_1[\dot{_G}] - \bar{c}[\dot{_G}]\} \\
&= \Lambda_0^{\text{cor}}\{\bar{c}\sigma_1\overline{c\sigma_1}^{-1}(\overline{c\sigma_1}[\dot{_G}]) - \bar{c}[\dot{_G}]\} \\
&= \bar{c}\sigma_1\overline{c\sigma_1}^{-1}[\dot{_H}] - [\dot{_H}] \\
&\qquad \text{(since } \Lambda_0 \text{ is an } H\text{-homomorphism)} \\
&= \partial_1^H[\bar{c}\sigma_1\overline{c\sigma_1}^{-1}]
\end{aligned}$$

so that Λ_1^{cor} should be defined by

$$\Lambda_1^{\text{cor}}(\bar{c}[\sigma_1]) = [\bar{c}\sigma_1\overline{c\sigma_1}^{-1}]$$

In the next step, we must have

$$\begin{aligned}
\partial_2^H\{\Lambda_2^{\text{cor}}(\bar{c}[\sigma_1, \sigma_2])\} &= \Lambda_1^{\text{cor}}\{\partial_2^G(\bar{c}[\sigma_1, \sigma_2])\} \\
&= \Lambda_1^{\text{cor}}(\bar{c}\partial_2^G[\sigma_1, \sigma_2]) \\
&= \Lambda_1^{\text{cor}}\{\bar{c}\sigma_1\overline{c\sigma_1}^{-1}(\overline{c\sigma_1}[\sigma_2]) - \bar{c}[\sigma_1\sigma_2] + \bar{c}[\sigma_1]\} \\
&= \bar{c}\sigma_1\overline{c\sigma_1}^{-1}[\overline{c\sigma_1}\sigma_2\overline{\overline{c\sigma_1}\sigma_2}^{-1}] - [\bar{c}\sigma_1\sigma_2\overline{c\sigma_1\sigma_2}^{-1}] + [\bar{c}\sigma_1\overline{c\sigma_1}^{-1}] \\
&= \partial_2^H[\bar{c}\sigma_1\overline{c\sigma_1}^{-1}, \overline{c\sigma_1}\sigma_2\overline{c\sigma_1\sigma_2}^{-1}]
\end{aligned}$$

so that Λ_2^{cor} should be defined by

$$\Lambda_2^{\text{cor}}(\bar{c}[\sigma_1, \sigma_2]) = [\bar{c}\sigma_1\overline{c\sigma_1}^{-1}, \overline{c\sigma_1}\sigma_2\overline{c\sigma_1\sigma_2}^{-1}]$$

It may now be left to the reader to show inductively that Λ_n^{cor}, for $n \geqslant 1$, may be taken as

$$\begin{aligned}
&\Lambda_n^{\text{cor}}(\bar{c}[\sigma_1, \ldots, \sigma_n]) \\
&\quad = [\bar{c}\sigma_1\overline{c\sigma_1}^{-1}, \overline{c\sigma_1}\sigma_2\overline{c\sigma_1\sigma_2}^{-1}, \ldots, \overline{c\sigma_1\sigma_2\cdots\sigma_{n-1}}\sigma_n\overline{c\sigma_1\cdots\sigma_n}^{-1}]
\end{aligned}$$

From (#) and the fact that $G = \bigcup \bar{c}^{-1}H$ is a left coset decomposition, so that $S_{H\to G} = \sum_{c\in C} \bar{c}^{-1}$, we conclude that :

2-5-2. Proposition. For positive dimensions, the transfer (or corestriction) is determined by a standard complex cochain map of form

$$(\text{cor } u)([\dot{_G}]) = \sum_{c\in C} \bar{c}^{-1}u([\dot{_H}])$$

$$\begin{aligned}
&(\text{cor } u)([\sigma_1, \ldots, \sigma_n]) \\
&\quad = \sum_{c\in C} \bar{c}^{-1}u([\bar{c}\sigma_1\overline{c\sigma_1}^{-1}, \overline{c\sigma_1}\sigma_2\overline{c\sigma_1\sigma_2}^{-1}, \ldots, \overline{c\sigma_1\cdots\sigma_{n+1}}\sigma_n\overline{c\sigma_1\cdots\sigma_n}^{-1}])
\end{aligned}$$

It remains to consider negative dimensions. Inflation does not exist for negative dimensions, so it does not enter into the discussion. As for restriction and corestriction we have already defined, for $n \geqslant 0$, H-homomorphisms

$$\Lambda_n^{\text{res}} : X_n^H \longrightarrow X_n^G \qquad \text{and} \qquad \Lambda_n^{\text{cor}} : X_n^G \longrightarrow X_n^H$$

which are associated with the commutative diagrams

$$\begin{array}{ccccccc} \cdots \longrightarrow & X_1^H & \xrightarrow{\partial_1^H} & X_0^H & & & \\ & \Big\downarrow \Lambda_1^{\text{res}} & & \Big\downarrow \Lambda_0^{\text{res}} & \searrow^{\varepsilon^H} & & \\ & & & & & \mathbf{Z} & \\ & & & & \nearrow_{\varepsilon^G} & & \\ \cdots \longrightarrow & X_1^G & \xrightarrow{\partial_1^G} & X_0^G & & & \end{array} \tag{$*$}$$

and

$$\begin{array}{ccccccc} \cdots \longrightarrow & X_1^G & \xrightarrow{\partial_1^G} & X_0^G & & & \\ & \Big\downarrow \Lambda_1^{\text{cor}} & & \Big\downarrow \Lambda_0^{\text{cor}} & \searrow^{\varepsilon G} & & \\ & & & & & \mathbf{Z} & \\ & & & & \nearrow_{\varepsilon H} & & \\ \cdots \longrightarrow & X_1^H & \xrightarrow{\partial_1^H} & X_1^H & & & \end{array} \tag{$**$}$$

Using the notation of Section 1-4 we have the dual mappings

$$\widehat{\Lambda_n^{\text{res}}} : \widehat{X_n^G} \longrightarrow \widehat{X_n^H} \qquad \text{and} \qquad \widehat{\Lambda_n^{\text{cor}}} : \widehat{X_n^H} \longrightarrow \widehat{X_n^G}.$$

Now putting

$$\Lambda_n^{\text{res}} = \widehat{\Lambda_{-(n+1)}^{\text{cor}}} \qquad \Lambda_n^{\text{cor}} = \widehat{\Lambda_{-(n+1)}^{\text{res}}} \qquad n < 0$$

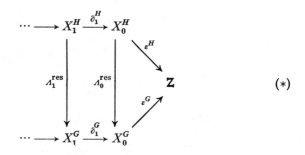

we have H-homomorphisms

$$\Lambda_n^{\mathrm{res}} : X_n^H \longrightarrow X_n^G \qquad \text{and} \qquad \Lambda_n^{\mathrm{cor}} : X_n^G \longrightarrow X_n^H \text{ for } n < 0$$

which are associated with the commutative diagrams

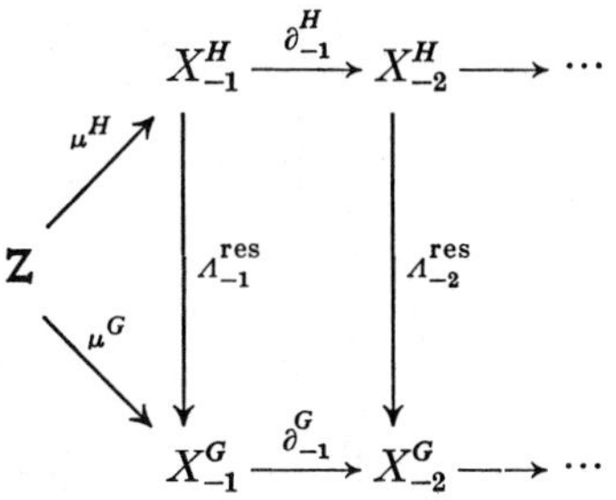

(***)

and

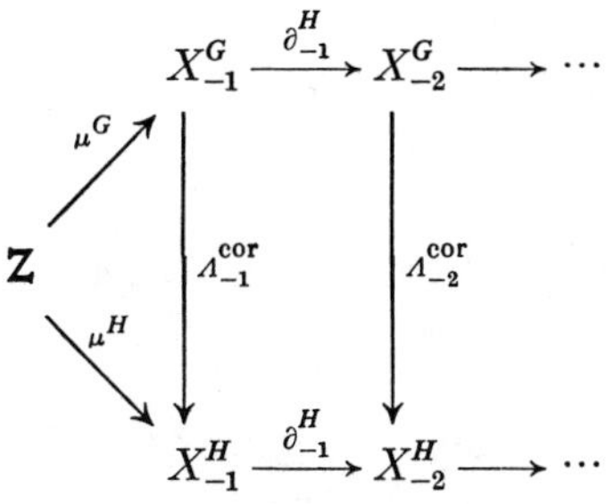

(****)

When (*) is combined with (***) and (**) is combined with (****) we have commutativity for the rectangles connecting dimensions 0 and -1—for example,

$$\Lambda_{-1}^{\mathrm{res}} \circ \partial_0^H = \Lambda_{-1}^{\mathrm{res}} \circ \mu^H \circ \varepsilon^H = \mu^G \circ \varepsilon^H = \mu^G \circ \varepsilon^G \circ \Lambda_0^{\mathrm{res}} = \partial_0^G \circ \Lambda_0^{\mathrm{res}}$$

This completes the construction of the negative dimensional Λ's, and it remains to determine their action explicitly.

It suffices to describe $\Lambda_{-n}^{\mathrm{res}}$ on an H-basis of X_n^H. We assert, first of all, that

$$\Lambda_{-1}^{\mathrm{res}}(\langle\dot{}_H\rangle) = \sum_{c \in \mathcal{C}} \langle\dot{}_G\rangle^{\bar{c}}$$

To verify this, we note that $\Lambda_{-1}^{\mathrm{res}}(\langle \dot{_G} \rangle) \in \mathrm{Hom}\,(\Gamma_G[\dot{_G}], \mathbf{Z})$ so that it is determined by its behavior on a $\mathbf{Z}$-basis of $\Gamma_G[\dot{_G}]$—that is, on $\{\sigma[\dot{_G}] \mid \sigma \in G\}$. Now

$$\begin{aligned}(\Lambda_{-1}^{\mathrm{res}}(\langle \dot{_H} \rangle))(\sigma[\dot{_G}]) &= (\widehat{\Lambda_0^{\mathrm{cor}}(\langle \dot{_H} \rangle)})(\sigma[\dot{_G}]) \\ &= (\langle \dot{_H} \rangle \circ \Lambda_0^{\mathrm{res}})(\rho\bar{c}[\dot{_G}]) \qquad \text{where} \quad \sigma = \rho\bar{c} \quad \rho \in H \\ &= \langle \dot{_H} \rangle(\rho[\dot{_H}]) \\ &= \begin{cases} 1 & \text{if} \quad \sigma = \text{some } \bar{c} \\ 0 & \text{otherwise} \end{cases}\end{aligned}$$

and it is clear that $\sum_c \langle \dot{_G} \rangle^{\bar{c}}$ takes on the same values on $\{\sigma[\dot{_G}]\}$. This proves the assertion. By the same procedure one may show that for $n \geqslant 2$, $\rho_1, \ldots, \rho_{n-1} \in H$

$$\Lambda_{-n}^{\mathrm{res}}(\langle \rho_1, \ldots, \rho_{n-1} \rangle) = \sum_{c_0, \ldots, c_{n-1}} \langle \bar{c}_0^{-1}\rho_1\bar{c}_1, \bar{c}_1^{-1}\rho_2\bar{c}_2, \ldots, \bar{c}_{n-2}^{-1}\rho_{n-1}\bar{c}_{n-1} \rangle^{\bar{c}_0}$$

As for $\Lambda_{-n}^{\mathrm{cor}}$, it is sufficient to give its action on an H-basis of X_{-n}^G. We assert first that

$$\Lambda_{-1}^{\mathrm{cor}}(\langle \dot{_G} \rangle^{\bar{c}}) = \begin{cases} \langle \dot{_H} \rangle^{\bar{c}} & \text{if} \quad c = H \\ 0 & \text{otherwise} \end{cases}$$

In fact, for $\rho \in H$

$$(\Lambda_{-1}^{\mathrm{cor}}(\langle \dot{_G} \rangle^{\bar{c}}))(\rho[\dot{_H}]) = \langle \dot{_G} \rangle^{\bar{c}}(\rho[\dot{_G}]) = \begin{cases} 1 & \text{if} \quad \bar{c} = \rho \\ 0 & \text{otherwise} \end{cases}$$

and the assertion follows immediately. In the same way, one may show that for $n \geqslant 2$, $\sigma_1, \ldots, \sigma_{n-1} \in G$

$$\Lambda_{-n}^{\mathrm{cor}}(\langle \sigma_1, \ldots, \sigma_{n-1} \rangle^{\bar{c}}) = \begin{cases} \langle \sigma_1, \ldots, \sigma_{n-1} \rangle^{\bar{c}} & \text{if} \quad c = H, \quad \sigma_1, \ldots, \sigma_{n-1} \in H \\ 0 & \text{otherwise} \end{cases}$$

From all this we conclude

2-5-3. Proposition. For negative dimensions, restriction and corestriction are determined by standard complex cochain maps of form:

$$(\text{res } u)(\langle\dot{_H}\rangle) = \sum_c \bar{c}(u(\langle\dot{_G}\rangle))$$

$$(\text{res } u)(\langle\rho_1, \ldots, \rho_{n-1}\rangle) = \sum_{c_0, \ldots, c_{n-1}} \bar{c}_0 u(\langle\bar{c}_0^{-1}\rho_1\bar{c}_1, \ldots, \bar{c}_{n-2}^{-1}\rho_{n-1}\bar{c}_{n-1}\rangle)$$

$$n \geqslant 2, \quad \rho_1, \ldots, \rho_n \in H$$

$$(\text{cor } u)(\langle\dot{_G}\rangle) = u(\langle\dot{_H}\rangle)$$

$$(\text{cor } u)(\langle\sigma_1, \ldots, \sigma_{n-1}\rangle) = \begin{cases} u(\langle\sigma_1, \ldots, \sigma_{n-1}\rangle) \text{ if all } \sigma_i \in H \\ 0 \quad \text{otherwise} \end{cases} \quad \begin{matrix} n \geqslant 2 \\ \text{all } \sigma_i \in G \end{matrix}$$

2-5-4. Proposition. For negative dimensions, conjugation is determined by a standard complex cochain map of form

$$(\text{con}_\sigma u)(\langle\dot{_{H^\sigma}}\rangle) = \sigma u(\langle\dot{_H}\rangle)$$

$$(\text{con}_\sigma u)(\langle\rho_1, \ldots, \rho_{n-1}\rangle) = \sigma u(\langle\sigma^{-1}\rho_1\sigma, \ldots, \sigma^{-1}\rho_n\sigma\rangle) \quad n \geqslant 1, \quad \rho_1, \ldots, \rho_{n-1} \in H^\sigma$$

Proof: In connection with the homomorphism of pairs $(\sigma^{-1}, \sigma) : (H, B) \longrightarrow (H^\sigma, \sigma B)$ we have for $n \geqslant 0$ the maps $\Lambda_n^\sigma : X_n^{H^\sigma} \longrightarrow X_n^H$ used for the computation of con_σ. They are given by the formulas (see (2-5-1))

$$\Lambda_0^\sigma([\dot{_{H^\sigma}}]) = [\dot{_H}]$$

$$\Lambda_n^\sigma([\rho_1, \ldots, \rho_n]) = [\sigma^{-1}\rho_1\sigma, \ldots, \sigma^{-1}\rho_n\sigma] \qquad n \geqslant 1, \quad \rho_1, \ldots, \rho_n \in H^\sigma$$

In the same way, for $(\sigma, \sigma^{-1}) : (H^\sigma, \sigma B) \longrightarrow (H, B)$ we have $\Lambda_n^{\sigma^{-1}} : X_n^H \longrightarrow X_n^{H^\sigma}$ for $n \geqslant 0$. According to the dualization procedure used earlier we may define

$$\Lambda_n^\sigma = \widehat{\Lambda_{-(n+1)}^{\sigma^{-1}}} : X_n^{H^\sigma} \longrightarrow X_n^H \qquad n < 0$$

It follows easily that

$$\Lambda^{\sigma}_{-1}(\langle \dot{\ }_{H^{\sigma}} \rangle) = \langle \dot{\ }_{H} \rangle$$

$$\Lambda^{\sigma}_{-n}(\langle \rho_1, \ldots, \rho_{n-1} \rangle) = \langle \sigma^{-1}\rho_1\sigma, \ldots, \sigma^{-1}\rho_{n-1}\sigma \rangle \qquad n \geqslant 2, \quad \rho_1, \ldots, \rho_{n-1} \in H^{\sigma}$$

so that, in negative dimensions, $\mathrm{con}_{\sigma} = (\Lambda^{\sigma}, \sigma)$ is as described. ▮

In virtue of these explicit formulas our mappings of cohomology groups can, for dimensions 0 and -1, be described in terms of mappings of modules—more precisely,

2-5-5. Proposition. The following diagrams commute :

$$\begin{array}{ccc} A^G & \xrightarrow{\ i\ } & A^H \\ {\scriptstyle \kappa^G}\downarrow & & \downarrow{\scriptstyle \kappa^H} \\ H^0(G, A) & \xrightarrow{\mathrm{res}_{G\to H}} & H^0(H, A) \end{array} \qquad \begin{array}{ccc} A_{S_G} & \xrightarrow{\sum_{c} \bar{c}} & A_{S_H} \\ {\scriptstyle \eta^G}\downarrow & & \downarrow{\scriptstyle \eta^H} \\ H^{-1}(G, A) & \xrightarrow{\mathrm{res}_{G\to H}} & H^{-1}(H, A) \end{array}$$

$$\begin{array}{ccc} A^H & \xrightarrow{S_{H\to G}} & A^G \\ {\scriptstyle \kappa^H}\downarrow & & \downarrow{\scriptstyle \kappa^G} \\ H^0(H, A) & \xrightarrow{\mathrm{cor}_{H\to G}} & H^0(G, A) \end{array} \qquad \begin{array}{ccc} A_{S_H} & \xrightarrow{\ i\ } & A_{S_G} \\ {\scriptstyle \eta^H}\downarrow & & \downarrow{\scriptstyle \eta^G} \\ H^{-1}(H, A) & \xrightarrow{\mathrm{cor}_{H\to G}} & H^{-1}(G, A) \end{array}$$

$$\begin{array}{ccc} B^H & \xrightarrow{\ \sigma\ } & (\sigma B)^{H^{\sigma}} = \sigma B^H \\ {\scriptstyle \kappa^H}\downarrow & & \downarrow{\scriptstyle \kappa^{H^{\sigma}}} \\ H^0(H, B) & \xrightarrow{\ \sigma_*\ } & H^0(H^{\sigma}, \sigma B) \end{array} \qquad \begin{array}{ccc} B_{S_H} & \xrightarrow{\ \sigma\ } & (\sigma B)_{S_{H^{\sigma}}} = \sigma B_{S_H} \\ {\scriptstyle \eta^H}\downarrow & & \downarrow{\scriptstyle \eta^{H^{\sigma}}} \\ H^{-1}(H, B) & \xrightarrow{\ \sigma_*\ } & H^{-1}(H^{\sigma}, \sigma B) \end{array}$$

CHAPTER

III

Some Properties of Cohomology Groups

This chapter is devoted to several special and unrelated topics about cohomology groups. The only thread connecting them is their applicability in the development of class field theory.

As usual, G will refer to a finite group; however, we shall be careless about the notation $H^n(G, A)$. It will be clear from the context whether we are dealing with concrete cohomology groups computed with respect to some G-complex or with abstract ones.

3-1. MISCELLANEOUS FACTS

This section deals with some simple properties of cohomology groups—for example, special conditions under which the cohomology groups vanish, and also the relations between the cohomology of a group and the cohomology of its Sylow subgroups.

3-1-1. Lemma. Let $f : A \longrightarrow B$ be a G-homomorphism of G-modules. If f is a trace (meaning that $f = Sg$ for some $g \in \mathrm{Hom}\,(A, B)$), then $f_* = 0$.

Proof: Let X be a G-complex, and let D be a contracting homotopy for it—so $D\partial + \partial D = 1$. For any n, let u be an n-cocycle of G in A with respect to X; in other words, $u \in \mathrm{Hom}_G(X_n, A)$ with $\delta u = 0$. Therefore, $u = u \cdot 1 = uD\partial + u\partial D = uD\partial$. Now, making use of (2-4-1) we have

$$(1, f)\, u = fu = S(g)\, u = S(gu) = S(guD\partial) = S(guD)\, \partial = \delta[S(guD)]$$

so that the image of an n-cocycle is a coboundary, and $f_* = 0$. ▮

3-1-2. Corollary. If the identity map of A, $1 : A \longrightarrow A$ is the trace of an endomorphism of A, then $H^n(G, A) = (0)$ for all n.

Proof: From above $1_* = 0$; but $1_* : H^n(G, A) \longrightarrow H^n(G, A)$ is the identity. ▮

3-1-3. Corollary. If the module A is G-regular, then $H^n(G, A) = (0)$ for all n.

Proof: By (2-1-4) the identity map of A is a trace. ▮

3-1-4. Corollary. Suppose that K/F is a finite Galois extension with Galois group G, then

$$H^n(G, K^+) = (0) \qquad n \in \mathbf{Z}$$

Proof: According to the normal basis theorem—see [3, p. 66] or [41, pp. 57, 61]—there exists $\alpha \in K$ such that

$$K^+ = \sum_{\sigma \in G} \oplus F^+(\sigma\alpha)$$

in other words, K^+ is G-regular. ▮

3-1-5. Corollary. If $G = \{1\}$, then $H^n(G, A) = (0)$ for all n.

Proof: The identity is a trace. ▮

3-1-6. Proposition. If G has order m, then for any n every element of $H^n(G, A)$ has m as a period—that is

$$m \cdot H^n(G, A) = (0) \qquad n \in \mathbf{Z}$$

Proof: Let $m : A \longrightarrow A$ be the map which takes $a \longrightarrow ma$. Since m is the trace of the identity, we have $m_* = 0$. On the other hand, $m_* : H^n(G, A) \longrightarrow H^n(G, A)$ is the map which takes $\alpha \longrightarrow m\alpha$. ▮

3-1-7. Corollary. Suppose that G has order m and that r is an integer for which $(m, r) = 1$ and $rA = (0)$. Then $H^n(H, A) = (0)$ for all n and all subgroups H of G.

Proof: Let $x, y \in \mathbf{Z}$ be such that $xm + yr = 1$. Then $xm : A \longrightarrow A$ is the identity map, and

$$xm = (xm)_* : H^n(G, A) \longrightarrow H^n(G, A)$$

is the identity. Because $mH^n(G, A) = (0)$, we have $H^n(G, A) = (0)$. The same argument applies to any $H \subset G$ because its order is prime to r. ▮

3-1-8. Corollary. Let V be a vector space over the field F of characteristic $p \neq 0$. If V is a G-module and $(p, m) = 1$, where $m = \#(G)$, then $H^n(H, A) = (0)$ for all n and all subgroups H of G.

Proof: Apply (3-1-7). ▮

3-1-9. Proposition. If A is a finitely generated G-module, then $H^n(G, A)$ is finite for every n.

Proof: The result holds if A is finite, because then, for any G-complex X, there are only a finite number of cochains of G in A. In the general case, consider the exact G-sequence

$$0 \longrightarrow A \xrightarrow{m} A \xrightarrow{\pi} \frac{A}{mA} \longrightarrow 0 \qquad m = \#(G)$$

where π is the canonical map. For each n, it determines an exact sequence

$$H^n(G, A) \xrightarrow{m_* = m} H^n(G, A) \xrightarrow{\pi_*} H^n\left(G, \frac{A}{mA}\right)$$

Now, π_* is a monomorphism because $m_* = 0$, and $H^n(G, A/(mA))$ is finite because $A/(mA)$ is finite. This completes the proof. ∎

3-1-10. Corollary. $H^n(G, \mathbf{Z})$ is a finite group for every n.

The G-module A is said to be **uniquely divisible by r** when then map $r : A \longrightarrow A$ is a G-isomorphism onto. The inverse map is denoted by $1/r$, and we have $(1/r) \circ (r) = (r) \circ (1/r) = 1_A$, the identity map of A.

3-1-11. Proposition. If A is uniquely divisible by the order m of G, then $H^n(G, A) = (0)$ for all n.

Proof: We have $m_* = 0$ and also $(m)_*(1/m)_* = (1_A)_* = 1$. ∎

3-1-12. Proposition. Suppose that G acts trivially on a field F of characteristic 0 (viewed additively), then

$$H^n\left(G, \frac{F}{\mathbf{Z}}\right) \approx H^{n+1}(G, \mathbf{Z})$$

Proof: Consider the exact additive G-sequence

$$0 \longrightarrow \mathbf{Z} \longrightarrow F \longrightarrow \frac{F}{\mathbf{Z}} \longrightarrow 0,$$

where the action of G is trivial. Since F is uniquely divisible by $m = \#(G)$, the desired isomorphism follows from the exactness of

$$H^n(G,F) \longrightarrow H^n\left(G, \frac{F}{\mathbf{Z}}\right) \xrightarrow{\delta_*} H^{n+1}(G, \mathbf{Z}) \longrightarrow H^{n+1}(G,F) \quad ∎$$

3-1-13. Corollary. For all n, we have

$$H^n\left(G, \frac{\mathbf{Q}}{\mathbf{Z}}\right) \approx H^{n+1}(G, \mathbf{Z}) \approx H^n\left(G, \frac{\mathbf{R}}{\mathbf{Z}}\right)$$

3-1-14. Proposition. If G is a group of automorphisms of a finite field K, then

$$H^0(G, K^*) = (1)$$

Proof: Let $F = K^G$ (the fixed field of G),

$$q = \#(F), \qquad m = \#(G) = [K : F], \qquad \text{and} \qquad q^m = \#(K).$$

Of course, G is cyclic with generator $\sigma : \alpha \longrightarrow \alpha^q$, $\alpha \in K$, and according to (1-5-8), $H^0(G, K^*) \approx F^*/(N_{K \to F}K^*)$. Now,

$$N_{K \to F} : K^* \longrightarrow F^*$$

is a homomorphism given by

$$N_{K \to F}(\alpha) = \alpha^{1+q+\cdots+q^{m-1}} = \alpha^{(q^m-1)/(q-1)}$$

so that $\#(\ker N_{K \to F}) \leqslant (q^m - 1)/(q - 1)$. On the other hand, $\#(K^*)/\#(\ker N_{K \to F}) \leqslant \#(F^*)$ leads to

$$(q^m - 1)/(q - 1) \leqslant \#(\ker N_{K \to F}).$$

Hence equality holds, the norm map is onto, and

$$H^0(G, K^*) = (1).$$

∎

By combining this result with (1-5-4) and (3-2-1) it will follow that $H^n(G, K^*) = (1)$ for all n, when K is finite.

Consider the group G of order m. For each prime p, let G_p be a p-Sylow subgroup of G, so that, according to the definition of p-Sylow subgroup, $\#(G_p) = p^r$ where $m = p^r m'$, $(m', p) = 1$. (The basic properties of Sylow groups are given, for example, in [83, Chapter 4]; we shall require no information about them other than their existence.) Now, consider a G-module A. For each, n, $mH^n(G, A) = (0)$, which implies that $H^n(G, A)$ is an abelian torsion group. (For facts about such groups, see [42, Section 3].) Let $H^n(G, A)_p$ denote the set of all elements of $H^n(G, A)$ whose order is a power of p. Then $H^n(G, A)_p$ is a subgroup (known as the p-primary part) of $H^n(G, A)$ and

$$H^n(G, A) = \sum_p \oplus H^n(G, A)_p \qquad n \in \mathbf{Z}$$

Of course, if $p \nmid m$, then $H^n(G, A)_p = (0)$, so that the direct sum is a finite one, and, in particular, if G is a p-group, then $H^n(G, A) = H^n(G, A)_p$. The cohomology of G_p is related to that of G by the following result:

3-1-15. Proposition. Let A be a G-module, and let G_p be a p-Sylow subgroup of G, then for all n

(1) $0 \longrightarrow H^n(G, A)_p \xrightarrow{\text{res}} H^n(G_p, A)$ is exact.

(2) $H^n(G_p, A) \xrightarrow{\text{cor}} H^n(G, A)_p \longrightarrow 0$ is exact.

(3) $H^n(G_p, A) = \text{im (res)} \oplus \ker \text{(cor)}$.

Proof: By res we mean $\text{res}_{G \to G_p}$ restricted to $H^n(G, A)_p$, while cor is $\text{cor}_{G_p \to G}$ which indeed maps $H^n(G_p, A)$ into $H^n(G, A)_p$ because every element of $H^n(G_p, A)$ has order a power of p.

Write $m = p^r m'$, $(p^r, m') = 1$, so $\#(G_p) = p^r$, $(G : G_p) = m'$. Choose integers x, y such that $xp^r + ym' = 1$. For any $\alpha \in H^n(G, A)_p$ we have

$$\alpha = (xp^r + ym')\,\alpha = ym'\alpha = y \text{ cor res } \alpha = \text{cor}(y \text{ res } \alpha)$$

which proves both (1) and (2).

Given $\beta \in H^n(G_p, A)$ we may write

$$\beta = y \text{ res cor } \beta + (\beta - y \text{ res cor } \beta)$$

with y res cor $\beta \in \text{im (res)}$ and $\beta - y \text{ res cor } \beta \in \ker \text{(cor)}$. As for directness, if $\beta = \text{res } \alpha$ and $\text{cor } \beta = 0$, then $0 = \text{cor (res } \alpha) = m'\alpha$ so that $ym'\alpha = 0$, $\alpha = 0$, and $\beta = 0$. This completes the proof. ∎

3-1-16. Corollary. For each n, the map

$$\alpha = \sum_{p|m} \alpha_p \longrightarrow \sum_{p|m} \text{res}_{G \to G_p}(\alpha_p)$$

of

$$H^n(G, A) = \sum_{p|m} \oplus H^n(G, A)_p \longrightarrow \sum_{p|m} \oplus H^n(G_p, A)$$

is a monomorphism, and the image is a direct summand. Thus, if $H^n(G_p, A)$ is finite for each $p \mid m$, then $H^n(G, A)$ is finite, and its order divides the product of the orders of the $H^n(G_p, A)$.

3-2. CYCLIC GROUPS AND THE HERBRAND QUOTIENT

3-2-1. Proposition. Suppose that G is a cyclic group, then for any G-module A

$$H^n(G, A) \approx H^{n+2}(G, A) \qquad n \in \mathbf{Z}$$

Moreover, if σ is a generator of G, then

$$H^0(G, A) \approx \frac{A_{\sigma-1}}{SA} \qquad H^{-1}(G, A) \approx \frac{A_S}{(\sigma - 1) A}$$

Proof: Because G is cyclic, we can construct an especially simple G-complex. For each $n \in \mathbf{Z}$ choose an indeterminate x_n, and let X_n be the free module over $\Gamma = \mathbf{Z}[G]$ with the single free generator x_n—$X_n = \Gamma x_n$. Define the boundary operator $\partial = \sum \oplus \partial_n$ by putting for $\gamma \in \Gamma$

$$\partial(\gamma x_n) = \partial_n(\gamma x_n) = \begin{cases} S(\gamma x_{n-1}) & n \text{ even} \\ (\sigma - 1)(\gamma x_{n-1}) & n \text{ odd} \end{cases}$$

Since Γ is commutative and $S \cdot (\sigma - 1) = (\sigma - 1) \cdot S = 0$, we see that ∂ is a G-homomorphism with $\partial^2 = 0$.

To show that the finite free chain complex $(X, \partial, -1)$ is acyclic, let us write $G = \{1, \sigma, \ldots, \sigma^{m-1}\}$ (where G has order m) so that $S = \sum_0^{m-1} \sigma^i$ and any element $\gamma \in \Gamma$ is of the form

$$\gamma = \sum_0^{m-1} c_i \sigma^i, \qquad c_i \in \mathbf{Z}.$$

Since $\partial^2 = 0$, it suffices to show that $\ker \partial \subset \operatorname{im} \partial$. If n is odd, then

$$\gamma \in \ker(\sigma - 1) \Longrightarrow \sigma\gamma = \gamma \Longrightarrow \text{all } c_i \text{ are equal, say } c_i = c$$
$$\Longrightarrow \gamma = cS \Longrightarrow \gamma \in \operatorname{im} S,$$

so that $\operatorname{im} S = \ker(\sigma - 1)$. If n is even, then

$$\gamma \in \ker S \Longrightarrow S\left(\sum c_i\sigma^i\right) = 0 \Longrightarrow \left(\sum c_i\right) S = 0 \Longrightarrow \sum c_i = 0$$
$$\Longrightarrow \gamma = \sum c_i\sigma^i = \sum c_i(\sigma^i - 1) \Longrightarrow \gamma \in \operatorname{im}(\sigma - 1),$$

so that $\operatorname{im}(\sigma - 1) = \ker S$.

One may also prove acyclicity by verifying that the following $D = \sum \oplus D_n$, which is defined on the **Z**-generators and extended linearly, is a contracting homotopy :

$$n \text{ odd}: \quad D_n(\sigma^i x_n) = \begin{cases} x_{n+1} & i = m - 1 \\ 0 & \text{otherwise} \end{cases}$$

$$n \text{ even}: \quad D_n(\sigma^i x_n) = S^{(i)} x_{n+1}$$

where for $i = 1, \ldots, m$ we let $S^{(i)} = 1 + \sigma + \cdots + \sigma^{i-1}$ and $S^{(0)} = 0$.

As for the augmentation, put

$$\varepsilon(x_0) = 1 \qquad \mu(1) = Sx_{-1}$$

and extend them to Γ-homomorphisms $\varepsilon : X_0 \twoheadrightarrow \mathbf{Z}$, $\mu : \mathbf{Z}u \rightarrowtail X_{-1}$ which then satisfy $\mu \circ \varepsilon = \partial_0$. Thus, $(X, \partial, -1)$ is indeed a G-complex, and because it is periodic, it follows immediately from the definition of cohomology groups of G in A that

$$H^n(G, A) \approx H^{n+2}(G, A) \qquad n \in \mathbf{Z}$$

Finally, denoting the kernel of $(\sigma - 1) : A \longrightarrow A$ by $A_{\sigma-1}$, we have $A^G = A_{\sigma-1}$ and $IA = \sum_1^{m-1} (\sigma^i - 1)A = (\sigma - 1)A$, so that from (1-5-6) and (1-5-9),

$$H^0(G, A) \approx \frac{A_{\sigma-1}}{SA} \qquad H^{-1}(G, A) \approx \frac{A_S}{(\sigma - 1)A}$$

Of course, these isomorphisms, which may be given in terms of the maps κ and η, respectively (see Section 2-2), are independent of the choice of generator σ of G. ∎

3-2-2. Remark. One may be somewhat more explicit, at this stage, about the isomorphisms $H^n(G, A) \approx H^{n+2}(G, A)$; more detailed information about these isomorphisms will be given in (4-5-10). Define a map $\Lambda : X \longrightarrow X$ by putting $\Lambda(x_n) = x_{n+2}$ for all n and then extending by G-linearity. Thus, Λ commutes with ∂, and the map

$$(\Lambda, 1) : \mathrm{Hom}_G (X, A) \longrightarrow \mathrm{Hom}_G (X, A)$$

which takes n-dimensional cochains to $(n-2)$-dimensional cochains, commutes with $\delta = (\partial, 1)$ and induces an isomorphism

$$\Lambda_* = (\Lambda, 1)_* : H^n(G, A) \rightarrowtail\!\!\!\twoheadrightarrow H^{n-2}(G, A) \qquad n \in \mathbf{Z}$$

The location of the augmentation in X is obviously not significant. It is easy to see that Λ_* has the "proper" cohomological behavior. In particular, it commutes with induced homomorphisms and with coboundaries arising from exact sequences. Therefore, it follows that if $0 \longrightarrow A \xrightarrow{i} B \xrightarrow{j} C \longrightarrow 0$ is an exact G-sequence, then we have an exact hexagon

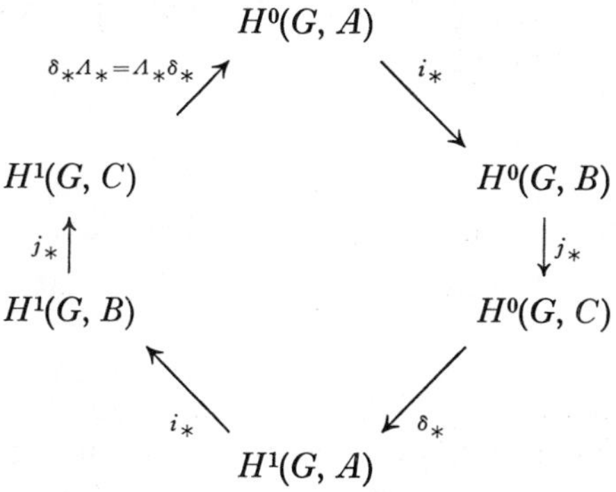

and can treat the Herbrand quotient by the method used in Section 1-2. However, we prefer to do so by a slightly different procedure.

For any finite group G and any G-module A we denote the order of the nth cohomology group (in a manner consistent with the notation introduced in (1-2-4)) by

$$h_n(A) = h_n(G, A) = \#(H^n(G, A))$$

If G is cyclic, we define

$$h(A) = \frac{h_2(A)}{h_1(A)} = \frac{h_0(A)}{h_1(A)}$$

provided both terms on the right side are finite, and call it the **Herbrand quotient** of the G-module A.

Now, for any $\sigma \in G$, A is an abelian group with endomorphisms $\sigma - 1$ and S such that $(\sigma - 1) \circ S = S \circ (\sigma - 1) = 0$. In particular, if σ is a generator of the cyclic group G, then

$$h(A) = Q_{\sigma-1,S}(A) \qquad (*)$$

where $Q_{\sigma-1,S}(A)$ is the Herbrand quotient of A with respect to the endomorphisms $\sigma - 1$ and S as defined in Section 1-2. Consequently, (1-2-2) and (1-2-3) can be carried over to h, and we have :

3-2-3. Proposition. Suppose that G is a cyclic group and that $0 \longrightarrow A \longrightarrow B \longrightarrow C \longrightarrow 0$ is an exact G-sequence such that two of the three quotients $h(A)$, $h(B)$, $h(C)$ are defined, then the third one is defined and

$$h(B) = h(A)\, h(C)$$

Moreover, if the G-module A is finite then

$$h(A) = 1$$

3-2-4. Corollary. If G is cyclic and $f: A \longrightarrow B$ is a G-homomorphism of G-modules such that the kernel $f^{-1}(0)$ and the cokernel $B/(f(A))$ are both finite, then $h(A)$ is defined $\Longleftrightarrow$ $h(B)$ is defined, and in this situation

$$h(A) = h(B)$$

Proof: Apply the preceding result to the exact G-sequences $0 \longrightarrow \ker f \longrightarrow A \longrightarrow A/\ker f \longrightarrow 0$, $0 \longrightarrow A/\ker f \longrightarrow \operatorname{im} f \longrightarrow 0$, and $0 \longrightarrow \operatorname{im} f \longrightarrow B \longrightarrow B/\operatorname{im} f \longrightarrow 0$. ∎

Any G-module A may also be viewed as a G-module for which the action of G is trivial. If $G = \{1, \sigma, \ldots, \sigma^{m-1}\}$ is cyclic, then there is a Herbrand quotient defined in this case too; we denote it by $h^0(A) = h^0(G, A)$. Because the action of G is trivial, $\sigma - 1$ is the 0-map on A and S is multiplication by m; so according to (*) we have

$$h^0(A) = Q_{0,m}(A)$$

In other words,

$$h^0(A) = \frac{(A : mA)}{(A_m : (0))}$$

provided both terms on the right are finite, and where A_m is the kernel of the map $a \longrightarrow ma$. It is surprising, perhaps, that for certain G there is a connection between $h(A)$ and $h^0(A)$. This connection is given by our next result, which is an extension of a result that appears in [17, Theorem 10.3], and whose primary application is to give an efficient proof of the first inequality of global class field theory.

3-2-5. Proposition. (Tate) Suppose that

$$G = \{1, \sigma, \ldots, \sigma^{p-1}\}$$

is a cyclic group of order p, p prime, and that A is a G-module. If $h^0(A)$ is defined, then so are $h^0(A^G)$ and $h(A)$, and then

$$\{h(A)\}^{p-1} = \frac{\{h^0(A^G)\}^p}{h^0(A)}$$

Proof: We prepare for the proof by proving three lemmas.

3-2-6. Lemma. Let $B \subset A \subset E$, $C \subset E$, D be abelian groups, let $\varphi : A \longrightarrow D$ be a homomorphism and write, as usual, A^φ for the image of A under φ and A_φ for the kernel of φ on A; then

$$(A : B) = (A^\varphi : B^\varphi)(A_\varphi : B_\varphi)$$

and

$$(A : B) = \left(\frac{AC}{C} : \frac{BC}{C}\right)(A \cap C : B \cap C)$$

Proof: Start from $(A : B) = (A : BA_\varphi)(BA_\varphi : B)$, and observe that the natural map $A \longrightarrow\!\!\!\!\rightarrow A^\varphi/B^\varphi$ has kernel BA_φ, while $(BA_\varphi)/B \approx A_\varphi/(B \cap A_\varphi) = A_\varphi/B_\varphi$.

For the second part, consider the natural map $\varphi : A \longrightarrow\!\!\!\!\rightarrow (AC)/C$, and apply the first part. ∎

3-2-7. Lemma. Let φ, ψ be commuting endomorphisms of the abelian group A, then

$$Q_{0,\varphi\psi}(A) = Q_{0,\varphi}(A)\, Q_{0,\psi}(A)$$

with the understanding that if the left side is defined then so is the right side, and conversely.

Proof: The assertion is that

$$\frac{(A : A^{\varphi\psi})}{(A_{\varphi\psi} : (0))} = \frac{(A : A^\varphi)(A : A^\psi)}{(A_\varphi : (0))(A_\psi : (0))}$$

To prove it, note first that

$$(A^\varphi)^\psi = A^{\psi\varphi} = A^{\varphi\psi} \subset A^\varphi, \qquad (A^\psi)_\varphi = \{\psi(\alpha) \mid \varphi\psi(a) = 0\} = (A_{\varphi\psi})^\psi$$

and

$$(A_{\varphi\psi})_\psi = A_\psi .$$

Now consider

$$(A : A^{\varphi\psi}) = (A : A^\varphi)(A^\varphi : A^{\varphi\psi})$$

From (3-2-6) we have

$$(A : A^\psi) = (A^\varphi : (A^\psi)^\varphi)(A_\varphi : (A^\psi)_\varphi) = (A^\varphi : A^{\varphi\psi})(A_\varphi : (A_{\varphi\psi})^\psi)$$

and also

$$(A_{\varphi\psi} : (0)) = ((A_{\varphi\psi})^{\psi} : (0))((A_{\varphi\psi})_{\psi} : (0)) = ((A_{\varphi\psi})^{\psi} : (0))(A_{\psi} : (0))$$

Therefore,

$$\frac{(A : A^{\varphi\psi})}{(A_{\varphi\psi} : (0))} = \frac{(A : A^{\varphi})}{1} \cdot \frac{(A : A^{\psi})}{(A_{\varphi} : (A_{\varphi\psi})^{\psi})} \cdot \frac{1}{((A_{\varphi\psi})^{\psi} : (0))(A_{\psi} : (0))}$$

which is the desired result. The finiteness conditions are easily checked. ∎

3-2-8. Lemma. Let $G = \{1, \sigma, \ldots, \sigma^{p-1}\}$ be a cyclic group of prime order p, and let A be a G-module. Then, on A, the maps $(\sigma - 1)^{p-1}$ and p (that is, multiplication by p) have the same image and the same kernel.

Proof: We work in the ring $\mathbf{Z}[G]$. Since σ is essentially a primitive pth root of unity, it follows that

$$X^{p-1} + \cdots + X + 1 = \prod_{1}^{p-1} (X - \sigma^i),$$

so that

$$p = \prod_{1}^{p-1} (1 - \sigma^i)$$

For any $i = 1, \ldots, p - 1$, $(1 - \sigma^i) = (1 - \sigma)(1 + \sigma + \cdots + \sigma^{i-1})$ so that $\operatorname{im}(1 - \sigma^i) \subset \operatorname{im}(1 - \sigma)$ and $\ker(1 - \sigma) \subset \ker(1 - \sigma^i)$. But the roles of σ and σ^i may be interchanged, because σ^i is also a generator of G; consequently, for $i = 1, \ldots, p - 1$

$$\operatorname{im}(1 - \sigma^i) = \operatorname{im}(1 - \sigma) \qquad \ker(1 - \sigma^i) = \ker(1 - \sigma)$$

These relations hold on A and on any G-submodule of A; in particular, they hold on $A^{(\sigma-1)^i}$ and $A_{(\sigma-1)^i}$. Therefore,

$$\operatorname{im}(p) = \operatorname{im}\left(\prod_{1}^{p-1} (1 - \sigma^i)\right) = \operatorname{im}((1 - \sigma)^{p-1}) = \operatorname{im}((\sigma - 1)^{p-1})$$

and similarly $\ker(p) = \ker((\sigma - 1)^{p-1})$. ∎

Now let us turn to the proof of (3-2-5). Consider the exact G-sequence

$$0 \longrightarrow A^G = A_{\sigma-1} \longrightarrow A \xrightarrow{\sigma-1} A^{\sigma-1} \longrightarrow 0$$

The modules and homomorphisms are fixed, and the sequence is exact for each of the actions of G.

If $h^0(A) = (A : mA)/(A_m : (0))$ is defined, then so is

$$h^0(A^{\sigma-1}) = (A^{\sigma-1} : mA^{\sigma-1})/((A^{\sigma-1})_m : (0)),$$

since

$$(A : mA) = (A^{\sigma-1} : (mA)^{\sigma-1})(A_{\sigma-1} : (mA)_{\sigma-1}),$$

$(mA)^{\sigma-1} = mA^{\sigma-1}$, and $(A^{\sigma-1})_m \subset A_m$. Then, according to (3-2-3), $h^0(A^G)$ is defined, and $h^0(A) = h^0(A^G)\, h^0(A^{\sigma-1})$. Furthermore, $h(A^G) = h^0(A^G) < \infty$, and making use of (3-2-8) and (3-2-7) we have for $B = A^{\sigma-1}$

$$h^0(B) = \frac{(B : pB)}{(B_p : (0))} = \frac{(B : B^{(\sigma-1)^{p-1}})}{(B_{(\sigma-1)^{p-1}} : (0))}$$

$$= \frac{(B : B^{(\sigma-1)})^{p-1}}{(B_{\sigma-1} : (0))^{p-1}} = \frac{1}{\{h(B)\}^{p-1}} \qquad (\text{since} \quad SB = (0))$$

In particular, $h(A^{\sigma-1})$ is defined; so according to (3-2-3), $h(A)$ is defined and $h(A) = h(A^G)\, h(A^{\sigma-1})$. Therefore

$$\begin{aligned}\{h(A)\}^{p-1} &= \{h^0(A^G)\}^{p-1}\{h(A^{\sigma-1})\}^{p-1}\\ &= \frac{\{h^0(A^G)\}^{p-1}}{h^0(A^{\sigma-1})}\\ &= \{h^0(A^G)\}^{p-1}\frac{h^0(A^G)}{h^0(A)}\\ &= \frac{\{h^0(A^G)\}^p}{h^0(A)}\end{aligned}$$

This completes the proof. ▮

3-3. DIMENSION SHIFTING

The G-module P is said to be **projective** if whenever we have G-homomorphisms of G-modules $f : P \longrightarrow C$ and $g : B \twoheadrightarrow C$, there exists a G-homomorphism $h : P \longrightarrow B$ such that $f = g \circ h$. This is usually expressed by the requirement that every diagram

$$\begin{array}{ccccc} & & P & & \\ & \swarrow^{h} & \downarrow f & & \\ B & \longrightarrow & C & \longrightarrow & 0 \end{array}$$

of G-modules and G-homomorphisms in which the row is exact can be completed to a commutative diagram. It is clear that P is projective if and only if whenever $f : P \longrightarrow C$ and $g : B \longrightarrow C$ are G-homomorphisms of G-modules with $f(P) \subset g(B)$, then there exists a G-homomorphism $h : P \longrightarrow B$ with $f = g \circ h$.

3-3-1. Proposition. A G-free module is projective.

Proof: Let $\{x_\alpha\}$ be a $\Gamma = \mathbf{Z}[G]$ basis for the G-free module X. For each α, choose $b_\alpha \in B$ such that $g(b_\alpha) = f(x_\alpha)$, and then define h by putting $h(x_\alpha) = b_\alpha$. ∎

3-3-2. Proposition. Let X be an arbitrary G-module, then :

(1) If $A \xrightarrow{i} B \xrightarrow{j} C \longrightarrow 0$ is an exact G-sequence, then the following sequence of abelian groups is exact :

$$0 \longrightarrow \mathrm{Hom}_G(C, X) \xrightarrow{(j,1)} \mathrm{Hom}_G(B, X) \xrightarrow{(i,1)} \mathrm{Hom}_G(A, X)$$

(2) If $0 \longrightarrow A \xrightarrow{i} B \xrightarrow{j} C$ is an exact G-sequence, then the following sequence of abelian groups is exact :

$$0 \longrightarrow \mathrm{Hom}_G(X, A) \xrightarrow{(1,i)} \mathrm{Hom}_G(X, B) \xrightarrow{(1,j)} \mathrm{Hom}_G(X, C)$$

(3) Let $0 \longrightarrow A \longrightarrow B \longrightarrow C \longrightarrow 0$ be an exact G-sequence, then

$$0 \longrightarrow \mathrm{Hom}_G(X, A) \longrightarrow \mathrm{Hom}_G(X, B) \longrightarrow \mathrm{Hom}_G(X, C) \longrightarrow 0$$

is exact for all such short exact sequences $\Longleftrightarrow$ X is projective.

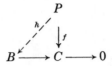

Proof: (1) To show that $(j, 1)$ is a monomorphism, consider $f \in \mathrm{Hom}_G(C, X)$, then

$$(j, 1)f = f \circ j = 0 \Longrightarrow fj(B) = 0 \Longrightarrow f(C) = 0 \Longrightarrow f = 0.$$

Since

$$(i, 1) \circ (j, 1) = (ji, 1) = (0, 1) = 0,$$

we have $\mathrm{im}\,(j, 1) \subset \ker\,(i, 1)$. To prove the reverse inclusion, suppose $f \in \mathrm{Hom}_G(B, X)$ satisfies $(i, 1)f = f \circ i = 0$. Then $f(iA) = 0$ and we may define a G-homomorphism $\bar{f} : B/(iA) \longrightarrow X$ by putting $\bar{f}(b + iA) = f(b)$. Furthermore, j induces a G-isomorphism $\bar{j} : B/(iA) \rightarrowtail\!\!\!\!\twoheadrightarrow C$ and the map $g = \bar{f} \circ \bar{j}^{-1} \in \mathrm{Hom}_G(C, X)$ satisfies $(j, 1)\,g = \bar{f} \circ \bar{j}^{-1} \circ j = f$. This completes the proof of (1). The proof of (2) involves similar arguments. As for (3), exactness at $\mathrm{Hom}_G(X, C)$ is precisely equivalent to the assertion that X is projective. ▮

3-3-3. Remark. To every ordered pair of G-modules A and B, we may assign the $\mathbf{Z}$-module (that is, abelian group) $\mathrm{Hom}_G(A, B)$. Furthermore, given G-homomorphisms of G-modules $f : B \longrightarrow B'$ and $g : A' \longrightarrow A$, there exist homomorphisms of groups

$$(1, f) : \mathrm{Hom}_G(A, B) \longrightarrow \mathrm{Hom}_G(A, B')$$

$$(g, 1) : \mathrm{Hom}_G(A, B) \longrightarrow \mathrm{Hom}_G(A', B)$$

We summarize these facts by saying, loosely, that the **functor** Hom_G is contravariant in the first variable and covariant in the second variable. (A precise formulation of these terms is given in [15, Chapter 2].) We also say that the functor Hom_G is **left exact** which is, by definition, the assertion that parts (1) and (2) of (3-3-2) hold. An exact G-sequence $0 \longrightarrow A \xrightarrow{i} B \xrightarrow{j} C \longrightarrow 0$ is said to **split** (with respect to G, or over G) when iA is a direct summand of B. In other words, there exists a G-submodule B' of B such that $B = iA \oplus B'$; note that then B' is an isomorphic copy of C, since $j : B' \rightarrowtail\!\!\!\!\twoheadrightarrow C$.

3-3-4. Exercise. The following conditions on a G-module P are equivalent :

(1) P is projective.
(2) P is a direct summand of a free G-module.
(3) Every exact G-sequence $0 \longrightarrow A \longrightarrow B \longrightarrow P \longrightarrow 0$ splits.
(4) P is $\mathbf{Z}$-free and the identity map of P is a trace.

3-3-5. Proposition. Let $0 \longrightarrow A \xrightarrow{i} B \xrightarrow{j} C \longrightarrow 0$ be an exact G-sequence, and let X be any G-module, then

(1) If iA is a direct summand of B as a G-module, the sequence

$$0 \longrightarrow \mathrm{Hom}_G(C, X) \xrightarrow{(j,1)} \mathrm{Hom}_G(B, X) \xrightarrow{(i,1)} \mathrm{Hom}_G(A, X) \longrightarrow 0$$

of abelian groups is exact and splits.

(2) If iA is a direct summand of B as a $\mathbf{Z}$-module, then

$$0 \longrightarrow \mathrm{Hom}(C, X) \xrightarrow{(j,1)} \mathrm{Hom}(B, X) \xrightarrow{(i,1)} \mathrm{Hom}(A, X) \longrightarrow 0$$

is an exact G-sequence which splits as a $\mathbf{Z}$-sequence.

Proof: (1) Using the notation introduced in (3-3-3), we have

$$\mathrm{Hom}_G(B, X) = \mathrm{Hom}_G(iA \oplus B', X) \approx \mathrm{Hom}_G(iA, X) \oplus \mathrm{Hom}_G(B', X)$$

and when the appropriate identifications are made the result follows immediately.

(2) Apply part (1) and (1-4-3). Note that we do not assert that the sequence splits as a G-sequence. ∎

3-3-6. Proposition. If A is G-regular, then, for any G-module X, so are $\mathrm{Hom}(X, A)$ and $\mathrm{Hom}(A, X)$.

Proof: Let B be a $\mathbf{Z}$-module such that $A = \sum_{\sigma \in G} \oplus\, \sigma B$. Then

$$\begin{aligned}\mathrm{Hom}(X, A) = \mathrm{Hom}(X, \sum_\sigma \oplus\, \sigma B) &\approx \sum_\sigma \oplus\, \mathrm{Hom}(X, \sigma B)\\ &\approx \sum_\sigma \oplus\, (\mathrm{Hom}(X, B))^\sigma.\end{aligned}$$

A similar proof applies for $\mathrm{Hom}(A, X)$. ∎

3-3-7. Proposition. If A is G-regular and H is a normal subgroup of G, then A^H is (G/H)-regular.

Proof: Write $A = \sum_{\sigma \in G} \oplus \sigma B$ and let $G = \bigcup_i \sigma_i H = \bigcup_i H\sigma_i$ be a coset decomposition. Any element of A has a unique expression $a = \sum_{\sigma \in G} \sigma b_\sigma$, $b_\sigma \in B$. Thus, $a \in A^H \iff$ for each $\sigma \in G$, $b_{\rho\sigma} = b_\sigma$ for all $\rho \in H$. In other words, if $S_H = \sum_{\rho \in H} \rho$ is the H-trace, then $a \in A^H \iff a$ is of the form $a = \sum_i (S_H \sigma_i) b_{\sigma_i}$, $b_{\sigma_i} \in B$. Now, $S_H B \subset A^H$ and because the action of G/H on A^H is given completely by the action of the σ_i, it follows that

$$A^H = \sum_i \oplus (S_H \sigma_i) B = \sum_i \oplus \sigma_i (S_H B) = \sum_{\mu \in (G/H)} \oplus \mu(S_H B)$$

which completes the proof. ∎

3-3-8. Proposition. In connection with the G-module $\Gamma = \mathbf{Z}[G]$ we have exact G-sequences

$$0 \longrightarrow I \longrightarrow \Gamma \xrightarrow{\varepsilon} \mathbf{Z} \longrightarrow 0$$

and

$$0 \longrightarrow \mathbf{Z} \xrightarrow{\mu} \Gamma \longrightarrow J \longrightarrow 0$$

where $\varepsilon(1) = 1$, $\mu(1) = S$, and

$$I = \sum_{\sigma \in G} \oplus \mathbf{Z}(\sigma - 1) \qquad J = \frac{\Gamma}{\mathbf{Z}S}$$

Moreover, I and J are direct $\mathbf{Z}$-summands of Γ (that is, the exact sequences split as $\mathbf{Z}$-sequences) and we have isomorphisms of G-modules

$$\operatorname{Hom}(I, \mathbf{Z}) \approx J \qquad \operatorname{Hom}(J, \mathbf{Z}) \approx I$$

Proof: As usual, the action of G on $\mathbf{Z}$ is trivial. Since the element 1 is a G-basis for Γ we may put $\varepsilon(1) = 1 \in \mathbf{Z}$ and extend G-linearly. Thus, ε maps an element $\gamma = \sum_{\sigma \in G} n_\sigma \sigma \in \Gamma$ to $\sum_{\sigma \in G} n_\sigma \in \mathbf{Z}$. Clearly, ε is onto, and its kernel is

$$I = \{\gamma \in \Gamma \mid \sum_{\sigma \in G} n_\sigma = 0\} = \sum_{\sigma \in G} \oplus \mathbf{Z}(\sigma - 1) = \sum_{\sigma \neq 1} \oplus \mathbf{Z}(\sigma - 1)$$

As for the second sequence, define $\mu(1) = S \in \Gamma$ and extend linearly—so $\mu(n) = nS$ for $n \in \mathbf{Z}$, and μ is a G-homomorphism. Consequently, if we put $J = \Gamma/(\mathbf{Z}S)$, then the second sequence is exact.

Note that if we define a G-homomorphism $S : \Gamma \longrightarrow \Gamma$ by putting $S(1) = S = \sum_{\sigma\in G} \sigma \in \Gamma$ and extending G-linearly from the Γ-basis $\{1\}$, then we have a commutative triangle

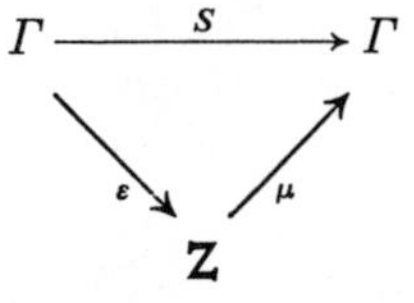

which is essentially the augmentation of the standard G-complex.

Because $\mathbf{Z} \approx \Gamma/I$ and $J \approx \Gamma/(\mu(\mathbf{Z}))$ are $\mathbf{Z}$-free, it follows from (3-3-4) that our sequences split over $\mathbf{Z}$. However, these assertions may also be proved directly. Thus, it is clear that

$$\Gamma = I \oplus \mathbf{Z} \cdot 1 \approx I \oplus \mathbf{Z}$$

In fact, $\gamma = \sum_{\sigma\in G} n_\sigma \sigma \in \Gamma$ is of form

$$\gamma = \sum_{\sigma\in G} n_\sigma(\sigma - 1) + \Big(\sum_{\sigma\in G} n_\sigma\Big) 1$$

As for $J = \Gamma/(\mathbf{Z}S)$, let $\gamma \longrightarrow \bar{\gamma}$ denote the natural map of $\Gamma \longrightarrow\!\!\!\!\rightarrow J$. It is clear that $\bar{1} = -\sum_{\sigma\neq 1} \bar{\sigma}$ and that $J = \sum_{\sigma\neq 1} \oplus \mathbf{Z}\bar{\sigma}$; so

$$\Gamma = \mathbf{Z}S \oplus \sum_{\sigma\neq 1} \oplus \mathbf{Z}\sigma \approx \mathbf{Z} \oplus J$$

We note in passing that the map $\Gamma \longrightarrow J$ is given by

$$\gamma = \sum_{\sigma\in G} n_\sigma \sigma \longrightarrow \bar{\gamma} = \sum_{\sigma\in G} n_\sigma \bar{\sigma} = \sum_{\sigma\neq 1} (n_\sigma - n_1)\, \bar{\sigma}$$

The uniqueness of this last expression implies that if

$$\sum_{\sigma\in G} n_\sigma \bar{\sigma} = \sum_{\sigma\in G} n'_\sigma \bar{\sigma}$$

then $n_\sigma - n_1 = n'_\sigma - n'_1$ for all $\sigma \neq 1$.

To prove the final assertion, let us first define a map $\mathrm{Hom}(I, \mathbf{Z}) \longrightarrow J$ by

$$f \longrightarrow \sum_{\sigma\neq 1} f(\sigma - 1)\,\bar{\sigma}$$

(and the right side equals $\sum_{\sigma\in G} f(\sigma - 1)\bar{\sigma}$, since $f(0) = 0$). This is clearly an isomorphism of abelian groups—and the inverse map takes $\sum_{\sigma\neq 1} n_\sigma \bar{\sigma} \in J$ to the $f \in \mathrm{Hom}(I, \mathbf{Z})$ for which $f(\sigma - 1) = n_\sigma$, $\sigma \neq 1$. To prove that this is a G-isomorphism, we must show that for $\tau \in G$

$$\sum_{\sigma\neq 1} f^\tau(\sigma - 1)\,\bar{\sigma} = \tau\left[\sum_{\sigma\neq 1} f(\sigma - 1)\,\bar{\sigma}\right]$$

—but $\sum_{\sigma\neq 1} f^\tau(\sigma - 1)\bar{\sigma} = \sum_{\sigma\neq 1} f(\tau^{-1}\sigma - \tau^{-1})\bar{\sigma}$, while

$$\begin{aligned}\tau\left[\sum_{\sigma\neq 1} f(\sigma - 1)\,\bar{\sigma}\right] &= \tau\sum_{\sigma\in G} f(\sigma - 1)\,\bar{\sigma} = \sum_{\sigma\in G} f(\sigma - 1)\,\overline{\tau\sigma}\\ &= \sum_{\sigma\in G} f(\tau^{-1}\sigma - 1)\,\bar{\sigma} = \sum_{\sigma\neq 1}\,[f(\tau^{-1}\sigma - 1) - f(\tau^{-1} - 1)]\,\bar{\sigma}\\ &= \sum_{\sigma\neq 1} f(\tau^{-1}\sigma - \tau^{-1})\,\bar{\sigma}\end{aligned}$$

Finally, define a map $\mathrm{Hom}(J, \mathbf{Z}) \longrightarrow I$ by

$$f \longrightarrow \sum_{\sigma\neq 1} f(\bar{\sigma})(\sigma - 1)$$

(and the right side equals $\sum_{\sigma\in G} f(\bar{\sigma})(\sigma - 1)$, where, of course, $f(\bar{1}) = -\sum_{\sigma\neq 1} f(\bar{\sigma})$). This is clearly an isomorphism of abelian groups—and the inverse map takes $\sum_{\sigma\neq 1} n_\sigma(\sigma - 1)$ to the $f \in \mathrm{Hom}(J, \mathbf{Z})$ for which $f(\bar{\sigma}) = n_\sigma$, $\sigma \neq 1$. To prove that this is a G-isomorphism, we must show that for $\tau \in G$

$$\sum_{\sigma\neq 1} f^\tau(\bar{\sigma})(\sigma - 1) = \tau\left[\sum_{\sigma\neq 1} f(\bar{\sigma})(\sigma - 1)\right]$$

—but $\sum_{\sigma\neq 1} f^{\tau}(\bar{\sigma})(\sigma - 1) = \sum_{\sigma\in G} f(\overline{\tau^{-1}\sigma})(\sigma - 1) = \sum_{\sigma\in G} f(\bar{\sigma})(\tau\sigma - 1)$, while

$$\tau\left[\sum_{\sigma\neq 1} f(\bar{\sigma})(\sigma - 1)\right] = \sum_{\sigma\in G} f(\bar{\sigma})(\tau\sigma - \tau)$$

$$\sum_{\sigma\in G} f(\bar{\sigma})[(\tau\sigma - 1) - (\tau - 1)]$$

$$\sum_{\sigma\in G} f(\bar{\sigma})(\tau\sigma - 1)$$

This completes the proof. ∎

For any G-module A we get, from (3-3-8) and (3-3-5) the exact G-sequences

$$0 \longrightarrow \operatorname{Hom}(\mathbf{Z}, A) \xrightarrow{(\varepsilon,1)} \operatorname{Hom}(\Gamma, A) \longrightarrow \operatorname{Hom}(I, A) \longrightarrow 0$$

$$0 \longrightarrow \operatorname{Hom}(J, A) \longrightarrow \operatorname{Hom}(\Gamma, A) \xrightarrow{(\mu,1)} \operatorname{Hom}(\mathbf{Z}, A) \longrightarrow 0$$

and these sequences split as $\mathbf{Z}$-sequences. According to (1-4-5), $\operatorname{Hom}(\mathbf{Z}, A) \approx A$ as G-modules, and according to (3-3-6), $\operatorname{Hom}(\Gamma, A)$ is G-regular; thus we have proved :

3-3-9. Proposition. Given a G-module A, there exist two exact G-seqeunces

$$0 \longrightarrow A \longrightarrow M \longrightarrow A^{-} \longrightarrow 0$$

$$0 \longrightarrow A^{+} \longrightarrow M \longrightarrow A \longrightarrow 0$$

which split as $\mathbf{Z}$-sequences, and where M is G-regular. In fact, $M = \operatorname{Hom}(\Gamma, A)$, $A^{+} = \operatorname{Hom}(J, A)$, $A^{-} = \operatorname{Hom}(I, A)$.

3-3-10. Theorem. (Dimension Shifting). Given a G-module A, there exist G-modules A^{+} and A^{-} such that for any subgroup H of G and any integer n we have

$$H^{n-1}(H, A^{-}) \approx H^{n}(H, A) \approx H^{n+1}(H, A^{+})$$

Proof: For any $H \subset G$ the sequences of (3-3-9) are exact

H-sequences, and it is easy to check that M is H-regular. From (2-2-4) and (3-1-3) we arrive at exact sequences

$$(0) = H^{n-1}(H, M) \longrightarrow H^{n-1}(H, A^-) \xrightarrow{\delta_*} H^n(H, A) \longrightarrow H^n(H, M) = (0)$$

$$(0) = H^n(H, M) \longrightarrow H^n(H, A) \xrightarrow{\delta_*} H^{n+1}(H, A^+) \longrightarrow H^{n+1}(H, M) = (0)$$

and the proof is complete. ▮

Because of these isomorphisms, the exact sequences of (3-3-9) are known as **dimension shifters.**

3-3-11. Corollary. Given a G-module A and an integer d, there exists a G-module B such that for any subgroup H of G and any integer n we have

$$H^n(H, A) \approx H^{n+d}(H, B)$$

Proof: Iterate the preceding result, using A^+ or A^- according as d is positive or negative. ▮

3-4. THE INFLATION-RESTRICTION SEQUENCE

In this section, we suppose that A is a G-module and that H is a normal subgroup of G. Then, using the results of Section 2-3 we have for each $n \geqslant 1$ the sequence

$$0 \longrightarrow H^n(G/H, A^H) \xrightarrow{\text{inf}} H^n(G, A) \xrightarrow{\text{res}} H^n(H, A)$$

which is known as the **inflation-restriction sequence** in dimension n.

3-4-1. Proposition. For each $n \geqslant 1$, res $\circ$ inf $= 0$.

Proof: Applying (2-3-7) with $H_1 = H_2 = H$ gives the commutative diagram

$$\begin{array}{ccc} H^n(G/H, A^H) & \xrightarrow{\text{inf}} & H^n(G, A) \\ \text{res}\downarrow & & \downarrow\text{res} \\ H^n(H/H, A^H) & \xrightarrow{\text{inf}} & H^n(H, A) \end{array}$$

—and by (3-1-5), $H^n(H/H, A^H) = (0)$. ▮

3-4-2. Proposition. In dimension 1, the inflation–restriction sequence

$$0 \longrightarrow H^1(G/H, A^H) \xrightarrow{\text{inf}} H^1(G, A) \xrightarrow{\text{res}} H^1(H, A)$$

is exact.

Proof: Let $\sigma \longrightarrow \bar{\sigma}$ denote the canonical map of $G \twoheadrightarrow G/H$, and for each of the groups G/H, G, H, use its standard complex for computation of the cohomology groups. As in Section 2-5 we use inf and res to denote the cochain maps and are thus in a position to use the appropriate formulas in (2-5-1).

exactness at $H^1(G/H, A^H)$: Let u be a 1-cocycle of G/H in A^H such that the cohomology class of $\inf u$ is 0. Since $\inf u$ is a coboundary and $(\inf u)[\sigma] = u[\bar{\sigma}]$, $\sigma \in G$, there exists $a \in A$ such that

$$u[\bar{\sigma}] = (\sigma - 1)\, a \qquad \forall \sigma \in G$$

For $\rho \in H$ we have $(\rho - 1)a = u[\bar{\rho}] = 0$—as may be seen by substituting ρ for τ in the cocycle identity $\bar{\sigma}u[\bar{\tau}] - u[\bar{\sigma}\bar{\tau}] + u[\bar{\sigma}] = 0$, $\bar{\sigma}, \bar{\tau} \in G/H$. This means that $a \in A^H$ and that

$$u[\bar{\sigma}] = (\bar{\sigma} - 1)\, a \qquad \forall \bar{\sigma} \in G/H$$

Thus, u is a coboundary, and its cohomology class is 0.

exactness at $H^1(G, A)$: In virtue of (3-4-1), it suffices to show that if u is a 1-cocycle of G in A for which $\operatorname{res} u$ is a coboundary, then there exists a 1-cocycle v of G/H in A^H such that $\inf v$ is cohomologous to u. According to (2-5-1) and the hypothesis on u there exists $a \in A$ such that

$$u[\rho] = (\operatorname{res} u)[\rho] = (\rho - 1)\, a \qquad \forall \rho \in H$$

Now put

$$u'[\sigma] = u[\sigma] - (\sigma - 1)\, a \qquad \forall \sigma \in G$$

so that u' is a 1-cocycle of G in A and $u' \sim u$. Then $u'[\rho] = 0$ for all $\rho \in H$, and from the cocycle identity $\sigma u'[\tau] - u'[\sigma\tau] + u'[\sigma] = 0$ we get (by taking $\tau = \rho \in H$) $u'[\sigma\rho] = u'[\sigma]\ \forall \sigma \in G, \rho \in H$ and (by taking $\sigma = \rho \in H$) $\rho u'[\tau] = u'[\rho\tau] \; \forall \tau \in G, \; \rho \in H$. Then, upon defining a 1-cochain v of G/H in A^H by

$$v[\bar{\sigma}] = u'[\sigma] \qquad \forall \bar{\sigma} \in G/H$$

it is immediate that v is a cocycle with $\inf v = u' \sim u$. This completes the proof. ▮

3-4-3. Theorem. Let A be a G-module, let H be a normal subgroup of G, and suppose that for an integer $n \geqslant 1$ we have

$$H^r(H, A) = (0) \qquad r = 1, 2, \ldots, n-1$$

Then, in dimension n, the inflation–restriction sequence

$$0 \longrightarrow H^n(G/H, A^H) \xrightarrow{\text{inf}} H^n(G, A) \xrightarrow{\text{res}} H^n(H, A)$$

is exact.

Proof: By induction on n. For $n = 1$, the hypotheses are vacuous and the result is given by (3-4-2). Suppose then that the result is true for $n - 1 \geqslant 1$. According to (3-3-9), A can be imbedded in an exact G-sequence

$$0 \longrightarrow A \longrightarrow B \longrightarrow C \longrightarrow 0 \qquad (*)$$

in which B is G-regular. Because this is also an exact H-sequence and $H^1(H, A) = (0)$ it follows from (2-2-4) (or, more precisely, from the comments following (2-2-4), in the case where cohomology is done with respect to the half-complex) that the sequence

$$0 \longrightarrow A^H \longrightarrow B^H \longrightarrow C^H \longrightarrow 0$$

is exact. Moreover, this is clearly an exact (G/H)-sequence, and according to (3-3-7), B^H is (G/H)-regular. Therefore, in the commutative diagram

$$\begin{array}{ccccccc} 0 \longrightarrow & H^{n-1}(G/H, C^H) & \xrightarrow{\text{inf}} & H^{n-1}(G, C) & \xrightarrow{\text{res}} & H^{n-1}(H, C) \\ & \downarrow{\scriptstyle \delta_*} & & \downarrow{\scriptstyle \delta_*} & & \downarrow{\scriptstyle \delta_*} \\ 0 \longrightarrow & H^n(G/H, A^H) & \xrightarrow{\text{inf}} & H^n(G, A) & \xrightarrow{\text{res}} & H^n(H, A) \end{array}$$

the three coboundary maps are isomorphisms onto. Thus, to prove our result, it suffices to show that the top row is exact. Since B is also H-regular, we conclude from the exact H-sequence $(*)$ that $H^r(H, C) = H^{r+1}(H, A)$ for all r. Now, the hypothesis on the cohomology of H in A (for $r = 2,..., n-1$) implies that $H^r(H, C) = (0)$ for $r = 1, 2,..., n-2$. Consequently, the induction hypothesis applies for C, and the top row is exact. This completes the proof. ∎

The case $n = 2$ of this result will play an important role in abstract class field theory (see Chapter VI) when we discuss the Brauer group. This case $n = 2$ was first proved in [34] for Galois cohomology and applied to local class field theory; its formulation for global class field theory was given in [54]. Our general result, considered from the point of view of spectral sequences, appears in [38].

3-4-4. Remark. It may be noted that it is possible to define a **deflation map** (see [79]) in negative dimensions, which is dual to inflation, and for which an analog of (3-4-3) holds (with inf replaced by defl and res replaced by cor). The sole fact of interest to us (since it will be used in Section 3-6) is that if H is a normal subgroup of G and A is any G-module, then we may define a deflation map in dimension 0 by the requirement that the diagram

$$\begin{array}{ccc} A^G & \xrightarrow{1} & H^G = (A^H)^{G/H} \\ \kappa\downarrow & & \downarrow\kappa \\ H^0(G, A) & \xrightarrow{\text{defl}} & H^0(G/H, A^H) \end{array}$$

be commutative—so that defl is defined by

$$\text{defl}\,(\kappa^G a) = \kappa^{G/H} a \qquad a \in A^G$$

Moreover, the sequence

$$H^0(H, A) \xrightarrow{\text{cor}} H^0(G, A) \xrightarrow{\text{defl}} H^0(G/H, A^H) \longrightarrow 0$$

is exact. To see this consider the diagram

$$\begin{array}{ccccccc}
\dfrac{A^H}{S_H A} & \xrightarrow{\bar{S}_{H\to G}} & \dfrac{A^G}{S_G A} & \xrightarrow{\bar{1}} & \dfrac{A^G}{S_{G/H} A^H} & \longrightarrow & 0 \\
\bar{\kappa}\downarrow & & \bar{\kappa}\downarrow & & \bar{\kappa}\downarrow & & \\
H^0(H, A) & \xrightarrow{\text{cor}} & H^0(G, A) & \xrightarrow{\text{defl}} & H^0(G/H, A^H) & \longrightarrow & 0
\end{array}$$

Here, the top row is induced from $A^H \xrightarrow{S_{H\to G}} A^G \xrightarrow{1} A^G \longrightarrow 0$ (which is not exact). It is exact since

$$S_G A = S_{H\to G}(S_H A) \subset S_{H\to G} A^H = S_{G/H} A^H$$

and

$$\ker \bar{1} = \frac{S_{G/H} A^H}{S_G A} = \bar{S}_{H\to G}\left(\frac{A^H}{S_H A}\right)$$

By (2-5-5) and the definition of defl the diagram commutes; and because the $\bar{\kappa}$ maps are isomorphisms onto, the bottom row is exact.

3-5. THE GROUP $H^{-2}(G, \mathbf{Z})$

In this section, we examine the group $H^{-2}(G, \mathbf{Z})$, which will play a key role in our formulation of the reciprocity law (see Section 6-5).

3-5-1. Proposition. For any group G we have an isomorphism

$$\frac{G}{G^c} \approx \frac{I}{I^2}$$

given by

$$\sigma G^c \longleftrightarrow (\sigma - 1) + I^2 \qquad \sigma \in G$$

Proof: Here, as usual, G^c is the commutator subgroup of G, while I^2 (which is viewed additively) is the square of the ideal $I = \sum_{\sigma \neq 1} \oplus \mathbf{Z}(\sigma - 1)$ of $\Gamma = \mathbf{Z}[G]$.

Let us map $G \longrightarrow I/I^2$ by the rule $\sigma \longrightarrow (\sigma - 1) + I^2$. This is a homomorphism since

$$(\sigma\tau - 1) + I^2 = (\sigma - 1)(\tau - 1) + (\sigma - 1) + (\tau - 1) + I^2$$
$$= \{(\sigma - 1) + I^2\} + \{(\tau - 1) + I^2\}.$$

Because I/I^2 is commutative, it follows that G^c is in the kernel, so that we have an induced homomorphism of $G/G^c \longrightarrow I/I^2$ given explicitly by : $\sigma G^c \longrightarrow (\sigma - 1) + I^2$. To show that this is an isomorphism, we construct its inverse. Define a map $I \longrightarrow G/G^c$ by mapping $(\sigma - 1) \longrightarrow \sigma G^c$, $\sigma \in G$ and extending linearly from the $\mathbf{Z}$-basis. This is clearly a homomorphism; moreover, I^2 is in the kernel because $\{(\sigma - 1)(\tau - 1) \mid \sigma, \tau \in G\}$ is a set of $\mathbf{Z}$-generators for I^2 and

$$(\sigma - 1)(\tau - 1) = (\sigma\tau - 1) - (\tau - 1) - (\sigma - 1)$$
$$\longrightarrow (\sigma\tau G^c)(\tau^{-1}G^c)(\sigma^{-1}G^c) = G^c$$

Thus, we have an induced homomorphism of $I/I^2 \longrightarrow G/G^c$ given by : $(\sigma - 1) + I^2 \longrightarrow \sigma G^c$. This completes the proof. ▌

3-5-2. Remark. Let us show how $H^{-2}(G, \mathbf{Z})$ can be interpreted as the group G/G^c. From the exact G-sequence $0 \longrightarrow I \longrightarrow \Gamma \longrightarrow \mathbf{Z} \longrightarrow 0$ we conclude (since Γ is G-regular) that the coboundary $\delta_* : H^{-2}(G, \mathbf{Z}) \longrightarrow H^{-1}(G, I)$ is an isomorphism onto. Furthermore, according to (2-2-6) we have (since $I_S = I$) a map $\eta : I \twoheadrightarrow H^{-1}(G, I)$ with kernel $I \cdot I = I^2$, which determines the isomorphism $\bar{\eta} : I/I^2 \rightarrowtail\!\!\!\twoheadrightarrow H^{-1}(G, I)$. Combining these with (3-5-1) leads to the chain of isomorphisms

$$H^{-2}(G, \mathbf{Z}) \xrightarrow{\delta_*} H^{-1}(G, I) \xleftarrow{\bar{\eta}} \frac{I}{I^2} \approx \frac{G}{G^c}$$

and provides a canonical procedure for identifying $H^{-2}(G, \mathbf{Z})$ with G/G^c. In particular, if we denote by ζ_σ the element of $H^{-2}(G, \mathbf{Z})$ corresponding to σG^c (so that $\sigma \longrightarrow \zeta_\sigma$ maps G onto $H^{-2}(G, \mathbf{Z})$ with kernel G^c), then

$$\delta_* \zeta_\sigma = \eta(\sigma - 1)$$

Our next objective is then to describe the usual cohomological maps of $H^{-2}(G, \mathbf{Z})$ as reflected in G/G^c.

3-5-3. Proposition. Let H be a subgroup of G. Then the map $\mathrm{cor} : H^{-2}(H, \mathbf{Z}) \longrightarrow H^{-2}(G, \mathbf{Z})$ corresponds to the natural map $\bar{\imath} : H/H^c \longrightarrow G/G^c$ which is induced by the inclusion map $i : H \longrightarrow G$.

Proof: Let $\Gamma_G = \mathbf{Z}[G]$, $I_G = \sum_{\sigma \neq 1 \in G} \oplus \mathbf{Z}(\sigma - 1)$, $\Gamma_H = \mathbf{Z}[H]$, $I_H = \sum_{\rho \neq 1 \in H} \oplus \mathbf{Z}(\rho - 1)$ and consider the commutative diagram

$$\begin{array}{ccccccccc}
0 & \longrightarrow & I_H & \xrightarrow{i} & \Gamma_H & \xrightarrow{\varepsilon^H} & \mathbf{Z} & \longrightarrow & 0 \\
 & & {\scriptstyle i}\downarrow & & {\scriptstyle i}\downarrow & & {\scriptstyle 1}\downarrow & & \\
0 & \longrightarrow & I_G & \xrightarrow{i} & \Gamma_G & \xrightarrow{\varepsilon^G} & \mathbf{Z} & \longrightarrow & 0 \\
 & & {\scriptstyle 1}\downarrow & & {\scriptstyle 1}\downarrow & & {\scriptstyle 1}\downarrow & & \\
0 & \longrightarrow & I_G & \xrightarrow{i} & \Gamma_G & \xrightarrow{\varepsilon^G} & \mathbf{Z} & \longrightarrow & 0
\end{array}$$

where i denotes inclusion and ε^H, ε^G are the respective augmentations (as in (3-3-8)). The top row is an exact H-sequence, the middle row is an exact G-sequence which we view as an exact H-sequence, and the bottom row is an exact G-sequence. This leads to the diagram

$$\begin{array}{ccccccc}
H^{-2}(H, \mathbf{Z}) & \xrightarrow{\delta_*} & H^{-1}(H, I_H) & \xleftarrow{\bar\eta^H} & \dfrac{I_H}{I_H^2} & \approx & \dfrac{H}{H^c} \\
{\scriptstyle 1_*=1}\downarrow & & {\scriptstyle i_*}\downarrow & & {\scriptstyle \bar{\imath}}\downarrow & & \\
H^{-2}(H, \mathbf{Z}) & \xrightarrow{\delta_*} & H^{-1}(H, I_G) & \xleftarrow{\bar\eta^H} & \dfrac{(I_G)_{S_H}}{I_H \cdot I_G} & & \\
{\scriptstyle \mathrm{cor}}\downarrow & & {\scriptstyle \mathrm{cor}}\downarrow & & {\scriptstyle \bar{\imath}}\downarrow & & \\
H^{-2}(G, \mathbf{Z}) & \xrightarrow{\delta_*} & H^{-1}(G, I_G) & \xleftarrow{\bar\eta^G} & \dfrac{I_G}{I_G^2} & \approx & \dfrac{G}{G^c}
\end{array}$$

The δ_* maps are isomorphisms because Γ_H is H-regular, and Γ_G is both G-regular and H-regular. The inclusion map $i : (I_G)_{S_H} \longrightarrow I_G$ maps $I_H \cdot I_G \rightarrowtail I_G^2$ and induces the map $\bar{\imath}$ of the factor groups. All the squares commute—by (2-2-5), (2-2-7), (2-4-7), and (2-5-5).

In virtue of this, we may compute $\text{cor} \circ 1_*$ (which is indeed the map we seek for standard complexes) by using the third column. Consequently, for $\rho \in H$

$$\rho H^c \longrightarrow (\rho - 1) + I_H^2 \longrightarrow (\rho - 1) + I_H \cdot I_G$$
$$\longrightarrow (\rho - 1) + I_G^2 \longrightarrow \rho G^c$$

and this completes the proof. ▮

3-5-4. Remark. As preparation for our next result, we need to know the definition of the group theoretical transfer (or **verlagerung**)

$$\bar{V} = \bar{V}_{G\to H} : G/G^c \longrightarrow H/H^c$$

The definition, along with the basic properties of this map, may be found in [83, Chapter 5] or [31, p. 202]; however, we shall arrive at the definition through cohomological considerations.

Let $\hat{G}$ denote the character group of G—that is, $\hat{G}$ is the group of all continuous homomorphisms of G into $\mathbf{Q}/\mathbf{Z}$. Because G is finite and the action of G on $\mathbf{Q}/\mathbf{Z}$ is trivial, we may identify

$$H^1(G, \mathbf{Q}/\mathbf{Z}) = \text{Hom}\,(G, \mathbf{Q}/\mathbf{Z}) = \hat{G} = \widehat{(G/G^c)}$$

If H is a normal subgroup of G the identification enables us to carry the map $\inf : H^1(G/H, \mathbf{Q}/\mathbf{Z}) \longrightarrow H^1(G, \mathbf{Q}/\mathbf{Z})$ over to a map

$$\inf : \widehat{(G/H)} \longrightarrow \hat{G}$$

which, in virtue of (2-5-1), is given by

$$(\inf \chi)(\sigma) = \chi(\bar{\sigma}) \qquad \chi \in \widehat{(G/H)}, \quad \sigma \in G$$

In other words, inflation of a (standard) cocycle corresponds to inflation (or lifting) of a character of G/H to G.

Similarly, if H is any subgroup of G, then, according to (2-5-1), restriction of a cocycle corresponds to restriction of a character of G to H—in other words, the map

$$\text{res} : \hat{G} \longrightarrow \hat{H}$$

corresponding to res : $H^1(G, \mathbf{Q}/\mathbf{Z}) \longrightarrow H^1(H, \mathbf{Q}/\mathbf{Z})$ is given by

$$(\text{res } \chi)(\rho) = \chi(\rho) \qquad \chi \in \hat{G}, \quad \rho \in H$$

As for the corestriction or transfer; here we have a map

$$\text{cor} : \hat{H} \longrightarrow \hat{G}$$

corresponding to cor : $H^1(H, \mathbf{Q}/\mathbf{Z}) \longrightarrow H^1(G, \mathbf{Q}/\mathbf{Z})$. According to (2-5-2), if $G = \bigcup c = \bigcup H\bar{c}$ is a right coset decomposition, this map is given by

$$(\text{cor } \chi)(\sigma) = \sum_c \chi(\bar{c}\sigma\overline{c\sigma}^{-1}) = \chi\left(\prod_c \bar{c}\sigma\overline{c\sigma}^{-1}\right) \qquad \chi \in \hat{H}, \quad \sigma \in G$$

Since $\hat{H} = \widehat{(H/H^c)}$ and since $\widehat{\widehat{(H/H^c)}}$ may be identified with H/H^c, there exists a dual map

$$\bar{V} = \bar{V}_{G \to H} : G/G^c \longrightarrow H/H^c$$

called the **reduced group theoretical transfer**, which satisfies the relation

$$\chi\{\bar{V}(\sigma G^c)\} = (\text{cor } \chi)(\sigma G^c) = \chi\left(\prod_c \bar{c}\sigma\overline{c\sigma}^{-1} H^c\right) \qquad \chi \in \widehat{(H/H^c)}, \quad \sigma \in G$$

Because this is true for all $\chi \in \widehat{(H/H^c)}$, it follows that

$$\bar{V}(\sigma G^c) = \prod_c \bar{c}\sigma\overline{c\sigma}^{-1} H^c$$

which is indeed the customary formula for the (reduced) group theoretical transfer. Of course, the usual group theoretical transfer is defined by this formula when G is infinite, provided that $(G : H) < \infty$. It may be left to the reader to show how, when cohomology is done for infinite groups, the cohomological formulation of the transfer still applies.

The preceding discussion has been concerned with the G-module $\mathbf{Q}/\mathbf{Z}$, but it is clear that everything (except the remarks about the reduced group theoretical transfer) may be carried over to the case of an arbitrary G-module A on which the action of G is trivial.

3-5-5. Proposition. Let H be a subgroup of G. Then the map $\text{res} : H^{-2}(G, \mathbf{Z}) \longrightarrow H^{-2}(H, \mathbf{Z})$ corresponds to the reduced group theoretical transfer $\bar{V} : G/G^c \longrightarrow H/H^c$.

Proof: The procedure used for the proof of (3-5-3) yields the diagram

$$\begin{array}{ccccc}
H^{-2}(H, \mathbf{Z}) & \xrightarrow{\delta_*} & H^{-1}(H, I_H) & \xleftarrow{\bar{\eta}} & \dfrac{I_H}{I_H^2} \approx \dfrac{H}{H^c} \\
{\scriptstyle 1_*=1}\downarrow & & {\scriptstyle i_*}\downarrow & & {\scriptstyle \bar{i}}\downarrow \\
H^{-2}(H, \mathbf{Z}) & \xrightarrow{\delta_*} & H^{-1}(H, I_G) & \xleftarrow{\bar{\eta}} & \dfrac{(I_G)S_H}{I_H \cdot I_G} \\
{\scriptstyle \text{res}}\uparrow & & {\scriptstyle \text{res}}\uparrow & & {\scriptstyle \Phi}\uparrow \\
H^{-2}(G, \mathbf{Z}) & \xrightarrow{\delta_*} & H^{-1}(G, I_G) & \xleftarrow{\bar{\eta}} & \dfrac{I_G}{I_G^2} \approx \dfrac{G}{G^c}
\end{array}$$

where (by (2-5-5)) Φ is induced by $\sum \bar{c}$ (where $G = \bigcup H\bar{c} = \bigcup c$ is a coset decomposition). Therefore, tracing res via the third column, we have for $\sigma \in G$

$$\sigma G^c \longrightarrow (\sigma - 1) + I_G^2 \longrightarrow \sum_c \bar{c}(\sigma - 1) + I_H I_G$$

Now, this element must be modified to enable us to recognize its pre-image under the isomorphism $\bar{i}$; thus

$$\begin{aligned}
\sum_c \bar{c}(\sigma - 1) + I_H I_G &= \sum_c \bar{c}\sigma - \sum_c \bar{c} + I_H I_G = \sum_c \bar{c}\sigma - \sum_c \overline{c\sigma} + I_H I_G \\
&= \sum_c (\underbrace{\bar{c}\sigma\overline{c\sigma}^{-1}}_{\in H}) + \sum_c \underbrace{(\bar{c}\sigma\overline{c\sigma}^{-1} - 1)(\overline{c\sigma} - 1)}_{\in I_H I_G} + I_H I_G \\
&= \sum_c (\bar{c}\sigma\overline{c\sigma}^{-1} - 1) + I_H I_G
\end{aligned}$$

and then

$$\sum_c (\bar{c}\sigma\overline{c\sigma}^{-1} - 1) + I_H I_G \xrightarrow{\bar{i}^{-1}} \sum_c (\bar{c}\sigma\overline{c\sigma}^{-1} - 1) + I_H^2 \longrightarrow \prod_c \bar{c}\sigma\overline{c\sigma}^{-1} H^c$$

This completes the proof. ∎

It is immediate from (3-5-4), (3-5-5), and the transitivity of restriction that the reduced transfer $\bar{V}$ is independent of the choice of coset representatives and is transitive. Of course, we may define a homomorphism $V = V_{G \to H} : G \longrightarrow H/H^c$ by $V(\sigma) = \bar{V}(\sigma G^c)$—this is what is usually called the **group theoretical transfer** of G into H.

3-5-6. Exercise. We sketch the standard discussion of the elementary properties of the group theoretical transfer. Let $G = \{\sigma, \tau, ...\}$ be an arbitrary multiplicative group, and let H be a subgroup of finite index—say, $(G : H) = n$. Fix a system of representatives $\sigma_1, ..., \sigma_n$ for the left cosets of H in G—$G = \bigcup_1^n \sigma_i H$. Consider any $\tau \in G$; for each $i = 1, ..., n$, we have $\tau\sigma_i \in G$, so there exist unique elements $\sigma_{\pi(i)}$ and $\rho_i \in H$ (depending on τ) such that $\tau\sigma_i = \sigma_{\pi(i)}\rho_i$. Define the **transfer** or **verlagerung** of G into H to be the map $V_{G \to H} : G \longrightarrow H/H^c$ given by

$$V_{G \to H} : \tau \longrightarrow \left(\prod_1^n \rho_i\right) H^c$$

It is not hard to verify that π is a permutation of the set $\{1, 2, ..., n\}$. Moreover, $V_{G \to H}$ is a homomorphism which is independent of the choice of the system of representatives. Since H/H^c is abelian, it follows that G^c is in the kernel of $V_{G \to H}$ and that we have an induced homomorphism $\bar{V}_{G \to H} : G/G^c \longrightarrow H/H^c$, called the **reduced transfer** of G into H. The reduced transfer is transitive—that is, if $G \supset H \supset K$ with $(G : K) < \infty$, then $\bar{V}_{G \to K} = \bar{V}_{H \to K} \circ \bar{V}_{G \to H}$.

The transfer may also be defined via the use of right cosets—thus, starting from $G = \bigcup_1^n H\sigma_i^{-1}$, we have $\sigma_i^{-1}\tau^{-1} = \rho_i^{-1}\sigma_{\pi(i)}^{-1}$ with the same π and ρ_i as before, and the map $\tau^{-1} \longrightarrow (\prod_1^n \rho_i^{-1})H^c$ is precisely $V_{G \to H}$. Of course, this procedure, too, is independent of the choice of representatives. Finally, if $G = \bigcup c = \bigcup H\bar{c}$ is a right coset decomposition (as in (3-5-4)), then this approach to the transfer yields $V_{G \to H}(\tau) = (\prod \bar{c}\tau\overline{\bar{c}\tau}^{-1})H^c$, which is the formula for the transfer arrived at by cohomological methods in (3-5-4).

3-5-7. Proposition. Let H be a subgroup of G. Then for $\sigma \in G$ the map $\sigma_* = \mathrm{con}_\sigma : H^{-2}(H, \mathbf{Z}) \longrightarrow H^{-2}(H^\sigma, \mathbf{Z})$ corresponds to the map of $H/H^c \longrightarrow H^\sigma/(H^\sigma)^c$ which is induced by conjugation with σ.

Proof: This is left to the reader. ▮

3-6. COHOMOLOGICAL EQUIVALENCE

The G-module A is said to be **cohomologically trivial** when $H^n(H, A) = (0)$ for all n and all subgroups H of G.

3-6-1. Theorem. (Tate [74]). If there exists an integer r such that

$$H^r(H, A) = H^{r+1}(H, A) = (0) \qquad \text{for all subgroups } H \text{ of } G$$

then A is cohomologically trivial.

Proof: It suffices to prove the following two statements :

(1) If $H^0(H, A) = H^1(H, A) = (0)\ \forall H \subset G$, then $H^2(H, A) = (0)\ \forall H \subset G$.

(2) If $H^1(H, A) = H^2(H, A) = (0)\ \forall H \subset G$, then $H^0(H, A) = (0)\ \forall H \subset G$.

To see this take G-modules B and C (which exist by (3-3-11)) such that

$$H^{r+i}(H, A) \approx H^i(H, B) \qquad i = 0, 1, 2, \quad \forall H \subset G$$

$$H^{r+i-1}(H, A) \approx H^i(H, C) \qquad i = 0, 1, 2, \quad \forall H \subset G$$

Applying (1) to B gives $H^{r+2}(H, A) = (0)\ \forall H \subset G$; and applying (2) to C gives $H^{r-1}(H, A) = (0)\ \forall H \subset G$. Thus, by moving up or down one dimension at a time, it follows that $H^n(H, A) = (0)\ \forall n, \forall H \subset G$.

Now, let us prove (1) and (2) by induction on $m = \#(G)$. If $m = 1$ the statements are trivial (by (3-1-5)), so suppose $m > 1$ and that (1) and (2) hold for all groups of order $<m$. If m is not a prime power then (1) and (2) hold for all Sylow subgroups G_p of G. Since, according to (3-1-16), $H^n(G_p, A) = (0)$ for all p implies $H^n(G, A) = (0)$, it follows that (1) and (2) hold.

It remains, therefore, to consider the case where m is a power of a prime, $m = p^s$. Then, according to a well-known property of

p-groups there exists a normal subgroup H such that G/H is cyclic of order p. Now, consider the diagram

$$\begin{array}{ccccccc} 0 & \longleftarrow & H^0(G/H, A^H) & \xleftarrow{\text{defl}} & H^0(G, A) & \xleftarrow{\text{cor}} & H^0(H, A) \\ & & \updownarrow & & & & \\ 0 & \longrightarrow & H^2(G/H, A^H) & \xrightarrow{\text{inf}} & H^2(G, A) & \xrightarrow{\text{res}} & H^2(H, A) \end{array}$$

with the top row exact (by (3-4-4)) and the vertical correspondence an isomorphism (because G/H is cyclic). The bottom row is exact by (3-4-3) because $H^1(H, A) = (0)$ in both cases (1) and (2).

To prove assertion (1), we note that $H^0(G, A) = (0)$ by hypothesis; hence, $H^0(G/H, A^H) = (0)$ and $H^2(G/H, A^H) = (0)$. By the induction hypothesis, $H^2(H, A) = (0)$, and therefore $H^2(G, A) = (0)$ —which suffices for the proof of (1). The proof of (2) goes the same way. ▮

This result has been generalized in [55] and [56] and then extended and the proofs simplified in [60] and [67, Chapter 10]. In particular, in order for the G-module A to be cohomologically trivial it suffices that for each prime p there exists an integer r (depending on p) such that $H^r(G_p, A) = H^{r+1}(G_p, A) = (0)$ for some p-Sylow subgroup G_p of G.

3-6-2. Proposition. Let $f : A \longrightarrow B$ be a G-homomorphism of G-modules; then there exists a G-module C such that for every subgroup H of G we have an exact cohomology sequence

$$\cdots \xrightarrow{\delta_*} H^n(H, A) \xrightarrow{f_*} H^n(H, B) \longrightarrow H^n(H, C) \xrightarrow{\delta_*} H^{n+1}(H, A) \longrightarrow \cdots$$

Proof: Since $f : A \longrightarrow B$ is an H-homomorphism of H-modules, we have $f_* : H^n(H, A) \longrightarrow H^n(H, B)$ for all n; the other cohomological maps will be described in the course of the proof.

By (3-3-9), A may be imbedded in a G-regular module M. Then $0 \longrightarrow A \xrightarrow{i} M$ is an exact G-sequence, and so is

$$0 \longrightarrow A \xrightarrow{f \oplus i} B \oplus M$$

(where G acts componentwise on $B \oplus M$). Let C be the factor module, j the canonical map of $B \oplus M \twoheadrightarrow C$, and π the projection of $B \oplus M$ on B. Then the diagram of G-modules and G-homomorphisms

$$\begin{array}{ccccccccc} & & & & (0) & & & & \\ & & & & \downarrow & & & & \\ & & & & M & & & & \\ & & & & \downarrow & & & & \\ (0) & \longrightarrow & A & \xrightarrow{f\oplus i} & B \oplus M & \xrightarrow{j} & C & \longrightarrow & (0) \\ & & & \searrow^{f} & \downarrow{\scriptstyle \pi} & & & & \\ & & & & B & & & & \\ & & & & \downarrow & & & & \\ & & & & (0) & & & & \end{array}$$

has the row and column exact and the triangle commutative. For any $H \subset G$ and any n we have, therefore,

$$\begin{array}{ccccccc} & & & H^n(H, M) = 0 & & & \\ & & & \downarrow & & & \\ \cdots \xrightarrow{\delta_*} H^n(H,A) & \xrightarrow{(f\oplus i)_*} & & H^n(H, B \oplus M) & \xrightarrow{j_*} H^n(H, C) & \xrightarrow{\delta_*} & H^{n+1}(H, A) \rightarrow \cdots \\ & \searrow^{f_*} & & \downarrow{\scriptstyle \pi_*} & & & \\ & & & H^n(H, B) & & & \\ & & & \downarrow & & & \\ & & & H^{n+1}(H, M) = (0) & & & \end{array}$$

with π_* an isomorphism onto, the row exact, and the triangle commutative. By using the map $j_* \circ \pi_*^{-1} : H^n(H, B) \longrightarrow H^n(H, C)$, the proof is completed. ▮

If instead of imbedding A in a G-regular module, one takes B

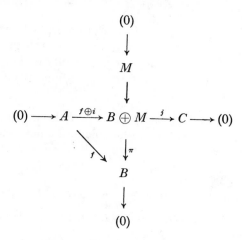

as the quotient of a G-regular module, then the same method of proof leads to an exact sequence

$$\cdots \xrightarrow{\delta_*} H^n(H, C) \longrightarrow H^n(H, A) \xrightarrow{f_*} H^n(H, B) \xrightarrow{\delta_*} H^{n+1}(H, C) \longrightarrow \cdots \quad (*)$$

If $f_* : H^n(H, A) \longrightarrow H^n(H, B)$ is an isomorphism onto for all n and all $H \subset G$, we say that A and B are **cohomologically** equivalent (with respect to f), or that f is a **cohomological equivalence**. It will be convenient to denote f_* in dimension n by $f_{(n)}$.

3-6-3. Proposition. Let $f : A \longrightarrow B$ be a G-homomorphism of G-modules. Then f is a cohomological equivalence between A and $B \Longleftrightarrow$ there exists an integer r such that for all $H \subset G$ we have

$$f_{(r-1)} : H^{r-1}(H, A) \twoheadrightarrow H^{r-1}(H, B) \quad \text{(epimorphism)}$$

$$f_{(r)} : H^r(H, A) \rightarrowtail\!\!\!\!\!\twoheadrightarrow H^r(H, B) \quad \text{(isomorphism)}$$

$$f_{(r+1)} : H^{r+1}(H, A) \rightarrowtail H^{r+1}(H, B) \quad \text{(monomorphism)}$$

Proof: By (3-6-2) we have (with obvious choice of notation) an exact sequence

$$\cdots \longrightarrow H^n(H, A) \xrightarrow{f_{(n)}} H^n(H, B) \xrightarrow{\varphi_{(n)}} H^n(H, C) \xrightarrow{\delta_{(n)}} H^{n+1}(H, A) \longrightarrow \cdots$$

Now, $f_{(r-1)}$ is an epimorphism

$$\Longrightarrow \operatorname{im} \varphi_{(r-1)} = (0) \Longrightarrow \ker \delta_{(r-1)} = (0) \Longrightarrow H^{r-1}(H, C) = (0),$$

since $\operatorname{im} \delta_{(r-1)} = \ker f_{(r)} = (0)$. Furthermore,

$$\ker f_{(r+1)} = (0) \Longrightarrow \operatorname{im} \delta_{(r)} = (0) \Longrightarrow$$

$$\operatorname{im} \varphi_{(r)} = \ker \delta_{(r)} = H^r(H, C) \Longrightarrow H^r(H, C) = (0)$$

since $H^r(H, B) = \operatorname{im} f_{(r)} = \ker \varphi_{(r)}$. Therefore, $H^{r-1}(H, C) = H^r(H, C) = (0)$ for all $H \subset G$, and by (3-6-1) C is cohomologically trivial. Returning to our exact sequence, we see that $f : A \longrightarrow B$ is indeed a cohomological equivalence—which completes the proof. ∎

The same result and proof are valid if the exact sequence $(*)$ is used, because one can only work to the right (left) from an epimorphism $f_{(r-1)}$ (monomorphism $f_{(r+1)}$). It is also known (see [26]) that this result holds when $f_{(r)}$ and $f_{(r+1)}$ are isomorphisms onto.

3-6-4. Theorem. If the G-module A has the properties

(1) $H^{-1}(H, A) = (0)$ for all $H \subset G$.

(2) $H^0(H, A)$ is cyclic of order $\#(H)$ for all $H \subset G$.

then $\mathbf{Z}$ and A are cohomologically equivalent.

Proof: Let α be a generator of $H^0(G, A)$, so α has order $\#(G)$. Fix any subgroup H of G. Take any positive integer $s < \#(H)$. Then

$$\operatorname{cor}_{H\to G}(s \operatorname{res}_{G\to H} \alpha) = s \operatorname{cor}_{H\to G} \operatorname{res}_{G\to H} \alpha = s(G:H)\,\alpha$$

$$s(G:H)\,\alpha \neq 0 \in H^0(G, A)$$

Therefore, $\operatorname{res}_{G\to H} \alpha \in H^0(H, A)$ has order $\geqslant \#(H)$; so from (2) $\operatorname{res}_{G\to H} \alpha$ has order $\#(H)$, and it is a generator of $H^0(H, A)$.

Now choose $a \in A^G$ such that $\kappa^G(a) = \alpha$, and using this a define the G-homomorphism $f : \mathbf{Z} \longrightarrow A$ by $f(1) = a$. We show that f is a cohomological equivalence (with $r = 0$). Clearly,

$$f_{(-1)} : H^{-1}(H, \mathbf{Z}) \twoheadrightarrow H^{-1}(H, A) \quad \text{since} \quad H^{-1}(H, A) = (0)$$

$$f_{(+1)} : H^1(H, \mathbf{Z}) \rightarrowtail H^1(H, A) \quad \text{since} \quad H^1(H, \mathbf{Z}) = (0)$$

and it remains to show that

$$f_{(0)} : H^0(H, \mathbf{Z}) \rightarrowtail\!\!\!\!\twoheadrightarrow H^0(H, A) \qquad \text{(isomorphism)}$$

In virtue of (2-5-5), $\kappa^H(a) = \operatorname{res}_{G\to H} \alpha$ is a generator of $H^0(H, A)$. On the other hand, $H^0(H, \mathbf{Z})$ is cyclic of order $\#(H)$ and $\kappa^H(1)$ is a generator. By (2-2-7), we have the commutative diagram

$$\begin{array}{ccc} \mathbf{Z} = \mathbf{Z}^H & \xrightarrow{\quad f \quad} & A^H \\ \kappa \downarrow & & \kappa \downarrow \\ H^0(H, \mathbf{Z}) & \xrightarrow{f_* = f_{(0)}} & H^0(H, A) \end{array}$$

which says that $f_{(0)}$ maps the generator $\kappa^H(1)$ to the generator $\kappa^H(a)$. Thus, $f_{(0)}$ is an isomorphism onto, and the proof is complete. ▮

3-6-5. Corollary. (Tate). If the G-module A has the properties

(1) $H^1(H, A) = (0)$ for all $H \subset G$,
(2) $H^2(H, A)$ is cyclic of order $\#(H)$ for all $H \subset G$,

then, for all $H \subset G$ and all n we have

$$H^n(H, \mathbf{Z}) \approx H^{n+2}(H, A)$$

Proof: By dimension shifting, there exists a G-module C with $H^n(H, C) \approx H^{n+2}(H, A)\ \forall n, \forall H \subset G$, and (3-6-4) may then be applied to C. ▮

This result is a weak form of Tate's theorem (see [74]). The full theorem also gives an explicit form for the isomorphisms $H^n(H, \mathbf{Z}) \approx H^{n+2}(H, A)$, and this is one of the major objectives of the next chapter. Roughly speaking, the arithmetic part of class field theory consists of the verification of the hypotheses of (3-6-5), and then the case $n = -2$ of Tate's theorem gives the reciprocity law isomorphism.

3-6-6. Remark. The proofs of (3-6-3), (3-6-4), and (3-6-5) depend on (3-6-1) and (3-6-2). Furthermore, it is easy to see that given the stronger version of (3-6-1) (mentioned at the end of that proof) the following versions of (3-6-3), (3-6-4), and (3-6-5) hold.

(i) Let $f : A \longrightarrow B$ be a G-homomorphism of G-modules. Suppose that for each prime p dividing $\#(G)$ we have a p-Sylow subgroup G_p and an integer r_p such that

$$f_{(r_p-1)} : H^{r_p-1}(G_p, A) \longrightarrow H^{r_p-1}(G_p, B) \quad \text{is an epimorphism}$$

$$f_{(r_p)} : H^{r_p}(G_p, A) \longrightarrow H^{r_p}(G_p, B) \quad \text{is an isomorphism}$$

$$f_{(r_p+1)} : H^{r_p+1}(G_p, A) \longrightarrow H^{r_p+1}(G_p, B) \quad \text{is a monomorphism}$$

—then f is a cohomological equivalence.

(ii) Consider the G-module A, $\alpha \in H^0(G, A)$, and p, G_p as above. If the following properties hold

(1) $H^{-1}(G_p, A) = (0) \; \forall p$,

(2) $H^0(G_p, A)$ is cyclic of order $\#(G_p)$, and $\text{res}_{G \to G_p}(\alpha)$ is a generator,

—then $\mathbf{Z}$ and A are cohomologically equivalent.

(iii) Suppose that $\alpha \in H^2(G, A)$ and

(1) $H^1(G_p, A) = (0) \; \forall p$,

(2) $H^2(G_p, A)$ is cyclic of order $\#(G_p)$, and $\text{res}_{G \to G_p}(\alpha)$ is a generator,

—then $H^n(H, \mathbf{Z}) \approx H^{n+2}(H, A) \; \forall H \subset G, \forall n$.

3-7. PROBLEMS AND SUPPLEMENTS

3-7-1. Given a homomorphism of groups $\lambda : G' \longrightarrow G$, we have constructed (at the beginning of Section 2-5) an explicit G'-homomorphism $\Lambda : X' \longrightarrow X$, suitable for λ, in dimensions $n \geqslant 0$ where X' and X are the standard G'- and G-complexes, respectively. If λ is a monomorphism, find the negative dimensional Λ's explicitly.

3-7-2. Suppose that $(\lambda, f) : (G, A) \longrightarrow (G', A')$ is a homomorphism of pairs. If $t : G \longrightarrow A$ is a crossed homomorphism (see (1-5-2)), then $(\lambda, f)t = ft\lambda : \sigma' \longrightarrow ft[\lambda\sigma']$ is a crossed homomorphism of G' in A' whose cohomology class is the image of the class of t under the map $(\lambda, f)_* : H^1(G, A) \longrightarrow H^1(G', A')$. Thus, inflation of crossed homomorphisms is analogous to lifting of homomorphisms from factor group to group, while restriction of crossed homomorphisms is analogous to restricting homomorphisms to a subgroup. What about corestriction of crossed homomorphisms.

3-7-3. Suppose that $(\lambda, f) : (G, A) \longrightarrow (G', A')$ is a homomorphism of pairs. What can be said about the diagram

$$\begin{array}{ccc} A_S & \longrightarrow & A'_{S'} \\ \eta \downarrow & & \eta \downarrow \\ H^{-1}(G, A) & \xrightarrow{(\lambda, f)_*} & H^{-1}(G', A') \end{array}$$

3-7-4. Induced homomorphisms f_* commute with inflation, restriction, corestriction, and conjugation; of course, the hypotheses need to be formulated accurately.

3-7-5. Suppose that G is a multiplicative group, finite or infinite. If $0 \longrightarrow A \longrightarrow B \longrightarrow C \longrightarrow 0$ is an exact G-sequence show that $0 \longrightarrow A^G \longrightarrow B^G \longrightarrow C^G$ is an exact sequence of abelian groups. Can the hypothesis be weakened?

3-7-6. (i) Suppose that H is a subgroup of G. If the G-module A is G-free, then it is also H-free. If A is G-regular, then it is also H-regular.

(ii) Let us call a G-module A **G-special** when its identity map is a trace; so G-regular implies G-special. Show that the converse is false.

Discuss the properties of G-special modules; in particular, show that a module which is either G-projective or G-injective is G-special. In general, the notion of G-specialness is easily seen to be sufficient for our theory—for example, for dimension shifting.

3-7-7. Consider the exact G-sequence

$$0 \longrightarrow A \xrightarrow{i} B \xrightarrow{j} C \longrightarrow 0$$

For $c \in C_S$, choose $b \in B$ and $a \in A$ such that $jb = c$ and $ia = Sb$. Then $a \in A^G$ and the map $c \longrightarrow a + SA$ of $C_S \longrightarrow A^G/(SA)$ is well-defined (that is, independent of the choice of b) and a homomorphism. Call it φ. Since $IC \subset \ker \varphi$, we have an induced homomorphism

$$\bar{\varphi} : \frac{C_S}{IC} \longrightarrow \frac{A^G}{SA}$$

Consider the sequence

$$\frac{A_S}{IA} \xrightarrow{\bar{\imath}} \frac{B_S}{IB} \xrightarrow{\bar{\jmath}} \frac{C_S}{IC} \xrightarrow{\bar{\varphi}} \frac{A^G}{SA} \xrightarrow{\bar{\imath}} \frac{B^G}{SB} \xrightarrow{\bar{\jmath}} \frac{C^G}{SC}$$

where $\bar{\imath}$ and $\bar{\jmath}$ are the natural induced maps, and show that it is exact. In virtue of the interpretation of cohomology groups in dimensions 0 and -1, this provides a direct proof of the exactness of

$$H^{-1}(A) \longrightarrow H^{-1}(B) \longrightarrow H^{-1}(C) \xrightarrow{\delta_*} H^0(A) \longrightarrow H^0(B) \longrightarrow H^0(C)$$

where $\delta_* = \bar{\varphi}$, the Tate linking (see (2-2-8) and (2-2-9)).

3-7-8. Suppose that A is a left G-module, and $\Gamma = \mathbf{Z}[G]$, then the additive group $M = \text{Hom}(\Gamma, A) = \text{Hom}_{\mathbf{Z}}(\Gamma, A)$ may be made into a left G-module in several ways.

First of all, Γ may be viewed as a left G-module, and as in (1-4-2), M then becomes a left G-module (which we denote by M_1) in which the action of G on M_1 is given by $(f^{\sigma})_1 = \sigma \circ f \circ \sigma^{-1}$, that is, $(f^{\sigma})_1(\tau) = \sigma f(\sigma^{-1}\tau)$, $\sigma, \tau \in G, f \in M$.

Let $\mathring{A}$ denote the additive group of A viewed as a G-module with trivial action. Then $M = \text{Hom}(\Gamma, A) = \text{Hom}(\Gamma, \mathring{A})$ can be made into a G-module (which we denote by M_2) in the same way. Thus, the action of G on M_2 is here given by $(f^{\sigma})_2 = \sigma \circ f \circ \sigma^{-1}$, that is, $(f^{\sigma})_2(\tau) = f(\sigma^{-1}\tau)$.

On the other hand, Γ may be viewed, in a natural way, as a right G-module, and A may be made into a right G-module by putting $a\sigma = \sigma^{-1}a$. Of course, $\mathring{A}$ is also a right G-module with trivial action. For $f \in M$, $\sigma \in G$ let us now define ${}^{\sigma}f = f^{\sigma^{-1}} = \sigma^{-1} \circ f \circ \sigma$—which means, act first by σ (on the right!), then by f, and then σ^{-1}. Then $M = \text{Hom}(\Gamma, A)$ becomes a left G-module M_3, with the action of G on M_3 given by $({}^{\sigma}f)_3(\tau) = (f(\tau\sigma))\sigma^{-1} = \sigma(f(\tau\sigma))$. Finally $M = \text{Hom}(\Gamma, \mathring{A})$ becomes a left G-module M_4 with action given by $({}^{\sigma}f)_4(\tau) = f(\tau\sigma)$. Show that the four G-modules M_1, M_2, M_3, M_4 are G-isomorphic.

3-7-9. (i) Let G be a multiplicative group, finite or infinite, and let $\Gamma = \mathbf{Z}[G]$. If A and B are G-modules then $A \otimes B$ (the usual tensor product over $\mathbf{Z}$—that is, the tensor product as abelian groups) may be made into a G-module in which $\sigma(a \otimes b) = \sigma a \otimes \sigma b$. In particular, $\Gamma \otimes B$ becomes a G-module in this way. If X is any abelian group, then X may be viewed as a G-module with trivial action, and we may speak of the G-module $A \otimes X$.

Let $\mathring{B}$ denote B viewed solely as an abelian group—so that $\mathring{B}$ is a G-module with trivial action. Then $\Gamma \otimes \mathring{B}$ is a G-module with $\sigma(\gamma \otimes b) = \sigma\gamma \otimes b$, $\gamma \in \Gamma$ and it is isomorphic to the G-module $\Gamma \otimes B$—in fact, the desired isomorphism of G-modules is given by $\sigma \otimes b \longrightarrow \sigma \otimes \sigma b$.

(ii) The G-module A is said to be **induced** when there exists a subgroup X of A such that $A = \sum_{\sigma \in G} \otimes\, \sigma X$—thus, induced means G-regular. The G-module A is induced $\Longleftrightarrow$ there exists an abelian group X such that $A \approx \Gamma \otimes X$. Any G-module A can be expressed as a quotient of an induced module—more precisely, there is a canonical epimorphism of $\Gamma \otimes \mathring{A} \longrightarrow A$ which maps $\sigma \otimes a \longrightarrow \sigma a$.

(iii) For any abelian group X we may consider the G-module $\operatorname{Hom}(\Gamma, X)$—namely, X is viewed as a G-module with trivial action, Γ is viewed as a (left) G-module, and (see (1-4-2)) $\operatorname{Hom}(\Gamma, X)$ becomes a G-module with $f^\sigma = \sigma \circ f \circ \sigma^{-1}$ and consequently $f^\sigma(\gamma) = f(\sigma^{-1}\gamma)$. The G-module A is said to be **co-induced** when there exists an abelian group X such that $A \approx \operatorname{Hom}(\Gamma, X)$. Any G-module A can be imbedded in a co-induced module—more precisely, there is a canonical monomorphism of $A \longrightarrow \operatorname{Hom}(\Gamma, \mathring{A})$ which maps $a \longrightarrow f_a$, where $f_a(\sigma) = \sigma^{-1}a$.

(iv) If G is finite, then for any abelian group X the G-modules $\operatorname{Hom}(\Gamma, X)$ and $\Gamma \otimes X$ are isomorphic; in particular, the notions of regular, induced, and co-induced modules then coincide.

3-7-10. Suppose that H is a subgroup of G and that $G = \bigcup c = \bigcup_c H\bar{c}$ is a right coset decomposition. Is the map $\gamma \longrightarrow \sum_c \bar{c}\gamma$ a homomorphism of $I_G \longrightarrow I_H$?

3-7-11. For a cyclic group G, we have defined a special G-

complex in (3-2-1). Using this complex, describe (for any G-module A) the cochains, cocycles, and coboundaries in dimensions 0 and -1, and the maps κ and η.

3-7-12. Suppose that H is a subgroup of G; show, directly and cohomologically, that the reduced transfer has the property

$$\overline{V}_{G\to\sigma H\sigma^{-1}}(\tau) = \sigma(\overline{V}_{G\to H}(\tau))\,\sigma^{-1} \qquad \sigma, \tau \in G$$

3-7-13. (i) If H and K are subgroups of G, then we have a disjoint decomposition into double cosets $G = \bigcup_{i=1}^{r} H\sigma_i K$. If we write $K^{\sigma_i} = \sigma_i K \sigma_i^{-1}$, then $(G:K) = \sum_{i=1}^{r} (H : H \cap K^{\sigma_i})$; in fact, if $H = \bigcup_j \rho_{ji} F_i$, $\rho_{ji} \in H$, $F_i = H \cap K^{\sigma_i}$ is a left coset decomposition for each i, then $G = \bigcup_{j,i} (\rho_{ji}\sigma_i)K$ is a left coset decomposition.

(ii) For any G-module A and any integer n, the map $\operatorname{res}_{G\to H} \circ \operatorname{cor}_{K\to G} : H^n(K, A) \longrightarrow H^n(H, A)$ is given by

$$\operatorname{res}_{G\to H} \circ \operatorname{cor}_{K\to G} = \sum_{i=1}^{r} \operatorname{cor}_{H\cap K^{\sigma_i}\to H} \circ \operatorname{res}_{K^{\sigma_i}\to H\cap K^{\sigma_i}} \circ (\sigma_i)_*$$

If, in addition, H is a normal subgroup of G, then

$$G = \bigcup H\sigma_i H = \bigcup \sigma_i H,$$

and for $\alpha \in H^n(H, A)$

$$\operatorname{res}_{G\to H} \operatorname{cor}_{H\to G} \alpha = \sum (\sigma_i)_* \alpha = S_{G/H}\,\alpha$$

(see (2-3-3) concerning the action of G/H on $H^n(H, A)$).

(iii) An element $\alpha \in H^n(H, A)$ is said to be **stable** when

$$\operatorname{res}_{H^\sigma\to H\cap H^\sigma} \sigma_*(\alpha) = \operatorname{res}_{H\to H\cap H^\sigma} \alpha$$

for every $\sigma \in G$. (In particular, if H is normal, then α is stable $\Longleftrightarrow$ $\sigma_*\alpha = \alpha$ for every $\sigma \in G$.) If $\beta \in H^n(G, A)$ and $\alpha = \operatorname{res}_{G\to H}\beta$ then α is stable. Finally, for any stable $\alpha \in H^n(H, A)$ we have

$$\operatorname{res}_{G\to H} \operatorname{cor}_{H\to G} (\alpha) = (G : H)\,\alpha$$

3-7-14. (i) Suppose that H is a subgroup of G and that B is an H-module. In the natural way, we may view $\Gamma = \mathbf{Z}[G]$ as a left G-module, and hence as an H-module, too. According to our standard procedure Hom (Γ, B) is a left H-module, with $f^\rho = \rho \circ f \circ \rho^{-1}, \rho \in H$. Let us put

$$B^* = (\operatorname{Hom}(\Gamma, B))^H = \operatorname{Hom}_H(\Gamma, B)$$

Since there is a 1–1 correspondence between Hom (Γ, B) and the set of all mappings $f : G \longrightarrow B$, it follows that under this identification

$$B^* = \{f : G \longrightarrow B \mid f(\rho\sigma) = \rho f(\sigma) \qquad \forall \rho \in H, \ \ \forall \sigma \in G\}$$

We make B^* into a G-module as follows : for $\tau \in G, f \in B^*$ define ${}^\tau f \in B^*$ by

$$ {}^\tau f(\sigma) = f(\sigma\tau) \qquad \sigma \in G$$

(ii) Define $\theta : B^* \longrightarrow B$ by

$$\theta f = f(1)$$

Then θ is an H-homomorphism, and if $i : H \longrightarrow G$ is the inclusion map, then $(i, \theta) : (G, B^*) \longrightarrow (H, B)$ is a homomorphism of pairs. In order to compute

$$(i, \theta)_* : H^n(G, B^*) \longrightarrow H^n(H, B) \qquad n \in \mathbf{Z}$$

choose a G-complex X and view it as an H-complex also. The map $\Lambda : X \longrightarrow X$ associated with $i : H \longrightarrow G$ may be taken as $\Lambda = 1$ (see (2-3-4)), and

$$(1, \theta) : \operatorname{Hom}_G(X, B^*) \longrightarrow \operatorname{Hom}_H(X, B)$$

then commutes with coboundaries and induces $(i, \theta)_*$.

Now, for any G-module A, the map

$$(1, \theta) : \operatorname{Hom}_G(A, B^*) \longrightarrow \operatorname{Hom}_H(A, B)$$

is an isomorphism onto. In fact, the inverse mapping is the following : for $\nu \in \operatorname{Hom}_H(A, B)$ define $\varphi_\nu \in \operatorname{Hom}_G(A, B^*)$ by

$$\varphi_\nu(a)(\sigma) = \nu(\sigma a) \qquad \sigma \in G, \quad a \in A$$

Thus, $(1, \theta)^{-1} : \nu \longrightarrow \varphi_\nu$.

It follows then that $(i, \theta)_*$ is an isomorphism onto in all dimensions—that is,

$$H^n(G, B^*) \approx H^n(H, B) \qquad n \in \mathbf{Z}$$

(This result is known as **Shapiro's lemma.**) Moreover, we have the commutative diagram

$$\begin{array}{ccc} H^n(G, B^*) & \xrightarrow{(i,1)_*=\text{res}} & H^n(H, B^*) \\ & \searrow^{(i,\theta)_*} & \downarrow{\scriptstyle (i,\theta)_*=\theta_*} \\ & & H^n(H, B) \end{array}$$

so that

$$(i, \theta)_* = \theta_* \circ \text{res}$$

(iii) Suppose further that B is a G-module. Define

$$\mu : B \longrightarrow B^*$$

by putting

$$(\mu b)(\sigma) = \sigma b \qquad b \in B, \quad \sigma \in G$$

Then μ is a G-homomorphism, $(1, \mu) : (G, B) \longrightarrow (G, B^*)$ is a homomorphism of pairs, and $\theta \circ \mu = 1$ (the identity map on B). Therefore, the following diagram commutes

$$\begin{array}{ccc} H^n(G, B^*) & & \\ \uparrow{\scriptstyle (i,\mu)_*=\mu_*} & \searrow^{(i,\theta)_*} & \\ H^n(G, B) & \xrightarrow[(i,1)_*=\text{res}]{} & H^n(H, B) \end{array} \qquad n \in \mathbf{Z}$$

so that

$$(i, \theta)_* \circ \mu_* = \text{res}$$

(iv) Suppose again that B is a G-module, and let $G = \bigcup_\tau \tau H$ be a left coset decomposition. Define

$$\pi : B^* \longrightarrow B$$

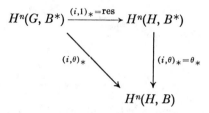

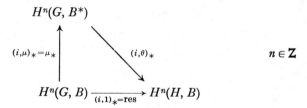

by

$$\pi f = \sum_{\tau} \tau f(\tau^{-1})$$

Then π is independent of the choice of coset representatives, and π is a G-homomorphism. Furthermore, π is an epimorphism; in fact, if we define

$$\lambda : B \longrightarrow B^*$$

by putting

$$(\lambda b)(\sigma) = \begin{cases} \sigma b & \sigma \in H \\ 0 & \sigma \notin H \end{cases}$$

then λ is an H-homomorphism with $\pi \circ \lambda = 1$ (the identity map on B).

For any G-module A, the following diagram commutes :

$$\begin{array}{ccc} \mathrm{Hom}_H(A, B) & \xrightarrow{(1,\theta)^{-1}} & \mathrm{Hom}_G(A, B^*) \\ & \searrow^{s_{H\to G}} & \downarrow{\scriptstyle (1,\pi)} \\ & & \mathrm{Hom}_G(A, B) \end{array}$$

It follows, therefore, that we have a commutative diagram

$$\begin{array}{ccc} H^n(H, B) & \xrightarrow{(i,\theta)_*^{-1}} & H^n(G, B^*) \\ & \searrow^{\mathrm{cor}} & \downarrow{\scriptstyle (1,\pi)_* = \pi_*} \\ & & H^n(G, B) \end{array} \qquad n \in \mathbf{Z}$$

and the relation

$$\pi_* \circ (i, \theta)_*^{-1} = \mathrm{cor}$$

(v) The mappings μ and π arise in a natural way; more precisely, suppose that B is a G-module—then the mapping of $B \longrightarrow \mathrm{Hom}_G(\Gamma, B)$ given by $b \longrightarrow f_b$ where $f_b(1) = b$, is an isomorphism onto (of G-modules), and we may identify the two G-modules. Now, μ and π are defined in order to make the following diagrams commute :

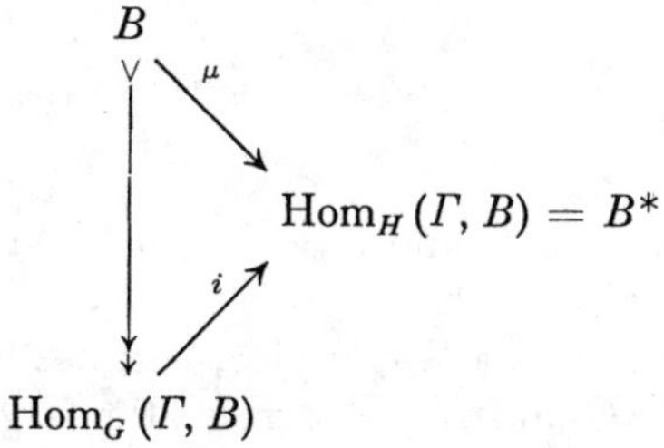

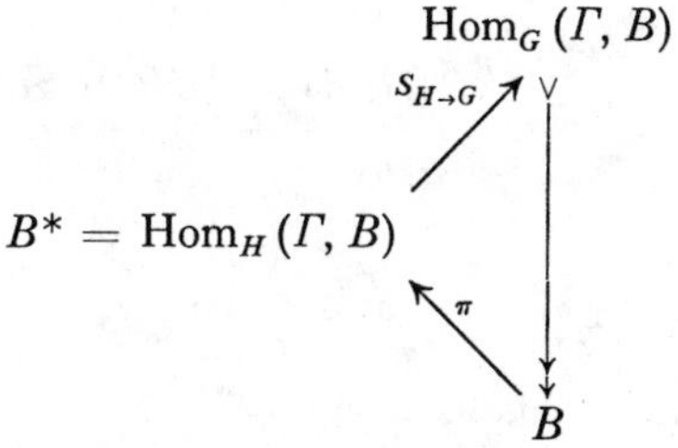

3-7-15. (i) Suppose that H is a subgroup of G, and let $G = \bigcup_1^m \tau_i H$ be a left coset decomposition (and for convenience, take $\tau_1 \in H$). Suppose that A is a G-module and A_1 is an H-submodule of G such that $A = \sum_1^m \oplus \tau_i A_1$. This kind of situation occurs naturally and often in class field theory, and leads us to say here that $\{G, H, A, A_1\}$ is **semilocal**. Note that the translates $\tau_i A_1$ are independent of the choice of coset representatives τ_i and that $H = \{\sigma \in G \mid \sigma A_1 = A_1\}$.

Suppose that A is a G-module and A_1' is an additive subgroup (i.e., **Z**-submodule) of A such that A is the direct sum of some of the translated subgroups $\sigma A_1'$, $\sigma \in G$—that is, $A = \sum_1^t \oplus \sigma_j A_1'$, for example. Consider the subgroup $H = \{\sigma \in G \mid \sigma A_1' = A_1'\}$ of G; are any additional hypotheses needed to guarantee that $\{G, H, A, A_1'\}$ is semilocal ?

(ii) Let X be any G-module and $\{G, H, A, A_1\}$ be semilocal; we seek to set up inverse isomorphisms

$$\mathrm{Hom}_G(X, A) \underset{\beta}{\overset{\alpha}{\rightleftarrows}} \mathrm{Hom}_H(X, A_1)$$

Let $\pi : A \longrightarrow A_1 = \tau_1 A_1$ be the projection map (there is no loss of generality in taking $\tau_1 = 1$), so π is an H-homomorphism, as is the inclusion map $i : A_1 \longrightarrow A$. Consider the diagram

$$\operatorname{Hom}_G(X, A) \underset{S_{H\to G}}{\overset{(1,1)}{\rightleftarrows}} \operatorname{Hom}_H(X, A) \underset{(1,i)}{\overset{(1,\pi)}{\rightleftarrows}} \operatorname{Hom}_H(X, A_1)$$

If we put $\alpha = (1, \pi) \circ (1, 1)$ and $\beta = S_{H\to G} \circ (1, i)$ then α and β are the desired isomorphisms.

Now, suppose that X is a G-complex, and use it to compute the cohomology of both G and H. We arrive at a diagram

$$H^n(G, A) \underset{\text{cor}}{\overset{\text{res}}{\rightleftarrows}} H^n(H, A) \underset{i_*}{\overset{\pi_*}{\rightleftarrows}} H^n(H, A_1) \qquad n \in \mathbf{Z}$$

and the maps

$$\pi_* \circ \text{res} \qquad \text{and} \qquad \text{cor} \circ i_*$$

provide inverse isomorphisms

$$H^n(G, A) \approx H^n(H, A_1) \qquad n \in \mathbf{Z}$$

3-7-16. We show that the hypotheses of Shapiro's lemma (see (3-7-14(ii))) and those of the semilocal situation (see (3-7-15)) are essentially equivalent—hence, each of the conclusions about the isomorphism of cohomology groups may be used to prove the other.

(i) Let B be an H-module, and form the G-module $B^* = \operatorname{Hom}_H(\Gamma, B) = \{f : G \longrightarrow B \mid f(\rho\sigma) = \rho f(\sigma)\ \rho \in H, \sigma \in G\}$. Fix a left coset decomposition $G = \bigcup_1^m \tau_i H$ and define homomorphisms

$$\pi_i : B^* \longrightarrow B^* \qquad i = 1, \ldots, m$$

by

$$(\pi_i f)(\sigma) = \begin{cases} f(\sigma\tau_i^{-1}) & \sigma \in H \\ 0 & \sigma \notin H \end{cases} \qquad f \in B^*$$

If we put

$$B_1 = \{f \in B^* \mid f(\sigma) = 0 \quad \forall \sigma \notin H\}$$

then B_1 is an H-module isomorphic to B; in fact, the map $f \longrightarrow f(1)$ is an H-isomorphism of $B_1 \rightarrowtail\!\!\!\twoheadrightarrow B$. Moreover, every $f \in B^*$ can be expressed uniquely in the form

$$f = \sum_1^m {}^{\tau_i}(\pi_i f)$$

—it follows that $B^* = \sum_1^m \oplus {}^{\tau_i}(B_1)$ and that $\{G, H, B^*, B_1\}$ is semilocal.

(ii) Suppose that $\{G, H, A, A_1\}$ is semilocal; thus, for a left coset decomposition $G = \bigcup_1^m \tau_i H$ we have $A = \sum_1^m \oplus \tau_i A_1$ and $H = \{\sigma \in G \mid \sigma A_1 = A_1\}$. Then the G-module

$$A_1{}^* = \operatorname{Hom}_H(\Gamma, A_1) = \{f : G \longrightarrow A_1 \mid f(\rho\sigma) = \rho f(\sigma) \quad \rho \in H, \sigma \in G\}$$

is isomorphic to A; in fact, the mapping

$$f \longrightarrow \sum_1^m \tau_i f(\tau_i^{-1}) \qquad f \in A_1^*$$

is a G-isomorphism of $A_1^* \rightarrowtail\!\!\!\twoheadrightarrow A$. (What is the inverse isomorphism?)

3-7-17. Give an example of a G-module with $H^n(G, A) = (0)$ for all n, but where A is not cohomologically trivial.

CHAPTER

IV

The Cup Product

The main objective of this chapter is to prove a more refined version of Tate's theorem (see (3-6-5)) and in this way to provide a hold on the reciprocity law isomorphism and the Nakayama map of class field theory. This will be accomplished by exploitation of the notion of cup product.

4-1. CUP PRODUCTS

In order to define the cup product it is necessary to prepare some background material about tensor products.

As usual, let G denote a finite group, and let $\Gamma = \mathbf{Z}[G]$. If A and B are G-modules, we denote by $A \otimes B$ the tensor product of A and B as $\mathbf{Z}$-modules; in general, all our tensor products will be taken over $\mathbf{Z}$. (The basic facts about tensor products, which we shall take for granted, may be found in [11] or in [18, §12].) Now, the additive group $A \otimes B$ may be viewed as a G-module in the following way. For each $\sigma \in G$, consider the bilinear map $\theta_\sigma : A \times B \longrightarrow A \otimes B$ given by $\theta_\sigma(a, b) = \sigma a \otimes \sigma b$. Then, by a standard property of the tensor product, there exists a unique linear map (which we denote by σ) of $A \otimes B \longrightarrow A \otimes B$ such that $\theta_\sigma(a, b) = \sigma(a \otimes b)$. In other words,

$$\sigma(a \otimes b) = \sigma a \otimes \sigma b \qquad \sigma \in G, \quad a \in A, \quad b \in B$$

and $A \otimes B$ is clearly a G-module.

Thus, the operator σ is an element of Hom $(A \otimes B, A \otimes B)$ and is sometimes denoted by $\sigma \otimes \sigma$. Of course, this $\sigma \otimes \sigma$ must be distinguished from $\sigma \otimes \sigma$, belonging to the tensor product of G-modules Hom $(A, A) \otimes$ Hom (B, B). As a matter of fact, the first of these maps $\sigma \otimes \sigma$ is the image of the second under the natural homomorphism which is a special case of the one given in the following statement.

4-1-1. Proposition. Let A, A', B, B' be G-modules. Then the mapping of G-modules

$$\text{Hom}\,(A, A') \otimes \text{Hom}\,(B, B') \longrightarrow \text{Hom}\,(A \otimes B, A' \otimes B')$$

which takes

$$f \otimes g \longrightarrow f \otimes g$$

is a G-homomorphism. Furthermore, if $f \in \text{Hom}_G\,(A, A')$ and $g \in \text{Hom}_G\,(B, B')$, then $f \otimes g \in \text{Hom}_G\,(A \otimes B, A' \otimes B')$.

Proof: Application of the standard procedure for defining a linear map of a tensor product via a bilinear map shows that $f \otimes g \longrightarrow f \otimes g$ determines an additive map of G-modules. As for the action of G, if $\sigma \in G, f \in \text{Hom}\,(A, A'), g \in \text{Hom}\,(B, B')$, then

$$(f \otimes g)^\sigma = f^\sigma \otimes g^\sigma \longrightarrow f^\sigma \otimes g^\sigma = (\sigma \otimes \sigma)(f \otimes g)(\sigma^{-1} \otimes \sigma^{-1}) = (f \otimes g)^\sigma$$

so that the map is a G-homomorphism.

Furthermore, if f and g are both G-homomorphisms, so that $f^\sigma = f$ and $g^\sigma = g$ for all $\sigma \in G$, then $(f \otimes g)^\sigma = f \otimes g$ and $f \otimes g$ is a G-homomorphism. It may be noted that this statement differs from the assertion that there is a homomorphism of

$$\text{Hom}_G\,(A, A') \otimes \text{Hom}_G\,(B, B') \longrightarrow \text{Hom}_G\,(A \otimes B, A' \otimes B'). \quad \blacksquare$$

The behavior of exact sequences under tensoring is given in part by the results which follow.

4-1-2. Proposition. If

$$A \xrightarrow{i} B \xrightarrow{j} C \longrightarrow 0 \quad \text{and} \quad A' \xrightarrow{i'} B' \xrightarrow{j'} C' \longrightarrow 0$$

are exact G-sequences, then so is

$$(A \otimes B') \oplus (B \otimes A') \xrightarrow{(i\otimes 1)\oplus(1\otimes i')} B \otimes B' \xrightarrow{j\otimes j'} C \otimes C' \longrightarrow 0$$

Proof: According to (4-1-1) this is indeed a G-sequence. Since j and j' are onto and $C \otimes C'$ is generated by the elements of the form $c \otimes c'$, it follows that $j \otimes j'$ is onto. In order to prove exactness at $B \otimes B'$, we observe that the image D of $(i \otimes 1) \oplus (1 \otimes i')$ is contained in $\ker (j \otimes j')$, so that $j \otimes j'$ induces a linear map $\beta : (B \otimes B')/D \twoheadrightarrow C \otimes C'$. On the other hand, the map $\gamma' : C \times C' \longrightarrow (B \otimes B')/D$ given by

$$\gamma'(c, c') = j^{-1}(c) \otimes j'^{-1}(c') + D \qquad c \in C, \quad c' \in C'$$

(where $j^{-1}(c)$ and $j'^{-1}(c')$ denote any preimages of c and c', respectively) is well-defined (since both $iA \otimes B'$ and $B \otimes i'A'$ are contained in D) and bilinear. Therefore, γ' induces a linear map $\gamma : C \otimes C' \longrightarrow (B \otimes B')/D$. It is clear that $\beta \circ \gamma =$ identity and $\gamma \circ \beta =$ identity—consequently, $D = \ker (j \otimes j')$. ▌

4-1-3. Corollary. The functor $\otimes$ is right exact; this means that if $0 \longrightarrow A \longrightarrow B \longrightarrow C \longrightarrow 0$ is any exact G-sequence and B' is any G-module, then the sequences

$$A \otimes B' \longrightarrow B \otimes B' \longrightarrow C \otimes B' \longrightarrow 0$$

$$B' \otimes A \longrightarrow B' \otimes B \longrightarrow B' \otimes C \longrightarrow 0$$

are exact G-sequences.

Proof: The first sequence arises from (4-1-2) with $A' = (0)$, $C' = B'$. A switch of notation leads to the second sequence. ▌

4-1-4. Corollary. If $0 \longrightarrow A \xrightarrow{i} B \xrightarrow{j} C \longrightarrow 0$ is an exact G-sequence which splits as a **Z**-sequence, then, for any G-module B',

$$0 \longrightarrow A \otimes B' \xrightarrow{i \otimes 1} B \otimes B' \xrightarrow{j \otimes 1} C \otimes B' \longrightarrow 0$$

is an exact G-sequence which splits as a **Z**-sequence.

Proof: By hypothesis, there exists an additive subgroup B'' of B with $B = iA \oplus B''$ as **Z**-modules, and then

$$B \otimes B' = (iA \otimes B') \oplus (B'' \otimes B')$$

as **Z**-modules. (Note that we are somewhat careless about distinguishing between $=$ and $\approx$.) This provides a **Z**-splitting and shows that $i \otimes 1$ is a monomorphism. ▮

4-1-5. Proposition. If the G-module A is G-regular, then so is $A \otimes A'$ for any G-module A'.

Proof: If A_0 is a subgroup of A with $A = \sum_{\sigma \in G} \oplus \sigma A_0$, then

$$A \otimes A' = \sum_{\sigma \in G} \oplus (\sigma A_0 \otimes A') = \sum_{\sigma \in G} \oplus (\sigma A_0 \otimes \sigma A')$$
$$= \sum_{\sigma \in G} \oplus \sigma(A_0 \otimes A'). \quad ▮$$

4-1-6. Remark. (**Dimension Shifting**). If A is a G-module, the bilinear map $(n, a) \longrightarrow na$ of $\mathbf{Z} \times A \longrightarrow A$ leads to a linear map of $\mathbf{Z} \otimes A \longrightarrow A$ which takes $n \otimes a \longrightarrow na$. This is an isomorphism; the inverse map is given by $a \longrightarrow 1 \otimes a$, so that $na \longrightarrow 1 \otimes na = n \otimes a$. Moreover, this is a G-isomorphism, since for $\sigma \in G$,

$$\sigma(n \otimes a) = \sigma n \otimes \sigma a = n \otimes \sigma a \longrightarrow n(\sigma a) = \sigma(na)$$

We shall often identify the G-modules $\mathbf{Z} \otimes A \approx A$. In particular, we have the identification of G-modules $\mathbf{Z} \otimes \mathbf{Z} \approx \mathbf{Z}$.

Now consider the exact G-sequences (see (3-3-8))

$$0 \longrightarrow I \longrightarrow \Gamma \xrightarrow{\varepsilon} \mathbf{Z} \longrightarrow 0$$

$$0 \longrightarrow \mathbf{Z} \xrightarrow{\mu} \Gamma \longrightarrow J \longrightarrow 0$$

Since these split as $\mathbf{Z}$-sequences, we have, by (4-1-4), exact G-sequences (which split as $\mathbf{Z}$-sequences)

$$0 \longrightarrow I \otimes A \xrightarrow{i \otimes 1} \Gamma \otimes A \xrightarrow{\varepsilon \otimes 1} \mathbf{Z} \otimes A \approx A \longrightarrow 0$$

$$0 \longrightarrow \mathbf{Z} \otimes A \approx A \xrightarrow{\mu \otimes 1} \Gamma \otimes A \longrightarrow J \otimes A \longrightarrow 0$$

for any G-module A. By (4-1-5), $\Gamma \otimes A$ is G-regular, and hence H-regular for every subgroup H of G. Therefore, for all n and all $H \subset G$, we have

$$H^{n-1}(H, J \otimes A) \approx H^n(H, A) \approx H^{n+1}(H, I \otimes A)$$

Of course, tensoring by A on the left leads to an analogous result.

4-1-7. Remark. Consider the G-complex $(X, \partial, \varepsilon, \mu)$; for the construction of the cup product it is necessary to look at $X \otimes X$. We recall that for all n there exist $\mathbf{Z}$-homomorphisms $D : X_n \longrightarrow X_{n+1}$ (see (1-4-6)) and $\pi : X_n \longrightarrow X_n$ (see (2-1-4)) such that $D\partial + \partial D = 1$ and the G-trace $S(\pi) = 1$. Then, for all p and q we may define G-homomorphisms ∂', ∂'', D', D'' in the diagram

$$\begin{array}{ccccc} & & X_p \otimes X_{q+1} & & \\ & & \partial'' \downarrow \uparrow D'' & & \\ X_{p-1} \otimes X_q & \underset{D'}{\overset{\partial'}{\rightleftarrows}} & X_p \otimes X_q & \underset{D'}{\overset{\partial'}{\rightleftarrows}} & X_{p+1} \otimes X_q \\ & & \partial'' \downarrow \uparrow D'' & & \\ & & X_p \otimes X_{q-1} & & \end{array}$$

so that $\partial'^2 = \partial''^2 = 0$, $\partial' \circ \partial'' = \partial'' \circ \partial'$, $\partial' D' + D' \partial' = 1$, $\partial'' D'' + D'' \partial'' = 1$. In fact, $\partial' = \partial \otimes 1$ and $\partial'' = 1 \otimes \partial$ are commuting G-homomorphisms of "square" zero. Furthermore, from the $\mathbf{Z}$-homomorphisms $D \otimes \pi : X_p \otimes X_q \longrightarrow X_{p+1} \otimes X_q$

and $\pi \otimes D : X_p \otimes X_q \longrightarrow X_p \otimes X_{q+1}$ we construct the G-homomorphisms $D' = S(D \otimes \pi)$ and $D'' = S(\pi \otimes D)$. Finally,

$$\begin{aligned}\partial' D' + D'\partial' &= (\partial \otimes 1) \circ S(D \otimes \pi) + S(D \otimes \pi) \circ (\partial \otimes 1) \\ &= S(\partial D \otimes \pi) + S(D\partial \otimes \pi) = S((\partial D + D\partial) \otimes \pi)) \\ &= S(1 \otimes \pi) = \sum_\sigma (1 \otimes \pi)^\sigma = \sum_\sigma (1 \otimes \pi^\sigma) = 1 \otimes \sum_\sigma \pi^\sigma = 1,\end{aligned}$$

and similarly $\partial'' D'' + D''\partial'' = 1$.

The cup product is to be an operation which, for any G-modules A and B, provides a pairing of

$$H^p(G, A) \times H^q(G, B) \longrightarrow H^{p+q}(G, A \otimes B)$$

for all p and q. In order to define it, one naturally goes back to cochains. Consider then $f \in \operatorname{Hom}_G(X_p, A)$ and $g \in \operatorname{Hom}_G(X_q, B)$; we seek to define $f \smile g \in \operatorname{Hom}_G(X_{p+q}, A \otimes B)$. An obvious procedure is to construct G-homomorphisms $\varphi_{p,q}$ in

$$X_{p+q} \xrightarrow{\varphi_{p,q}} X_p \otimes X_q \xrightarrow{f \otimes g} A \otimes B$$

in order to put $f \smile g = (f \otimes g) \circ \varphi_{p,q}$. If this cup product of cochains is to lead to a bilinear map of cohomology classes, it is necessary that the cup of two cocycles be a cocycle and that the cup of a cocycle and a coboundary be a coboundary. Since it is also desirable that the coboundary relation be symmetric, this suggests $\delta(f \smile g) = \delta f \smile g + f \smile \delta g$. In other words, we want $(f \otimes g)\, \varphi_{p,q}\, \partial = (\delta f \otimes g)\, \varphi_{p+1,q} + (f \otimes \delta g)\, \varphi_{p,q+1}$ which requires $\varphi_{p,q}\, \partial = \partial' \varphi_{p+1,q} + \partial'' \varphi_{p,q+1}$. In view of all this we prove:

4-1-8. Lemma. For any G-complex X there exist G-homomorphisms

$$\varphi_{p,q} : X_{p+q} \longrightarrow X_p \otimes X_q \qquad \text{all } p, q$$

such that

$$\text{I}: \quad \varepsilon = (\varepsilon \otimes \varepsilon) \circ \varphi_{0,0}$$

$$\text{II}_{p,q}: \varphi_{p,q}\partial = \partial' \varphi_{p+1,q} + (-1)^p\, \partial'' \varphi_{p,q+1} \qquad \text{all } p, q$$

Proof: Condition I is a normalization which expresses the commutativity of the diagram

$$\begin{array}{ccc} X_0 & \xrightarrow{\varphi_{0,0}} & X_0 \otimes X_0 \\ & {\scriptstyle\varepsilon}\searrow \quad \swarrow{\scriptstyle\varepsilon\otimes\varepsilon} & \\ & \mathbf{Z} \approx \mathbf{Z} \otimes \mathbf{Z} & \end{array}$$

It excludes, in particular, the choice of all $\varphi_{p,q} = 0$. The introduction of the factor $(-1)^p$ in $\mathrm{II}_{p,q}$ is a purely technical device. At the outset, we observe that operating on $\mathrm{II}_{p,q}$ on the left by ∂' gives

$$\mathrm{III}_{p,q} : \partial' \varphi_{p,q} \partial = (-1)^p \, \partial' \partial'' \varphi_{p,q+1}$$

The first step is then to construct $\varphi_{0,q}$ for all q such that I holds and $\mathrm{III}_{0,q}$ holds for all q. Choose $\xi \in X_0$ with $\varepsilon(\xi) = 1$ and for each q define a $\mathbf{Z}$-homomorphism $\psi_{0,q} : X_q \longrightarrow X_0 \otimes X_q$ by

$$\psi_{0,q}(x) = \xi \otimes \pi_q x \qquad x \in X_q$$

Then

$$\varphi_{0,q} = S(\psi_{0,q}) = \sum_{\sigma \in G} \psi_{0,q}^{\sigma}$$

is a G-homomorphism of $X_q \longrightarrow X_0 \otimes X_q$ whose action is given by

$$\varphi_{0,q}(x) = \sum_{\sigma \in G} (\sigma\xi \otimes \sigma\pi_q \sigma^{-1} x) \qquad x \in X_q$$

In particular, for $x \in X_q$ we have

$$(\varepsilon \otimes 1)\varphi_{0,q}(x) = \sum_{\sigma \in G} (1 \otimes \pi_q^{\sigma} x) = 1 \otimes x \in \mathbf{Z} \otimes X_q$$

Consequently, for $x \in X_0$,

$$(\varepsilon \otimes \varepsilon)\varphi_{0,0}(x) = (1 \otimes \varepsilon)(\varepsilon \otimes 1)\varphi_{0,0}(x) = 1 \otimes \varepsilon(x),$$

which upon identification of $\mathbf{Z} \otimes \mathbf{Z}$ with $\mathbf{Z}$ equals $\varepsilon(x)$. Thus I holds. As for $\mathrm{III}_{0,q}$, for $x \in X_{q+1}$, we have

$$\begin{aligned} \partial' \partial'' \varphi_{0,q+1}(x) &= \partial'' \partial' \varphi_{0,q+1}(x) = \partial''(\partial \otimes 1)\varphi_{0,q+1}(x) \\ &= \partial''(\mu \otimes 1)(\varepsilon \otimes 1)\varphi_{0,q+1}(x) = \mu(1) \otimes \partial x \end{aligned}$$

which is also equal to $\partial' \varphi_{0,q} \, \partial(x)$.

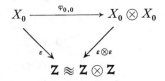

The next step is to prove the existence of all $\varphi_{p,q}$ with negative p by showing that if, for a given p and all q there exist G-homomorphisms $\varphi_{p,q}$ which satisfy $\mathrm{III}_{p,q}$, then we can define G-homomorphisms $\varphi_{p-1,q}$ for all q such that $\mathrm{II}_{p-1,q}$ holds. Since $\mathrm{II}_{p-1,q}$ implies $\mathrm{III}_{p-1\,q}$ the proof may proceed inductively.

To do this, put

$$\varphi_{p-1,q} = (-1)^p D''\partial'\varphi_{p,q-1} \qquad \text{all } q$$

so that

$$\begin{aligned}\varphi_{p-1,q}\partial &= (-1)^p D''[(-1)^p\partial'\partial''\varphi_{p,q}] = D''\partial''\partial'\varphi_{p,q} = (1 - \partial''D'')\partial'\varphi_{p,q} \\ &= \partial'\varphi_{p,q} - \partial''((-1)^p\varphi_{p-1,q+1}) = \partial'\varphi_{p,q} + (-1)^{p-1}\partial''\varphi_{p-1,q+1}\end{aligned}$$

and $\mathrm{II}_{p-1,q}$ holds.

Finally, to prove the existence of all $\varphi_{p,q}$ with positive p, we show that if for a given p and all q there exist G-homomorphisms $\varphi_{p\,q}$ which satisfy $\mathrm{III}_{p,q}$, then we can define G-homomorphisms $\varphi_{p+1,q}$ for all q such that $\mathrm{II}_{p,q}$ holds. Since, as one verifies easily, $\mathrm{II}_{p,q}$ and $\mathrm{II}_{p\,q+1}$ together imply $\mathrm{III}_{p+1,q}$, the proof may then proceed inductively. Now, using $\mathrm{III}_{p,q}$, we have

$$\begin{aligned}\varphi_{p,q}\partial &= (\partial'D' + D'\partial')\varphi_{p,q}\partial = \partial'D'\varphi_{p,q}\partial + (-1)^p D'\partial'\partial''\varphi_{p,q+1} \\ &= \partial'D'\varphi_{p,q}\partial + (-1)^p(1 - \partial'D')\partial''\varphi_{p,q+1} \\ &= \partial'[D'\varphi_{p,q}\partial + (-1)^{p+1}D'\partial''\varphi_{p,q+1}] + (-1)^p\partial''\varphi_{p,q+1}\end{aligned}$$

so that we should put

$$\varphi_{p+1,q} = D'\varphi_{p,q}\partial + (-1)^{p+1}D'\partial''\varphi_{p,q+1} \qquad \text{all } q$$

This completes the proof. ▮

A slightly different version of the construction of such $\varphi_{p,q}$'s is given in [15, Chapter 14].

4-1-9. Proposition. Suppose that A and B are G-modules. Then for every pair of integers p and q there exists a bilinear map, called the **cup product** (of dimension (p, q)), of

$$H^p(G, A) \times H^q(G, B) \longrightarrow H^{p+q}(G, A \otimes B)$$

and a homomorphism of

$$H^p(G, A) \otimes H^q(G, B) \longrightarrow H^{p+q}(G, A \otimes B)$$

Proof: Let X be any G-complex and consider

$$f, f' \in \operatorname{Hom}_G(X_p, A), \qquad g, g' \in \operatorname{Hom}_G(X_q, B).$$

From what has gone before, it is clear that

$$(f+f') \cup g = f \cup g + f' \cup g, \qquad f \cup (g+g') = f \cup g + f \cup g',$$

and

$$\delta(f \cup g) = \delta f \cup g + (-1)^p f \cup \delta g$$

Therefore, cupping two cocycles gives a cocycle, and cupping a cocycle with a coboundary gives a coboundary—and it follows that we have an induced bilinear cup product of cohomology classes. This, in turn, leads to a homomorphism of the tensor product. ∎

Strictly speaking, the cup product we have constructed depends on the G-complex X and maps

$$H^p(G, A, X) + H^q(G, B, X) \longrightarrow H^{p+q}(G, A \otimes B, X)$$

However, we shall continue to be careless with the notation; eventually we shall see that the cup product is indeed independent of the complex.

4-1-10. Exercise. Suppose that X is the standard G-complex; then the G-homomorphisms $\varphi_{p,q} : X_{p+q} \longrightarrow X_p \otimes X_q$ are completely determined by their action on the G-cells. Show that the $\varphi_{p,q}$'s can be constructed in such a way that

$$\varphi_{0,0}([\cdot]) = [\cdot] \otimes [\cdot] \qquad \varphi_{0,-1}(\langle\cdot\rangle) = [\cdot] \otimes \langle\cdot\rangle$$

$$\varphi_{0,q}([\sigma_1, \ldots, \sigma_q]) = [\cdot] \otimes [\sigma_1, \ldots, \sigma_q] \qquad q > 0$$

$$\varphi_{0,-q}(\langle\sigma_1, \ldots, \sigma_{q-1}\rangle) = [\cdot] \otimes \langle\sigma_1, \ldots, \sigma_{q-1}\rangle \qquad q > 1$$

$$\varphi_{-1,1}([\cdot]) = \sum_{\sigma \in G} (\langle\cdot\rangle^\sigma \otimes \sigma[\sigma^{-1}])$$

$$\begin{aligned} \varphi_{1,1}([\sigma, \tau]) &= \sigma[1] \otimes \sigma[\tau] - \sigma[\sigma^{-1}] \otimes \sigma[\tau] \\ &= \sigma\{([1] - [\sigma^{-1}]) \otimes [\tau]\} \end{aligned}$$

$$\varphi_{1,-1}([\cdot]) = \sum_{\sigma \in G} \{(\sigma[1] - \sigma[\sigma^{-1}]) \otimes \langle\cdot\rangle^\sigma\}$$

4-2. PROPERTIES OF THE CUP PRODUCT

In this section we describe the cup product in dimension (0, 0) and its behavior with respect to the coboundary arising from exact sequences. These properties serve to characterize the cup product completely, and the uniqueness of the cup product then implies its independence of the choice of a complex.

4-2-1. Proposition. (CP1). Let A and B be G-modules; then for $a \in A^G$ and $b \in B^G$ we have $a \otimes b \in (A \otimes B)^G$ and

$$\kappa a \smile \kappa b = \kappa(a \otimes b)$$

Proof: Of course, we must compute with respect to some G-complex $(X, \partial, \varepsilon, \mu)$. According to (2-2-10),

$$\kappa_X(a) = \omega_a \varepsilon + \mathscr{B}_X^0(G, A)$$

where $\omega_a : \mathbf{Z} \longrightarrow A$ is the unique G-homomorphism such that $\omega_a(1) = a$. Similarly, we have $\kappa_X(b) = \omega_b \varepsilon + \mathscr{B}_X^0(G, B)$. It is clear that $a \otimes b \in (A \otimes B)^G$, so that $\kappa_X(a \otimes b) = \omega_{a \otimes b}\varepsilon + \mathscr{B}_X^0(G, A \otimes B)$; and under the identification

$$\text{Hom}\,(\mathbf{Z} \otimes \mathbf{Z}, A \otimes B) \approx \text{Hom}\,(\mathbf{Z}, A \otimes B)$$

(which follows from that of $\mathbf{Z} \otimes \mathbf{Z}$ and $\mathbf{Z}$) we have $\omega_a \otimes \omega_b = \omega_{a \otimes b}$. Therefore,

$$\begin{aligned}
\kappa_X a \smile \kappa_X b &= (\omega_a \varepsilon \otimes \omega_b \varepsilon)\varphi_{0,0} + \mathscr{B}_X^0(G, A \otimes B)\\
&= (\omega_a \otimes \omega_b)(\varepsilon \otimes \varepsilon)\varphi_{0,0} + \mathscr{B}_X^0(G, A \otimes B)\\
&= \omega_{a \otimes b}\varepsilon + \mathscr{B}_X^0(G, A \otimes B)\\
&= \kappa_X(a \otimes b)
\end{aligned}$$

and the proof is complete. ▌

This result provides an explicit way of computing the cup product in dimension (0, 0)—namely, it is determined by the canonical map κ. More precisely, for $\alpha \in H^0(G, A)$, $\beta \in H^0(G, B)$

we compute $\alpha \cup \beta$ (all with respect to some G-complex X) by choosing $a \in A^G$, $b \in B^G$ such that $\kappa a = \alpha$, $\kappa b = \beta$ and noting that $\alpha \cup \beta = \kappa(a \otimes b)$.

4-2-2. Proposition. (CP2). Let A' be a G-module.

(i) If the G-sequences

$$0 \longrightarrow A \xrightarrow{i} B \xrightarrow{j} C \longrightarrow 0$$

$$0 \longrightarrow A \otimes A' \xrightarrow{i \otimes 1} B \otimes A' \xrightarrow{j \otimes 1} C \otimes A' \longrightarrow 0$$

are both exact, then, for all p and q, the following diagram commutes:

$$\begin{array}{ccc} H^p(G, C) \times H^q(G, A') & \xrightarrow{\cup} & H^{p+q}(G, C \otimes A') \\ \downarrow{\scriptstyle \delta_* \times 1} & & \downarrow{\scriptstyle \delta_*} \\ H^{p+1}(G, A) \times H^q(G, A') & \xrightarrow{\cup} & H^{p+q+1}(G, A \otimes A') \end{array}$$

In other words, for $\gamma \in H^p(G, C)$ and $\alpha' \in H^q(G, A')$ we have

$$\delta_*(\gamma \cup \alpha') = \delta_*\gamma \cup \alpha'$$

(ii) If the G-sequences

$$0 \longrightarrow A \xrightarrow{i} B \xrightarrow{j} C \longrightarrow 0$$

$$0 \longrightarrow A' \otimes A \xrightarrow{1 \otimes i} A' \otimes B \xrightarrow{1 \otimes j} A' \otimes C \longrightarrow 0$$

are both exact, then, for all p and q, the following diagram has commutativity character $(-1)^p$:

$$\begin{array}{ccc} H^p(G, A') \times H^q(G, C) & \xrightarrow{\cup} & H^{p+q}(G, A' \otimes C) \\ \downarrow{\scriptstyle 1 \times \delta_*} & (-1)^p & \downarrow{\scriptstyle \delta_*} \\ H^p(G, A') \times H^{q+1}(G, A) & \xrightarrow{\cup} & H^{p+q+1}(G, A' \otimes A) \end{array}$$

In other words, for $\alpha' \in H^p(G, A')$ and $\gamma \in H^q(G, C)$ we have

$$\delta_*(\alpha' \cup \gamma) = (-1)^p(\alpha' \cup \delta_*\gamma)$$

Proof: We prove (ii) in order to show how the factor $(-1)^p$ arises; the proof of (i) goes the same way. Let X be any G-complex. We observe that the coboundary maps arise (see (1-1-4) and (2-1-9)) from the exact sequence of differential graded groups

$$0 \longrightarrow \mathrm{Hom}_G(X, A) \xrightarrow{(1,i)} \mathrm{Hom}_G(X, B) \xrightarrow{(1,j)} \mathrm{Hom}_G(X, C) \longrightarrow 0$$

$$0 \longrightarrow \mathrm{Hom}_G(X, A' \otimes A) \xrightarrow{(1,1\otimes i)} \mathrm{Hom}_G(X, A' \otimes B) \xrightarrow{(1,1\otimes j)}$$

$$\mathrm{Hom}_G(X, A' \otimes C) \longrightarrow 0$$

Let $f' \in \mathrm{Hom}_G(X_p, A')$ and $h \in \mathrm{Hom}_G(X_q, C)$ be cocycles representing the cohomology classes α' and γ, respectively. Choose $g \in \mathrm{Hom}_G(X_q, B)$ such that $h = (1, j)g = jg$ and $f \in \mathrm{Hom}_G(X_{q+1}, A)$ such that $\delta g = (1, i)f = if$. Then $\delta_* \gamma$ is represented by f and $\alpha' \cup \delta_* \gamma$ is represented by $f' \cup f$.

On the other hand, $\alpha' \cup \gamma$ is represented by the cocycle $f' \cup h \in \mathrm{Hom}_G(X_{p+q}, A' \otimes C)$. We have, therefore,

$$\begin{aligned} f' \cup h &= (f' \otimes h)\varphi_{p,q} = (1 \otimes j)(f' \otimes g)\varphi_{p,q} \\ &= (1 \otimes j)(f' \cup g) = (1, 1 \otimes j)(f' \cup g) \end{aligned}$$

and

$$\begin{aligned} \delta(f' \cup g) &= \delta f' \cup g + (-1)^p (f' \cup \delta g) \\ &= (-1)^p (1 \otimes i)(f' \cup f) = (1, 1 \otimes i)((-1)^p (f' \cup f)), \end{aligned}$$

so that $\delta_*(\alpha' \cup \gamma)$ is represented by $(-1)^p (f' \cup f)$. This completes the proof. ∎

4-2-3. Remark. In virtue of the foregoing, the cup product $\cup$ is really a collection $\{\cup_{p,q} \mid p, q \in \mathbf{Z}\}$ of bilinear maps

$$\cup_{p,q} : H^p(G, A) \times H^q(G, B) \longrightarrow H^{p+q}(G, A \otimes B)$$

which are related to the coboundaries arising from exact sequences by the properties CP1 and CP2. It follows that for each pair (p, q) and any G-modules A and B we have a homomorphism of groups

$$\Omega_{p,q} : H^p(G, A) \otimes H^q(G, B) \longrightarrow H^{p+q}(G, A \otimes B)$$

given by

$$\Omega_{p,q}(\alpha \otimes \beta) = \alpha \cup_{p,q} \beta = \alpha \cup \beta$$

and these $\Omega_{p,q}$ satisfy CP1 and CP2 when the appropriate notational changes are made. It is clear that statements made about $\cup = \{\cup_{p,q}\}$ can be transferred to $\Omega = \{\Omega_{p,q}\}$, and conversely.

4-2-4. Theorem. The cup product is characterized by the fact that it is a bilinear map satisfying CP1 and CP2. More precisely, if $\Phi = \{\Phi_{p,q}\}$ is such that for any G-modules A and B

$$\Phi_{p,q} : H^p(G, A) \times H^q(G, B) \longrightarrow H^{p+q}(G, A \otimes B)$$

is a bilinear map, and the following conditions are satisfied:

CP1: If the hypotheses of CP1 hold, then

$$\Phi_{0,0}(\kappa a, \kappa b) = \kappa(a \otimes b).$$

CP2: If both hypotheses of CP2 hold, then we have

$$\delta_* \Phi_{p,q}(\gamma, \alpha') = \Phi_{p+1,q}(\delta_* \gamma, \alpha')$$

and

$$\delta_* \Phi_{p,q}(\alpha', \gamma) = (-1)^p \Phi_{p,q+1}(\alpha', \delta_* \gamma)$$

—then $\Phi = \cup$.

Proof: We work with respect to some fixed G-complex X (since the cup product is defined in terms of X—see (4-1-8) and (4-1-9)) even though it will not appear explicitly in the proof.

According to CP1, we have $\Phi_{0,0} = \cup_{0,0}$. To prove that $\Phi_{p,q} = \cup_{p,q}$ for all (p, q) we proceed inductively via the use of dimension shifters. Thus, suppose that for given (p, q) $\Phi_{p,q} = \cup_{p,q}$. Then it follows from (4-1-4) and (4-1-6) that for any fixed G-modules A and B the following are exact G-sequences which split as $\mathbf{Z}$-sequences:

$$0 \longrightarrow \mathbf{Z} \longrightarrow \Gamma \longrightarrow J \longrightarrow 0$$

$$0 \longrightarrow B \longrightarrow B \otimes \Gamma \longrightarrow B \otimes J \longrightarrow 0$$

$$0 \longrightarrow A \otimes B \longrightarrow A \otimes B \otimes \Gamma \longrightarrow A \otimes B \otimes J \longrightarrow 0$$

Since $B \otimes \Gamma$ and $A \otimes B \otimes \Gamma$ are both G-regular, the maps δ_* in the following diagram are isomorphisms onto:

$$\begin{array}{ccc} H^p(G, A) \times H^q(G, B \otimes J) & \longrightarrow & H^{p+q}(G, A \otimes B \otimes J) \\ {\scriptstyle 1\times\delta_*}\downarrow & (-1)^p & \downarrow{\scriptstyle \delta_*} \\ H^p(G, A) \times H^{q+1}(G, B) & \longrightarrow & H^{p+q+1}(G, A \otimes B) \end{array}$$

From CP2 we know that this diagram has commutativity character $(-1)^p$ when both horizontal maps are given by $\smile$ or by Φ. Consequently, $\Phi_{p,q} = \smile_{p,q}$ implies that $\Phi_{p,q+1} = \smile_{p,q+1}$.

The proof that $\Phi_{p,q} = \smile_{p,q}$ implies $\Phi_{p,q-1} = \smile_{p,q-1}$ is based on the exact G-sequences

$$0 \longrightarrow B \otimes I \longrightarrow B \otimes \Gamma \longrightarrow B \longrightarrow 0$$

$$0 \longrightarrow A \otimes B \otimes I \longrightarrow A \otimes B \otimes \Gamma \longrightarrow A \otimes B \longrightarrow 0$$

and the diagram

$$\begin{array}{ccc} H^p(G, A) \times H^{q-1}(G, B) & \longrightarrow & H^{p+q-1}(G, A \otimes B) \\ {\scriptstyle 1\times\delta_*}\downarrow & (-1)^p & \downarrow{\scriptstyle \delta_*} \\ H^p(G, A) \times H^{q}(G, B \otimes I) & \longrightarrow & H^{p+q}(G, A \otimes B \otimes I) \end{array}$$

while the proof that $\Phi_{p,q} = \smile_{p,q}$ implies $\Phi_{p-1,q} = \smile_{p-1,q}$ is based on the exact sequences

$$0 \longrightarrow I \otimes A \longrightarrow \Gamma \otimes A \longrightarrow A \longrightarrow 0$$

$$0 \longrightarrow I \otimes A \otimes B \longrightarrow \Gamma \otimes A \otimes B \longrightarrow A \otimes B \longrightarrow 0$$

and the diagram

$$\begin{array}{ccc} H^{p-1}(G, A) \times H^{q}(G, B) & \longrightarrow & H^{p+q-1}(G, A \otimes B) \\ {\scriptstyle \delta_*\times 1}\downarrow & & \downarrow{\scriptstyle \delta_*} \\ H^p(G, I \otimes A) \times H^{q}(G, B) & \longrightarrow & H^{p+q}(G, I \otimes A \otimes B) \end{array}$$

Finally, the details of the proof that $\Phi_{p,q} = \smile_{p,q}$ implies $\Phi_{p+1,q} = \smile_{p+1,q}$ are left to the reader. ▮

It is clear that this result can be generalized to a formal statement about conditions under which multilinear maps of cohomology groups are identical; however, we shall not do so. Instead, when the occasion arises, we shall, at best, provide a sketch of a proof based upon the principles used in the proof of (4-2-4).

4-2-5. Corollary. The cup product is independent of the choice of G-complex.

Proof: The statement means that if X and X' are G-complexes with associated functions $\{\varphi_{p,q}\}$ and $\{\varphi'_{p,q}\}$ in terms of which the cup products for cohomology groups $\smile$ and $\smile'$, respectively, are defined, then

$$\begin{array}{ccc} H_X^p(G, A) \times H_X^q(G, B) & \xrightarrow{\smile} & H_X^{p+q}(G, A \otimes B) \\ {\scriptstyle 1^p_{X,X'} \times 1^q_{X,X'}} \downarrow & & \downarrow {\scriptstyle 1^{p+q}_{X,X'}} \\ H_{X'}^p(G, A) \times H_{X'}^q(G, B) & \xrightarrow{\smile'} & H_{X'}^{p+q}(G, A \otimes B) \end{array}$$

is a commutative diagram for all p, q, A, B. To prove this, consider the maps $\Phi_{p,q} : H_X^p(G, A) \times H_X^q(G, B) \longrightarrow H_X^{p+q}(G, A \otimes B)$ given by $\Phi_{p,q} = 1^{p+q}_{X',X} \circ \smile' \circ 1^p_{X,X'} \times 1^q_{X,X'}$. It is easy to verify that $\Phi = \{\Phi_{p,q}\}$ is a bilinear map which satisfies CP1 and CP2; therefore, $\Phi = \smile$. ∎

It may also be noted that independence of the complex applies to the behavior of cup products with respect to coboundaries. More precisely, under the hypotheses of part (i) of CP2, if $\gamma \in H_X^p(G, C)$ and $\alpha' \in H_X^q(G, A')$, then $1_{X,X'}\delta_*(\gamma \smile \alpha') = \delta_*(1_{X,X'}\gamma \smile' 1_{X,X'}\alpha')$— and an analogous statement is valid in case (ii) of CP2.

4-2-6. Proposition. The cup product is associative. In other words, if $\alpha \in H^p(G, A)$, $\beta \in H^q(G, B)$, $\gamma \in H^r(G, C)$, then

$$(\alpha \smile \beta) \smile \gamma = \alpha \smile (\beta \smile \gamma)$$

Proof: Define maps

$$\Phi, \Psi : H^p(G, A) \times H^q(G, B) \times H^r(G, C) \longrightarrow H^{p+q+r}(G, A \otimes B \otimes C)$$

by $\Phi(\alpha, \beta, \gamma) = (\alpha \cup \beta) \cup \gamma$ and $\Psi(\alpha, \beta, \gamma) = \alpha \cup (\beta \cup \gamma)$. These maps are multilinear—that is, linear in each variable. For $a \in A^G$, $b \in B^G$, $c \in C^G$ we have $\Phi(\kappa a, \kappa b, \kappa c) = \kappa(a \otimes b \otimes c)$, so that Φ satisfies the analog of CP1. Of course, Ψ satisfies the same relation, and, in particular, $\Psi = \Phi$ in dimension (0, 0, 0). Furthermore, Φ satisfies the analog of CP2 which would have three parts in this situation. To be explicit, part (iii) would read as follows: if the G-sequences

$$0 \longrightarrow A \longrightarrow B \longrightarrow C \longrightarrow 0$$

$$0 \longrightarrow A' \otimes B' \otimes A \longrightarrow A' \otimes B' \otimes B \longrightarrow A' \otimes B' \otimes C \longrightarrow 0$$

are both exact, then the diagram

$$\begin{array}{ccc} H^p(G, A') \times H^q(G, B') \times H^r(G, C) & \xrightarrow{\Phi} & H^{p+q+r}(G, A' \otimes B' \otimes C) \\ {\scriptstyle 1\times 1\times\delta_*}\downarrow & (-1)^{p+q} & \downarrow{\scriptstyle \delta_*} \\ H^p(G, A') \times H^q(G, B') \times H^{r+1}(G, A) & \xrightarrow{\Phi} & H^{p+q+r+1}(G, A' \otimes B' \otimes A) \end{array}$$

has commutativity character $(-1)^{p+q}$ for all (p, q, r)—in other words, with obvious choice of notation,

$$\Phi(\alpha', \beta', \delta_*\gamma) = (-1)^{p+q}\delta_*\Phi(\alpha', \beta', \gamma).$$

As for parts (i) and (ii), their conclusions would be

$$\Phi(\delta_*\gamma, \alpha', \beta') = \delta_*\Phi(\gamma, \alpha', \beta') \quad \text{and} \quad \Phi(\alpha', \delta_*\gamma, \beta') = (-1)^p\delta_*\Phi(\alpha', \gamma, \beta'),$$

respectively. Again, Ψ satisfies the same relations. Therefore, by the method used to prove (4-2-4), we conclude that $\Phi = \Psi$. For example, the proof that $\Phi_{p,q,r} = \Psi_{p,q,r}$ implies $\Phi_{p\ q+1,r} = \Psi_{p,q+1,r}$ is based on the exact sequences

$$0 \longrightarrow B \longrightarrow B \otimes \Gamma \longrightarrow B \otimes J \longrightarrow 0$$

$$0 \longrightarrow A \otimes B \otimes C \longrightarrow A \otimes B \otimes \Gamma \otimes C \longrightarrow A \otimes B \otimes J \otimes C \longrightarrow 0$$

and the diagram

$$\begin{array}{ccc} H^p(G, A) \times H^q(G, B \otimes J) \times H^r(G, C) & \longrightarrow & H^{p+q+r}(G, A \otimes B \otimes J \otimes C) \\ {\scriptstyle 1 \times \delta_* \times 1}\downarrow & (-1)^p & \downarrow{\scriptstyle \delta_*} \\ H^p(G, A) \times H^{q+1}(G, B) \times H^r(G, C) & \longrightarrow & H^{p+q+r+1}(G, A \otimes B \otimes C) \end{array}$$

This completes the proof—with the missing details left to the reader. ∎

Let X be a G-complex and for every (p, q, r) consider the trilinear map

$$\mathrm{Hom}_G(X_p, A) \times \mathrm{Hom}_G(X_q, B) \times \mathrm{Hom}_G(X_r, C) \longrightarrow \mathrm{Hom}_G(X_{p+q+r}, A \otimes B \otimes C)$$

given by $(f, g, h) \longrightarrow (f \cup g) \cup h$. For notational convenience put $(f \cup g) \cup h = f \cup g \cup h$. Then

$$\delta(f \cup g \cup h) = \delta f \cup g \cup h + (-1)^p f \cup \delta g \cup h + (-1)^{p+q} f \cup g \cup \delta h$$

and this mapping induces the map of cohomology classes $(\alpha, \beta, \gamma) \longrightarrow (\alpha \cup \beta) \cup \gamma$. In the same way, we have

$$(f, g, h) \longrightarrow f \cup (g \cup h)$$

with corresponding coboundary formula and induced map of cohomology classes $(\alpha, \beta, \gamma) \longrightarrow \alpha \cup (\beta \cup \gamma)$. Of course, $(\alpha \cup \beta) \cup \gamma = \alpha \cup (\beta \cup \gamma)$ so that we may write $\alpha \cup \beta \cup \gamma$ without ambiguity. The question of equality of $(f \cup g) \cup h$ and $f \cup (g \cup h)$ is left to the reader.

For any sets A and B there is a pairing $\top : A \times B \longrightarrow B \times A$ called the **twist**, which is given by $\top(a, b) = (b, a)$. In particular, if A and B are G-modules, there is a canonical G-isomorphism (which we shall also call the **twist** and denote also by $\top$) $\top : A \otimes B \longrightarrow B \otimes A$ for which $\top(a \otimes b) = b \otimes a$. Then, for every integer n, $\top$ induces an isomorphism of cohomology groups

$$\top_* : H^n(G, A \otimes B) \rightarrowtail\!\!\!\!\rightarrow H^n(G, B \otimes A)$$

If the roles of A and B are reversed, we still use the same notation, and this will cause no confusion.

4-2-7. Proposition. Let A and B be G-modules. Then for any (p, q) and any $\alpha \in H^p(G, A)$, $\beta \in H^q(G, B)$ we have

$$\alpha \cup \beta = (-1)^{pq} \top_*(\beta \cup \alpha)$$

Proof: The map

$$\Phi : H^p(G, A) \times H^q(G, B) \longrightarrow H^{p+q}(G, A \otimes B)$$

given by

$$\Phi(\alpha, \beta) = (-1)^{pq} \top_*(\beta \cup \alpha)$$

(so that $\Phi = (-1)^{pq} \top_* \circ \cup \circ \top$) is bilinear and satisfies CP1 and CP2. The verification of CP1 makes use of the commutative diagram

$$\begin{array}{ccc} (B \otimes A) & \xrightarrow{\quad \top \quad} & (A \otimes B)^G \\ \kappa \downarrow & & \downarrow \kappa \\ H^0(G, B \otimes A) & \xrightarrow{\quad \top_* \quad} & H^0(G, A \otimes B) \end{array}$$

while the verification of CP2 makes use of $\top_* \circ \delta_* = \delta_* \circ \top_*$. ∎

4-2-8. Proposition. Suppose that $\varphi : A \longrightarrow A'$ and $\psi : B \longrightarrow B'$ are G-homomorphisms of G-modules, so that $\varphi \otimes \psi : A \otimes B \longrightarrow A' \otimes B'$ is a G-homomorphism. Then for $\alpha \in H^p(G, A)$ and $\beta \in H^q(G, B)$ we have

$$(\varphi \otimes \psi)_*(\alpha \cup \beta) = \varphi_*\alpha \cup \psi_*\beta$$

Proof: Fix a G-complex X, and let $f \in \operatorname{Hom}_G(X_p, A)$ and $g \in \operatorname{Hom}_G(X_q, B)$ be any cochains. Then

$$(\varphi \otimes \psi)(f \cup g) = \varphi f \cup \psi g$$

since

$$\varphi f \smile \psi g = (\varphi f \otimes \psi g) \circ \varphi_{p,q} = (\varphi \otimes \psi)(f \otimes g) \circ \varphi_{p,q} = (\varphi \otimes \psi)(f \smile g)$$

In other words, the following diagram commutes:

$$\begin{array}{ccc} \mathrm{Hom}_G(X_p, A) \times \mathrm{Hom}_G(X_q, B) & \xrightarrow{\smile} & \mathrm{Hom}_G(X_{p+q}, A \otimes B) \\ \downarrow{\scriptstyle (1,\varphi)\times(1,\psi)} & & \downarrow{\scriptstyle (1,\varphi\otimes\psi)} \\ \mathrm{Hom}_G(X_p, A') \times \mathrm{Hom}_G(X_q, B') & \xrightarrow{\smile} & \mathrm{Hom}_G(X_{p+q}, A' \otimes B') \end{array}$$

and the desired result follows easily. ▮

4-2-9. Exercise. Let X be a G-complex and let Δ denote the cochain cup product. In other words,

$$\Delta : \mathrm{Hom}_G(X_p, A) \times \mathrm{Hom}_G(X_q, B) \longrightarrow \mathrm{Hom}_G(X_{p+q}, A \otimes B)$$

is the bilinear map $\Delta(f, g) = f \smile g$. Then Δ, which induces the cup product $\smile$ of cohomology classes, satisfies the properties:

(CCP1) $\Delta(\dot{\kappa}_X a, \dot{\kappa}_X b) = \dot{\kappa}_X(a \otimes b)$ where $a \in A^G$, $b \in B^G$

(CCP2) $\delta\Delta(f, g) = \Delta(\delta f, g) + (-1)^p \Delta(f, \delta g)$

(CCP3) $(\varphi \otimes \psi)\Delta(f, g) = \Delta(\varphi f, \psi g)$ where $\varphi : A \longrightarrow A'$, $\psi : B \longrightarrow B'$ are G-homomorphisms.

Moreover, if Δ' is any bilinear map (more precisely, $\Delta' = \{\Delta'_{p,q}\}$ is a set of such maps) satisfying (CCP1), (CCP2), and (CCP3), then the bilinear mapping $H^p(G, A) \times H^q(G, B) \longrightarrow H^{p+q}(G, A \otimes B)$ which it induces on cohomology classes is precisely the cup product.

Strictly speaking, the symbol $\dot{\kappa}_X$ which appears in (CCP1) has never been defined. What is meant is $\dot{\kappa}_X a = \omega_a \varepsilon$, $a \in A^G$ (compare (2-2-10)), so that

$$\dot{\kappa}_X : A^G \longrightarrow \mathscr{Z}^0_X(G, A)$$

is a homomorphism whose image contains cocycle representatives of all the cohomology classes in $H^0(G, A)$.

4-3. CUP PRODUCTS FOR A PAIRING

The cup product discussed in the preceding section is not quite what we want; a slight extension is needed.

Suppose that A, B, C are G-modules and consider a **G-pairing** $\theta : A \times B \longrightarrow C$. By this we mean that θ is bilinear and that

$$\theta(\sigma a, \sigma b) = \sigma\theta(a, b) \qquad \sigma \in G, \quad a \in A, \quad b \in B$$

This latter condition may be expressed as $\theta \circ (\sigma \times \sigma) = \sigma \circ \theta$. If we let $\pi : (a, b) \longrightarrow a \otimes b$ denote the canonical map of $A \times B \longrightarrow A \otimes B$, then there exists a unique homomorphism

$$\bar{\theta} : A \otimes B \longrightarrow C$$

such that $\bar{\theta} \circ \pi = \theta$—or what is the same, $\bar{\theta}(a \otimes b) = \theta(a, b)$. Moreover, on the generators $a \otimes b$ of $A \otimes B$ we have, for $\sigma \in G$, $\bar{\theta}\sigma(a \otimes b) = \bar{\theta}(\sigma a \otimes \sigma b) = \theta(\sigma a, \sigma b) = \sigma\theta(a, b) = \sigma\bar{\theta}(a \otimes b)$, so that $\bar{\theta}$ is a G-homomorphism.

As natural examples of G-pairings we have:

(1) If A is any G-module, define $\theta : \mathbf{Z} \times A \longrightarrow A$ by $\theta(n, a) = na$. This G-pairing will often be denoted by θ_A.

(2) If A and B are any G-modules, let

$$\theta : \mathrm{Hom}\,(A, B) \times A \longrightarrow B$$

be given by $\theta(f, a) = f(a)$. Note that this is a G-pairing since $\theta(f^\sigma, \sigma a) = f^\sigma(\sigma a) = \sigma f(a) = \sigma\theta(f, a)$.

Of course, to every G-pairing $\theta : A \times B \longrightarrow C$ there is associated its **twist** $\theta^\top : B \times A \longrightarrow C$ which is the G-pairing given by

$$\theta^\top(b, a) = \theta(a, b)$$

—so that $\theta^\top = \theta \circ \top$ and $\theta^\top = \bar{\theta} \circ \top$.

It is also worth noting that a G-pairing $\theta : A \times B \longrightarrow C$ gives rise to G-homomorphisms of G-modules. Thus, for $a \in A^G$, define ${}^a\theta : B \longrightarrow C$ by

$${}^a\theta(b) = \theta(a, b)$$

Since ${}^a\theta(\sigma b) = \theta(a, \sigma b) = \theta(\sigma a, \sigma b) = \sigma\theta(a, b) = \sigma{}^a\theta(b)$, ${}^a\theta$ is indeed a G-homomorphism. In the same way, for $b \in B^G$, $\theta^b : A \longrightarrow C$ given by

$$\theta^b(a) = \theta(a, b)$$

is a G-homomorphism.

If $\theta : A \times B \longrightarrow C$ is a G-pairing we define the **cup product with respect to θ** (or simply the **θ-cup product)**

$$\cup_\theta : H^p(G, A) \times H^q(G, B) \longrightarrow H^{p+q}(G, C)$$

for all (p, q) by putting (with $\bar{\theta}_* = (\bar{\theta})_*$)

$$\alpha \cup_\theta \beta = \bar{\theta}_*(\alpha \cup \beta)$$

—in other words, $\cup_\theta = \bar{\theta}_* \circ \cup$; that is, $\cup_\theta$ is the composite map

$$H^p(G, A) \times H^q(G, B) \xrightarrow{\cup} H^{p+q}(G, A \otimes B) \xrightarrow{\bar{\theta}_*} H^{p+q}(G, C)$$

The properties of $\cup$ carry over to $\cup_\theta$, as will be seen in the next few results.

Of course, the θ-cup product $\cup_\theta$ on cohomology classes is induced from a θ-cup product (also denoted by $\cup_\theta$) on cochains. In more detail, let X be a G-complex, and suppose that $f \in \mathrm{Hom}_G(X_p, A)$, $g \in \mathrm{Hom}_G(X_q, B)$ are cochains. Then define $f \cup_\theta g \in \mathrm{Hom}_G(X_{p+q}, C)$ by

$$f \cup_\theta g = \bar{\theta} \circ (f \cup g) = \bar{\theta} \circ (f \otimes g) \circ \varphi_{p,q}$$

so that on cochains $\cup_\theta = \bar{\theta} \circ \cup$, and it follows easily that it induces $\cup_\theta$ on cohomology classes.

It should be noted that $\cup_\theta$ is a generalization of $\cup$; in more detail, the map $\theta : A \times B \longrightarrow A \otimes B$ given by $\theta(a, b) = a \otimes b$ is clearly a G-pairing for which $\bar{\theta}$ is the identity, so that $\cup_\theta = \cup$ on both cochains and cohomology classes.

4-3-1. Proposition. Consider the diagram

$$\left.\begin{array}{l} A' \xrightarrow{u} A \\ B' \xrightarrow{v} B \end{array}\right\} \xrightarrow{\theta} C \xrightarrow{w} C'$$

where u, v, w are G-homomorphisms and θ is a G-pairing. Then $\theta' : A' \times B' \longrightarrow C'$ given by $\theta' = w \circ \theta \circ (u \times v)$ is a G-pairing, and for $\alpha' \in H^p(G, A')$, $\beta' \in H^q(G, B')$ we have

$$\alpha' \smile_{\theta'} \beta' = w_*(u_*\alpha' \smile_\theta v_*\beta')$$

Proof: It is clear that θ' is a G-pairing, and that

$$\bar{\theta}' = w \circ \bar{\theta} \circ (u \otimes v) \qquad \text{and} \qquad \bar{\theta}'_* = w_* \circ \bar{\theta}_* \circ (u \otimes v)_*$$

Hence,

$$\begin{aligned} \smile_{\theta'} = \bar{\theta}'_* \circ \smile &= w_* \circ \bar{\theta}_* \circ (u \otimes v)_* \circ \smile \\ &= w_* \circ \bar{\theta}_* \circ \smile \circ (u_* \times v_*) \quad \text{(by (4-2-8))} \\ &= w_* \circ \smile_\theta \circ (u_* \times v_*) \end{aligned}$$

which is the desired result. ▌

4-3-2. Proposition. Let $\theta : A \times B \longrightarrow C$ be a G-pairing, then for $\alpha \in H^p(G, A)$, $\beta \in H^q(G, B)$ we have

$$\alpha \smile_\theta \beta = (-1)^{pq} \beta \smile_{\theta^\top} \alpha$$

Proof: In virtue of (4-2-7),

$$\begin{aligned} \smile_\theta = \bar{\theta}_* \circ \smile &= \bar{\theta}_* \circ (-1)^{pq} \top_* \circ \smile \circ \top \\ &= (-1)^{pq} \overline{\theta^\top}_* \circ \smile \circ \top = (-1)^{pq} \smile_{\theta^\top} \circ \top \end{aligned}$$

which is the desired result. ▌

4-3-3. Proposition. The cup product with respect to G-pairings is associative; more precisely, if

$$\theta_{1,2} : A_1 \times A_2 \longrightarrow A_{12}, \qquad \theta_{2,3} : A_2 \times A \longrightarrow A_{23},$$
$$\theta_{12,3} : A_{12} \times A_3 \longrightarrow A_{123}, \qquad \theta_{1,23} : A_1 \times A_{23} \longrightarrow A_{123}$$

are G-pairings such that $\theta_{12,3}(\theta_{1,2}(a_1, a_2), a_3) = \theta_{1,23}(a_1, \theta_{2,3}(a_2, a_3))$

for $a_1 \in A_1$, $a_2 \in A_2$, $a_3 \in A_3$ then for $\alpha_1 \in H^p(G, A_1)$, $\alpha_2 \in H^q(G, A_2)$, $\alpha_3 \in H^r(G, A_3)$ we have

$$(\alpha_1 \cup_{\theta_{1,2}} \alpha_2) \cup_{\theta_{12,3}} \alpha_3 = \alpha_1 \cup_{\theta_{1,23}} (\alpha_2 \cup_{\theta_{2,3}} \alpha_3)$$

Proof: The hypotheses imply that

$$\bar{\theta}_{12,3}[\bar{\theta}_{1,2}(a_1 \otimes a_2) \otimes a_3] = \bar{\theta}_{1,23}[a_1 \otimes \bar{\theta}_{2,3}(a_2 \otimes a_3)]$$

Define the G-pairings $\theta'_{12,3} : (A_1 \otimes A_2) \times A_3 \longrightarrow A_{123}$ and $\theta'_{1,23} : A_1 \times (A_2 \otimes A_3) \longrightarrow A_{123}$ by $\theta'_{12,3} = \theta_{12,3} \circ (\bar{\theta}_{1,2} \times 1)$ and $\theta'_{1,23} = \theta_{1,23} \circ (1 \times \bar{\theta}_{2,3})$, respectively. We have, therefore,

$$\bar{\theta}'_{12,3} = \bar{\theta}_{12,3} \circ (\bar{\theta}_{1,2} \otimes 1) = \bar{\theta}_{1,23} \circ (1 \otimes \bar{\theta}_{2,3}) = \bar{\theta}'_{1,23}$$

so that $(\bar{\theta}'_{12,3})_* = (\bar{\theta}'_{1,23})_*$. Now, by (4-3-1) and (4-2-8), it follows that

$$\begin{aligned}(\alpha_1 \cup_{\theta_{1,2}} \alpha_2) \cup_{\theta_{12,3}} \alpha_3 &= (\bar{\theta}_{12,3})_*[(\bar{\theta}_{1,2})_*(\alpha_1 \cup \alpha_2) \cup 1_*\alpha_3] \\ &= (\bar{\theta}'_{12,3})_*(\alpha_1 \cup \alpha_2 \cup \alpha_3) \\ &= (\bar{\theta}'_{1,23})_*(\alpha_1 \cup \alpha_2 \cup \alpha_3) \\ &= (\bar{\theta}_{1,23})_*[1_*\alpha_1 \cup (\bar{\theta}_{2,3})_*(\alpha_2 \cup \alpha_3)] \\ &= \alpha_1 \cup_{\theta_{1,23}} (\alpha_2 \cup_{\theta_{2,3}} \alpha_3)\end{aligned}$$

and the proof is complete. ∎

4-3-4. Proposition. Let $\theta : A \times B \longrightarrow C$ be a G-pairing; then for $a \in A^G$, $b \in B^G$ we have $\theta(a, b) \in C^G$ and

$$\kappa a \cup_\theta \kappa b = \kappa(\theta(a, b))$$

Proof: It is clear that $\theta(a, b) \in C^G$ and then from (4-2-1) and (2-2-10) it follows that

$$\kappa a \cup_\theta \kappa b = \bar{\theta}_*(\kappa a \cup \kappa b) = \bar{\theta}_*\kappa(a \otimes b) = \kappa\bar{\theta}(a \otimes b) = \kappa(\theta(a, b). \quad ∎$$

4-3-5. Exercise. The preceding result is a generalization of CP1 (see (4-2-1)) to θ-cup products. State and prove the appropriate generalization of CP2 (see (4-2-2)) for the case of cup products with respect to G-pairings.

4-3-6. Proposition. Cupping with respect to a G-pairing by a 0-dimensional cohomology class is the same as applying an induced map of cohomology groups. More precisely, if $\theta : A \times B \longrightarrow C$ is a G-pairing, then

(i) for $a \in A^G$, $\beta \in H^q(G, B)$ we have

$$\kappa a \cup_\theta \beta = ({}^a\theta)_*(\beta)$$

(ii) for $\alpha \in H^p(G, A)$, $b \in B^G$ we have

$$\alpha \cup_\theta \kappa b = (\theta^b)_*(\alpha)$$

Proof: Choose a G-complex $(X, \partial, \varepsilon, \mu)$ and let $g \in \mathrm{Hom}_G(X_q, B)$ be a cocycle representing β. According to (2-2-3), $(1, {}^a\theta)g$ is a cocycle representing $({}^a\theta)_*(\beta)$, and in virtue of (2-2-10) $\omega_a\varepsilon$ is a cocycle representing κa. Thus, $\omega_a\varepsilon \cup_\theta g$ is a cocycle representing $\kappa a \cup_\theta \beta$. Now, for $x \in X_q$ we have

$$(\omega_a\varepsilon \cup_\theta g)(x) = \bar{\theta}(\omega_a\varepsilon \otimes g)\varphi_{0,q}(x) = \bar{\theta}(\omega_a \otimes g)(\varepsilon \otimes 1)\varphi_{0,q}(x)$$

and

$$\begin{aligned}((1, {}^a\theta)g)(x) &= {}^a\theta(g(x)) = \theta(a, g(x)) = \bar{\theta}(a \otimes g(x)) \\ &= \bar{\theta}(\omega_a(1) \otimes g(x)) = \bar{\theta}(\omega_a \otimes g)(1 \otimes x)\end{aligned}$$

But, as seen in the proof of (4-1-8), $(\varepsilon \otimes 1)\,\varphi_{0,q}(x) = 1 \otimes x$ for the $\varphi_{0,q}$ defined there. Therefore, $\omega_a\varepsilon \cup_\theta g = (1, {}^a\theta)g$ and part (i) is proved. The proof of (ii) is now immediate:

$$\alpha \cup_\theta \kappa b = (-1)^0 \kappa b \cup_{\theta^\top} \alpha = ({}^b(\theta^\top))_*(\alpha) = (\theta^b)_*(\alpha). \quad \blacksquare$$

Now let us turn to the cohomology of an arbitrary subgroup H of G. Let X be a G-complex with associated G-homomorphisms $\varphi_{p,q} : X_{p+q} \longrightarrow X_p \otimes X_q$ in terms of which the cup product is defined. Then X is also an H-complex and the functions $\varphi_{p,q}$ are H-homomorphisms which may be used to define the cup product $\cup$ for the cohomology of H. Of course, all tensor products are over $\mathbf{Z}$, so they are not affected by a change from G to H. Furthermore, if $\theta : A \times B \longrightarrow C$ is a G-pairing, then it is also an H-pairing, and we can define $\smile_\theta$ (for both H-cochains and H-cohomology classes) in precisely the same way that was used for G.

4-3-7. Proposition. Let H be any subgroup of G, and suppose that $\theta : A \times B \longrightarrow C$ is a G-pairing. Then, writing $\text{res} = \text{res}_{G \to H}$ and $\text{cor} = \text{cor}_{H \to G}$ we have:

$$\text{(i) } \text{res}\,(\alpha \smile_\theta \beta) = \text{res}\,\alpha \smile_\theta \text{res}\,\beta \qquad \alpha \in H^p(G, A),\ \beta \in H^q(G, B)$$

$$\text{(ii) } \text{cor}\,(\text{res}\,\alpha \smile_\theta \beta) = \alpha \smile_\theta \text{cor}\,\beta \qquad \alpha \in H^p(G, A),\ \beta \in H^q(H, B)$$

$$\text{(iii) } \text{cor}\,(\alpha \smile_\theta \text{res}\,\beta) = \text{cor}\,\alpha \smile_\theta \beta \qquad \alpha \in H^p(H, A),\ \beta \in H^q(G, B)$$

Proof: (i) Choose a G-complex X with $\varphi_{p,q}$'s, and use X for the cohomology of H, too. Let $f \in \text{Hom}_G(X_p, A)$ and $g \in \text{Hom}_G(X_q, B)$ be cocycles representing α and β, respectively. According to (2-3-4), f represents $\text{res}\,\alpha$ and g represents $\text{res}\,\beta$. It follows that $\alpha \smile_\theta \beta$, $\text{res}\,(\alpha \smile_\theta \beta)$, and $\text{res}\,\alpha \smile_\theta \text{res}\,\beta$ are all represented by the cocycle $f \smile_\theta g = \bar{\theta} \circ (f \otimes g) \circ \varphi_{p,q}$.

(ii) Let $f \in \text{Hom}_G(X_p, A)$ represent α and $g \in \text{Hom}_H(X_q, B)$ represent β. Then f represents $\text{res}\,\alpha$ and $\bar{\theta} \circ (f \otimes g) \circ \varphi_{p,q}$ represents $\text{res}\,\alpha \smile_\theta \beta$. According to the definition of cor (see Section 2-4), $S_{H \to G}g$ represents $\text{cor}\,\beta$, so that $\alpha \smile_\theta \text{cor}\,\beta$ is represented by $\bar{\theta} \circ (f \otimes S_{H \to G}g) \circ \varphi_{p,q}$. On the other hand, if we write $G = \bigcup \sigma_i H$ and note that f, $\bar{\theta}$ and $\varphi_{p,q}$ are G-homomorphisms, then $\text{cor}\,(\text{res}\,\alpha \smile_\theta \beta)$ is represented by

$$\begin{aligned} S_{H \to G}(\bar{\theta} \circ (f \otimes g) \circ \varphi_{p,q}) &= \bar{\theta} \circ S_{H \to G}(f \otimes g) \circ \varphi_{p,q} \qquad \text{(see (2-4-11))} \\ &= \bar{\theta} \circ \sum_i (f \otimes g)^{\sigma_i} \circ \varphi_{p,q} \\ &= \bar{\theta} \circ \sum_i (f \otimes g^{\sigma_i}) \circ \varphi_{p,q} \\ &= \bar{\theta} \circ (f \otimes S_{H \to G}\, g) \circ \varphi_{p,q} \end{aligned}$$

This proves (ii). The proof of (iii) goes the same way. ∎

4-3-8. Theorem. Let $\theta : A \times B \longrightarrow C$ be a G-pairing and fix an element $\alpha \in H^p(G, A)$. For each subgroup H of G and each integer q, define a map

$$\Phi_H : H^q(H, B) \longrightarrow H^{p+q}(H, C)$$

by θ-cupping on the left by $\operatorname{res}_{G\to H} \alpha$—in other words,

$$\Phi_H : \beta \longrightarrow \operatorname{res}_{G\to H} \alpha \cup_\theta \beta$$

Suppose there exists an integer q_0 such that for all subgroups H of G we know that

(i) $\Phi_H : H^{q_0-1}(H, B) \longrightarrow H^{p+q_0-1}(H, C)$ is an epimorphism,

(ii) $\Phi_H : H^{q_0}(H, B) \longrightarrow H^{p+q_0}(H, C)$ is an isomorphism onto,

(iii) $\Phi_H : H^{q_0+1}(H, B) \longrightarrow H^{p+q_0+1}(H, C)$ is a monomorphism.

Then Φ_H is an isomorphism onto in all dimensions and for all subgroups H of G.

Proof: By induction on p. Consider first $p = 0$. If we choose $a \in A^G$ such that $\kappa^G(a) = \alpha$, then, according to (2-5-5), $\kappa^H(a) = \operatorname{res}_{G\to H} \alpha$. Now, θ is an H-pairing, and application of (4-3-6) to it yields $\Phi_H = ({}^a\theta)_*$. Furthermore, ${}^a\theta : B \longrightarrow C$ is a G-homomorphism which, in virtue of (3-6-3), is a cohomological equivalence. This proves our assertion for $p = 0$.

To pass from p to $p + 1$ we use dimension shifting techniques. Suppose the result holds for p. Consider the exact G-sequence (see (4-1-6))

$$0 \longrightarrow A \longrightarrow \Gamma \otimes A \longrightarrow J \otimes A \longrightarrow 0$$

and the commutative diagram

$$\begin{array}{ccccccccc}
0 \longrightarrow & A \otimes B & \longrightarrow & \Gamma \otimes A \otimes B & \longrightarrow & J \otimes A \otimes B & \longrightarrow 0 \\
& \downarrow{\scriptstyle \bar\theta} & & \downarrow{\scriptstyle 1\otimes\bar\theta} & & \downarrow{\scriptstyle 1\otimes\bar\theta} & \\
0 \longrightarrow & C & \longrightarrow & \Gamma \otimes C & \longrightarrow & J \otimes C & \longrightarrow 0
\end{array}$$

whose rows are exact G-sequences. Then for any $H \subset G$ we have the diagram

$$\begin{array}{ccccc}
H^p(H, J \otimes A) \times H^q(H, B) & \xrightarrow{\smile} & H^{p+q}(H, J \otimes A \otimes B) & \longrightarrow & \\
\downarrow{\scriptstyle \delta_* \times 1} & & \downarrow{\scriptstyle \delta_*} & & \\
H^{p+1}(H, A) \times H^q(H, B) & \xrightarrow{\smile} & H^{p+q+1}(H, A \otimes B) & \longrightarrow & \\
& & & \xrightarrow{(1 \otimes \bar{\theta})_*} & H^{p+q}(H, J \otimes C) \\
& & & & \downarrow{\scriptstyle \delta_*} \\
& & & \xrightarrow{(\bar{\theta})_*} & H^{p+q+1}(H, C)
\end{array}$$

The rectangles commute by (4-2-2) and (2-2-5), and the coboundary maps δ_* are isomorphisms onto since $\Gamma \otimes A$, $\Gamma \otimes C$ and $\Gamma \otimes A \otimes B$ are H-regular. Now, define the G-pairing $\theta' : (J \otimes A) \times B \longrightarrow J \otimes C$ by $\theta'(j \otimes a, b) = j \otimes \theta(a, b)$. It is clear that $\bar{\theta}' = 1 \otimes \bar{\theta}$ so that $(\bar{\theta}')_* = (1 \otimes \bar{\theta})_*$ and the composite map in the top row is $(1 \otimes \bar{\theta})_* \circ \smile = \smile_{\theta'}$.

To prove our result for $p + 1$, let $\alpha \in H^{p+1}(G, A)$ satisfy the hypotheses. Let $\alpha' \in H^p(G, J \otimes A)$ be the unique element for which $\delta_* \alpha' = \alpha$. Then for any subgroup H of G,

$$\mathrm{res}_{G \to H}\, \alpha = \delta_* \circ \mathrm{res}_{G \to H}\, \alpha'$$

and the following diagram commutes for all q,

$$\begin{array}{ccc}
H^q(H, B) & \xrightarrow{\Phi'_H} & H^{p+q}(H, J \otimes C) \\
{\scriptstyle 1}\,\wr\!\downarrow & & \wr\!\downarrow{\scriptstyle \delta_*} \\
H^q(H, B) & \xrightarrow{\Phi_H} & H^{p+q+1}(H, C)
\end{array}$$

where $\Phi_H(\beta) = \mathrm{res}_{G \to H}\, \alpha \smile_\theta \beta$ and $\Phi'_H(\beta) = \mathrm{res}_{G \to H}\, \alpha' \smile_{\theta'} \beta$. Since the hypotheses are satisfied for Φ_H (using α, $p + 1$, q_0, θ) they are satisfied for Φ'_H (using α', p, q_0, θ'). By the induction hypothesis, Φ'_H is an isomorphism onto in all dimensions and for all H; hence the same is true for Φ_H, and our result holds for $p + 1$.

The passage from p to $p - 1$ proceeds in the same manner. ∎

4-3-9. Exercise. Let $\theta : A \times B \longrightarrow C$ be a G-pairing of G-modules. If H is a normal subgroup of G then θ induces a (G/H)-pairing of (G/H)-modules $\theta' : A^H \times B^H \longrightarrow C^H$. Consider $a \in (A^H)^{G/H} = A^G$ and $\beta \in H^q(G/H, B^H)$ with $q \geqslant 1$; then

$$\inf_{G/H \to G} (\kappa^{G/H}(a) \cup_{\theta'} \beta) = \kappa^G(a) \cup_\theta \inf_{G/H \to G} \beta$$

A similar relation holds for the other variable. Furthermore, if $\alpha \in H^p(G/H, A^H)$ with $p \geqslant 1$, then

$$\inf_{G/H \to G} (\alpha \cup_{\theta'} \beta) = \inf_{G/H \to G} \alpha \cup_\theta \inf_{G/H \to G} \beta$$

4-4. THE DUALITY THEOREM

Let us consider any G-modules B and C. As was observed in the preceding section there is a canonical G-pairing

$$\theta : \mathrm{Hom}\,(B, C) \times B \longrightarrow C$$

such that for $\phi \in \mathrm{Hom}\,(B, C)$ and $b \in B$ we have $\theta(\phi, b) = \phi(b)$. If B' is another G-module, we shall also write

$$\theta : \mathrm{Hom}\,(B', C) \times B' \longrightarrow C$$

with $\theta(\phi', b') = \phi'(b')$ for the canonical G-pairing; this should cause no confusion. For any $\psi \in \mathrm{Hom}\,(B, B')$ we have an **adjoint** map, which is a homomorphism of groups,

$$\hat{\psi} = (\psi, 1) : \mathrm{Hom}\,(B', C) \longrightarrow \mathrm{Hom}\,(B, C)$$

Since for $\phi' \in \mathrm{Hom}\,(B', C)$ we have $\hat{\psi}\phi' = (\psi, 1)\,\phi' = \phi'\psi$, it follows that (with $b \in B$) $\theta(\hat{\psi}\phi', b) = \theta(\phi', \psi b)$.

Suppose further that ψ is a G-homomorphism, $\psi \in \mathrm{Hom}_G\,(B, B')$; then (see (1-4-3)) $\hat{\psi}$ is a G-homomorphism, and we may define a G-pairing

$$\theta_\psi : \mathrm{Hom}\,(B', C) \times B \longrightarrow C$$

by putting $\theta_\psi(\phi', b) = \theta(\phi', \psi b) = \theta(\hat{\psi}\phi', b) = \phi'(\psi b) = (\hat{\psi}\phi')(b)$; in other words,

$$\theta_\psi = \theta \circ (1 \times \psi) = \theta \circ (\hat{\psi} \times 1)$$

For $\sigma \in G$, $\theta_\psi(\phi'^\sigma, \sigma b) = \phi'^\sigma(\psi\sigma b) = \phi'^\sigma\sigma(\psi b) = \sigma\phi'(\psi b) = \sigma\theta_\psi(\phi', b)$, so that θ_ψ is indeed a G-pairing. Consequently, θ_ψ determines a cup product

$$\smile_{\theta_\psi} : H^p(G, \operatorname{Hom}(B', C)) \times H^q(G, B) \longrightarrow H^{p+q}(G, C)$$

which is induced by the corresponding cup product on cochains, and can be computed according to the following rule.

4-4-1. Proposition. If

$$\xi' \in H^p(G, \operatorname{Hom}(B', C)) \qquad \text{and} \qquad \beta \in H^q(G, B)$$

then

$$\xi' \smile_{\theta_\psi} \beta = \hat{\psi}_* \xi' \smile_\theta \beta = \xi' \smile_\theta \psi_* \beta$$

Proof: Fix a G-complex X with associated functions $\varphi_{p,q}$. It clearly suffices to prove that if $f' \in \operatorname{Hom}_G(X_p, \operatorname{Hom}(B', C))$ and $g \in \operatorname{Hom}_G(X_q, B)$ are p- and q-cochains, respectively, then

$$f' \smile_{\theta_\psi} g = \hat{\psi} f' \smile_\theta g = f' \smile_\theta \psi g$$

Now, let us observe that

$$f' \smile_{\theta_\psi} g = \bar{\theta}_\psi(f' \otimes g)\varphi_{p,q}, \qquad \hat{\psi} f' \smile_\theta g = \bar{\theta}(\hat{\psi} \otimes 1)(f' \otimes g)\varphi_{p,q}$$

and $f' \smile_\theta \psi g = \bar{\theta}(1 \otimes \psi)(f' \otimes g)\,\varphi_{p,q}$, so it suffices to show that

$$\bar{\theta}_\psi = \bar{\theta} \circ (\hat{\psi} \otimes 1) = \bar{\theta} \circ (1 \otimes \psi)$$

—in other words, that the following diagram commutes:

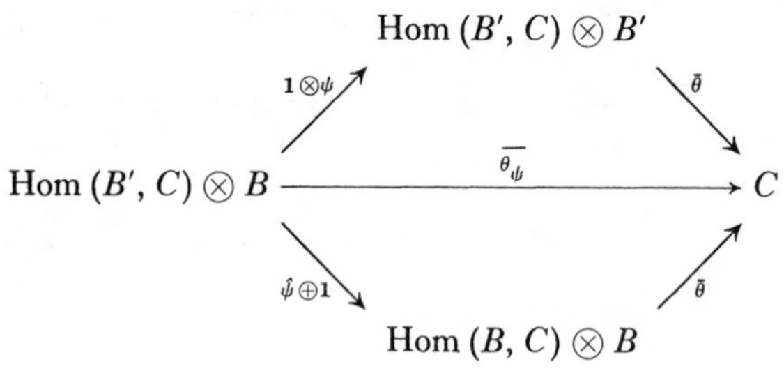

But this is immediate from the definition of θ_ψ. This completes the proof. ∎

For each pair of integers (p, q) let us consider the homomorphism

$$\Upsilon_{p,q} : H^p(G, \text{Hom}(B, C)) \longrightarrow \text{Hom}(H^q(G, B), H^{p+q}(G, C))$$

which is defined by taking the image of $\xi \in H^p(G, \text{Hom}(B, C))$ to be the homomorphism whose action on $H^q(G, B)$ is given by θ-cupping on the left by ξ—in other words,

$$(\Upsilon_{p,q}\xi)(\beta) = \xi \smile_\theta \beta \qquad \beta \in H^q(G, B)$$

4-4-2. Proposition. Suppose that

$$0 \longrightarrow B' \xrightarrow{i} B \xrightarrow{j} B'' \longrightarrow 0$$

is an exact G-sequence and that C is a G-module for which the adjoint G-sequence

$$0 \longrightarrow \text{Hom}(B'', C) \xrightarrow{\hat{j}=(j,1)} \text{Hom}(B, C) \xrightarrow{\hat{i}=(i,1)} \text{Hom}(B', C) \longrightarrow 0$$

is exact. Then the following diagram has commutativity character $(-1)^{p-1}$:

$$\begin{array}{ccc} H^p(G, \text{Hom}(B', C)) & \xrightarrow{\Upsilon_{p,q}} & \text{Hom}(H^q(G, B'), H^{p+q}(G, C)) \\ \delta_* \downarrow & (-1)^{p-1} & \downarrow (\delta_*, 1) \\ H^{p+1}(G, \text{Hom}(B'', C)) & \xrightarrow{\Upsilon_{p+1,q-1}} & \text{Hom}(H^{q-1}(G, B''), H^{p+q}(G, C)) \end{array}$$

Proof: We must show that for $\xi' \in H^p(G, \text{Hom}(B', C))$ and $\beta'' \in H^{q-1}(G, B'')$ we have

$$\delta_*\xi' \smile_\theta \beta'' = (-1)^{p-1}\xi' \smile_\theta \delta_*\beta''$$

Choose a G-complex X. Let $g'' \in \text{Hom}_G(X_{q-1}, B'')$ be a cocycle representing β''. Then there exist $g \in \text{Hom}_G(X_{q-1}, B)$ such that $g'' = (1, j)g = jg$ and $g' \in \text{Hom}_G(X_q, B')$ such that $\delta g = (1, i)g' = ig'$, and g' is a cocycle representing $\delta_*\beta''$. Similarly, if $f' \in \text{Hom}_G(X_p, \text{Hom}(B', C))$ is a cocycle representing ξ', then there exist $f \in \text{Hom}_G(X_p, \text{Hom}(B, C))$ such that

$$f' = (1, \hat{i})f = \hat{i}f$$

and

$$f'' \in \text{Hom}_G(X_{p+1}, \text{Hom}(B'', C)) \text{ such that } \delta f = (1, \hat{j})f'' = \hat{j}f'',$$

and f'' is a cocycle representing $\delta_* \xi'$. It now follows that $\delta_* \xi' \cup_\theta \beta'' + (-1)^p \xi' \cup_\theta \delta_* \beta'' = 0$ since it is represented by

$$\begin{aligned} & f'' \cup_\theta g'' + (-1)^p f' \cup_\theta g' \\ & = f'' \cup_\theta jg + (-1)^p \hat{\imath} f \cup_\theta g' \\ & = \hat{\jmath} f'' \cup_\theta g + (-1)^p f \cup_\theta ig' \qquad \text{(see proof of (4-4-1))} \\ & = \delta f \cup_\theta g + (-1)^p f \cup_\theta \delta g \\ & = \delta(f \cup_\theta g) \end{aligned}$$

which is a coboundary. This completes the proof. ∎

4-4-3. Proposition. Suppose that for a fixed pair of integers (p_0, q_0) and a fixed G-module C the mapping

$$\Upsilon_{p_0,q_0} : H^{p_0}(G, \operatorname{Hom}(B, C)) \longrightarrow \operatorname{Hom}(H^{q_0}(G, B), H^{p_0+q_0}(G, C))$$

is an isomorphism onto for every G-module B. Then for all (p, q) with $p + q = p_0 + q_0$, and all B, the map $\Upsilon_{p,q}$ is an isomorphism onto.

Proof: If the result is true for (p, q) then, using dimension shifting, we show that it is true for $(p + 1, q - 1)$. For any B,

$$0 \longrightarrow I \otimes B \longrightarrow \Gamma \otimes B \longrightarrow B \longrightarrow 0$$

is an exact G-sequence which splits as a $\mathbf{Z}$-sequence and has G-regular middle term (see (3-3-8), (4-1-4), and (4-1-5)). By (3-3-5) and (3-3-6), for any C

$$0 \longrightarrow \operatorname{Hom}(B, C) \longrightarrow \operatorname{Hom}(\Gamma \otimes B, C) \longrightarrow \operatorname{Hom}(I \otimes B, C) \longrightarrow 0$$

is an exact G-sequence with G-regular middle term. Thus the vertical maps in the following diagram (which, by (4-4-2), has commutativity character $(-1)^{p-1}$) are isomorphisms onto:

$$\begin{array}{ccc} H^p(G, \operatorname{Hom}(I \otimes B, C)) & \xrightarrow{\Upsilon_{p,q}} & \operatorname{Hom}(H^q(G, I \otimes B), H^{p+q}(G, C)) \\ {\scriptstyle \delta_*}\downarrow & (-1)^{p-1} & \downarrow{\scriptstyle (\delta_*, 1)} \\ H^{p+1}(G, \operatorname{Hom}(B, C)) & \xrightarrow{\Upsilon_{p+1,q-1}} & \operatorname{Hom}(H^{q-1}(G, B), H^{p+q}(G, C)) \end{array}$$

Consequently, if $\Upsilon_{p,q}$ is an isomorphism onto, then so is $\Upsilon_{p+1,q-1}$.
The details of the passage from (p, q) to $(p - 1, q + 1)$ are left to the reader. ∎

4-4-4. Proposition. Suppose that

$$f \in \mathrm{Hom}_G(B, C) = (\mathrm{Hom}(B, C))^G$$

so that $\kappa f \in H^0(G, \mathrm{Hom}(B, C))$. Then for all q,

$$\Upsilon_{0,q}(\kappa f) = f_* : H^q(G, B) \longrightarrow H^q(G, C)$$

Proof: If $\beta \in H^q(G, B)$, then

$$(\Upsilon_{0,q}(\kappa f))(\beta) = \kappa f \smile_\theta \beta = ({}^f\theta)_*(\beta) = f_*(\beta)$$

since $f = {}^f\theta$. ∎

4-4-5. Theorem. Suppose that C is a G-module on which the action of G is trivial and whose additive group is divisible (see (1-6-7)); then for all q and all G-modules B

$$\Upsilon_{-p,p-1} : H^{-p}(G, \mathrm{Hom}(B, C)) \longrightarrow \mathrm{Hom}(H^{p-1}(G, B), H^{-1}(G, C))$$

is an isomorphism onto.

Proof: By (4-4-3), it suffices to prove that $\Upsilon_{0,-1}$ is an isomorphism onto. For this it is useful to observe that any $\xi \in H^0(G, \mathrm{Hom}(B, C))$ can be written as $\xi = \kappa f$ for some $f \in (\mathrm{Hom}(B, C))^G = \mathrm{Hom}_G(B, C)$ and $\beta \in H^{-1}(G, B)$ can be written as $\beta = \eta_B(b)$ for some $b \in B_S$, so that the action of $\Upsilon_{0,-1}$ is described by

$$\begin{aligned}(\Upsilon_{0,-1}(\kappa f))(\eta_B b) &= f_*(\eta_B b) &&\text{(by (4-4-4))}\\ &= \eta_C(f(b)) &&\text{(by (2-2-7))}\end{aligned}$$

Moreover, because the action of G on C is trivial, the map $\eta_C : C_S \longrightarrow H^{-1}(G, C)$ is an isomorphism onto.
Suppose that $\Upsilon_{0,-1}(\kappa f) = 0$. Then $f(b) = 0$ for every $b \in B_S$,

and we may define an ordinary homomorphism $g_0 : SB \longrightarrow C$ by putting

$$g_0(Sb) = f(b) \qquad b \in B$$

Since C is divisible, g_0 may be extended (see (1-6-7)) to $g \in \text{Hom}(B, C)$. Now, for $b \in B$ we have

$$(Sg)(b) = \sum_\sigma g^\sigma b = \sum_\sigma g\sigma^{-1}b = g(Sb) = g_0(Sb) = f(b).$$

Therefore, $\kappa f = \kappa(Sg) = 0$ and $\Upsilon_{0,-1}$ is a monomorphism.

Finally, if $\nu \in \text{Hom}(H^{-1}(G, B), H^{-1}(G, C))$ is given, then we may define a homomorphism $f_0 : B_S \longrightarrow C_S$ by $f_0 = \eta_C^{-1} \circ \nu \circ \eta_B$ and extend f_0 to $f \in \text{Hom}(B, C)$. Then $f \in \text{Hom}_G(B, C)$ since $f(\sigma b) - \sigma f(b) = f(\sigma b) - f(b) = f((\sigma - 1)b) = 0$, and $\Upsilon_{0,-1}(\kappa f) = \nu$ since for $b \in B_S$ $(\Upsilon_{0,-1}(\kappa f))(\eta_B b) = \eta_C(f(b)) = \eta_C(f_0(b)) = \nu(\eta_B b)$. Thus, $\Upsilon_{0,-1}$ is an epimorphism, and the proof is complete. ∎

The preceding result may be applied to $C = \mathbf{Q}/\mathbf{Z}$. For any G-module B, we denote its **dual** by $\hat{B} = \text{Hom}(B, \mathbf{Q}/\mathbf{Z})$, and call the elements of $\hat{B}$ **characters** of B in $\mathbf{Q}/\mathbf{Z}$. If G has order n, then

$$\eta_{\mathbf{Q}/\mathbf{Z}} : \left(\frac{1}{n}\mathbf{Z}\right)\Big/\mathbf{Z} = (\mathbf{Q}/\mathbf{Z})_S \longrightarrow H^{-1}(G, \mathbf{Q}/\mathbf{Z})$$

is an isomorphism onto. Therefore, the map $(1, \eta_{\mathbf{Q}/\mathbf{Z}}^{-1})$ of $\text{Hom}(H^{p-1}(G, B), H^{-1}(G, \mathbf{Q}/\mathbf{Z}))$ into

$$\text{Hom}\left(H^{p-1}(G, B), \left(\frac{1}{n}\mathbf{Z}\right)\Big/\mathbf{Z}\right) = \text{Hom}(H^{p-1}(G, B), \mathbf{Q}/\mathbf{Z}) = \widehat{H^{p-1}(G, B)}$$

is also an isomorphism onto. We have therefore,

4-4-6. Corollary. (Duality Theorem). For any G-module B and any integer p the mapping

$$(1, \eta_{\mathbf{Q}/\mathbf{Z}}^{-1}) \circ \Upsilon_{-p,p-1} : H^{-p}(G, \hat{B}) \longrightarrow \widehat{H^{p-1}(G, B)}$$

is an isomorphism onto. This isomorphism associates with $\xi \in H^{-p}(G, \hat{B})$ the character of $H^{p-1}(G, B)$ in $\mathbf{Q}/\mathbf{Z}$ which takes

$\beta \longrightarrow \eta_{\mathbf{Q}/\mathbf{Z}}^{-1}(\xi \cup_\theta \beta)$ where $\theta : \hat{B} \times B \longrightarrow \mathbf{Q}/\mathbf{Z}$ is the canonical G-pairing.

4-4-7. Corollary. (Integral Duality Theorem). For every p, we have an isomorphism

$$H^{-p}(G, \mathbf{Z}) \approx \widehat{H^p(G, \mathbf{Z})}$$

More precisely, to $\xi \in H^{-p}(G, \mathbf{Z})$ this isomorphism associates the character of $H^p(G, \mathbf{Z})$ in $\mathbf{Q}/\mathbf{Z}$ whose action on $\zeta \in H^p(G, \mathbf{Z})$ is given by

$$\zeta \longrightarrow \eta_{\mathbf{Q}/\mathbf{Z}}^{-1}\delta_*^{-1}(\xi \cup_{\theta_\mathbf{Z}} \zeta)$$

where $\theta_\mathbf{Z} : \mathbf{Z} \times \mathbf{Z} \longrightarrow \mathbf{Z}$ is the natural G-pairing $\theta_\mathbf{Z}(m, n) = mn$ and $\delta_* : H^{-p-1}(G, \mathbf{Q}/\mathbf{Z}) \longrightarrow H^{-p}(G, \mathbf{Z})$ is the coboundary arising from the exact G-sequence (with trivial action of G)

$$0 \longrightarrow \mathbf{Z} \longrightarrow \mathbf{Q} \longrightarrow \mathbf{Q}/\mathbf{Z} \longrightarrow 0$$

Proof: According to (3-1-13), δ_* is an isomorphism onto. The identification of $\hat{\mathbf{Z}}$ and $\mathbf{Q}/\mathbf{Z}$ permits us to identify $H^{-p-1}(G, \mathbf{Q}/\mathbf{Z})$ and $H^{-p-1}(G, \hat{\mathbf{Z}})$. Therefore, by applying (4-4-6) with $B = \mathbf{Z}$ we conclude that

$$(1, \eta_{\mathbf{Q}/\mathbf{Z}}^{-1}) \circ Y_{-p-1,p} \circ \delta_*^{-1} : H^{-p}(G, \mathbf{Z}) \longrightarrow \widehat{H^p(G, \mathbf{Z})}$$

is an isomorphism onto, and the element of $\widehat{H^p(G, \mathbf{Z})}$ corresponding to $\xi \in H^{-p}(G, \mathbf{Z})$ maps

$$\zeta \longrightarrow \eta_{\mathbf{Q}/\mathbf{Z}}^{-1}(\delta_*^{-1}\xi \cup_\theta \zeta)$$

where θ denotes both the canonical G-pairing

$$\theta : \mathrm{Hom}\,(\mathbf{Z}, \mathbf{Q}/\mathbf{Z}) \times \mathbf{Z} \longrightarrow \mathbf{Q}/\mathbf{Z}$$

$\theta(f, a) = f(a)$, and the natural G-pairing $\theta : \mathbf{Q}/\mathbf{Z} \times \mathbf{Z} \longrightarrow \mathbf{Q}/\mathbf{Z}$ $\theta(a, n) = an$. Note that under the correspondence $f \longleftrightarrow f(1)$ of $\mathrm{Hom}\,(\mathbf{Z}, \mathbf{Q}/\mathbf{Z})$ and $\mathbf{Q}/\mathbf{Z}$ the two θ's correspond to each other.

Thus it remains to show that

$$\delta_*^{-1}\xi \cup_\theta \zeta = \delta_*^{-1}(\xi \cup_{\theta_{\mathbf{Z}}} \zeta)$$

—or, since δ_* is an isomorphism, that $\delta_*(\delta_*^{-1}\xi \cup_\theta \zeta) = \xi \cup_{\theta_{\mathbf{Z}}} \zeta$. Now, tensoring the exact G-sequence

$$0 \longrightarrow \mathbf{Z} \longrightarrow \mathbf{Q} \longrightarrow \mathbf{Q}/\mathbf{Z} \longrightarrow 0$$

with $\mathbf{Z}$ yields the same sequence (after identification), so that application of (4-3-5) gives the desired result. This completes the proof. ▮

4-4-8. Corollary. If $\xi_1, \xi_2 \in H^{-p}(G, \mathbf{Z})$ then

$$\xi_1 = \xi_2 \iff \xi_1 \cup_{\theta_{\mathbf{Z}}} \zeta = \xi_2 \cup_{\theta_{\mathbf{Z}}} \zeta$$

for all $\zeta \in H^p(G, \mathbf{Z})$.

4-5. THE NAKAYAMA MAP

In this section, we shall use the standard G-complex X exclusively (therefore, we shall omit all reference to it) and carry out some explicit computations. These revolve about the Nakayama map (which is sometimes referred to as the "Japanese homomorphism") and its connection with a certain cup product.

4-5-1. Proposition. For any G-module A, the map of

$$\mathrm{Hom}_G(I, A) \longrightarrow \mathscr{Z}^1(G, A)$$

which takes $f \longrightarrow u_f$ where

$$u_f([\sigma]) = f(\sigma - 1) \qquad \sigma \in G$$

is an isomorphism onto.

Proof: For $f \in \mathrm{Hom}(I, A)$, let us define the standard 1-cochain u_f by the given formula. Note that because f is a homomorphism

$u_f([1]) = f(0) = 0$. Since $\{(\sigma - 1) | \sigma \neq 1\}$ is a $\mathbf{Z}$-basis for I, it is clear that $f \longleftrightarrow u_f$ is an isomorphism between $\operatorname{Hom}(I, A)$ and the group of standard 1-cochains u for which $u([1]) = 0$. It should also be noted that if a 1-cochain u is a cocycle, then it is automatic that $u([1]) = 0$; this is immediate from the coboundary formula $\sigma u[\tau] - u[\sigma\tau] + u[\sigma] = 0$. Moreover,

$$\begin{aligned} u_f \text{ is a cocycle} &\Longleftrightarrow \sigma u_f[\tau] - u_f[\sigma\tau] + u_f[\tau] = 0 \quad \text{for all} \quad \sigma, \tau \in G \\ &\Longleftrightarrow \sigma f(\tau - 1) - f(\sigma\tau - 1) + f(\sigma - 1) = 0 \\ &\Longleftrightarrow f(\sigma(\tau - 1)) = \sigma f(\tau - 1) \\ &\Longleftrightarrow f \text{ is a } G\text{-homomorphism} \\ &\Longleftrightarrow f \in \operatorname{Hom}_G(I, A). \quad \blacksquare \end{aligned}$$

According to (3-3-9) we have, for any G-module A, the dimension shifting exact G-sequence

$$0 \longrightarrow A \approx \operatorname{Hom}(\mathbf{Z}, A) \xrightarrow{\hat{\varepsilon}=(\varepsilon,1)} \operatorname{Hom}(\Gamma, A)$$

$$\xrightarrow{\hat{\imath}=(i,1)} \operatorname{Hom}(I, A) \longrightarrow 0 \qquad (\#)$$

with middle term G-regular. It follows, in particular, that

$$\delta_* : H^0(G, \operatorname{Hom}(I, A)) \longrightarrow H^1(G, A) \approx H^1(G, \operatorname{Hom}(\mathbf{Z}, A))$$

is an isomorphism onto. We are now in a position to describe δ_* explicitly.

4-5-2. Proposition. Consider

$$f \in (\operatorname{Hom}(I, A))^G = \operatorname{Hom}_G(I, A)$$

and the corresponding

$$u_f \in \mathscr{Z}^1(G, A) \subset \operatorname{Hom}_G(X_1, A) \approx \operatorname{Hom}_G(X_1, \operatorname{Hom}(\mathbf{Z}, A))$$

If $\alpha \in H^1(G, A)$ is the cohomology class of u_f, then κf and $-\alpha$ correspond to each other under δ_*; in other words,

$$\delta_*(\kappa f) = -\alpha$$

Proof: To compute δ_* we must go back to the exact sequence of differential graded groups

$$0 \longrightarrow \mathrm{Hom}_G(X, A) \approx \mathrm{Hom}_G(X, \mathrm{Hom}(\mathbf{Z}, A)) \longrightarrow$$
$$\xrightarrow{(1,\hat{\varepsilon})} \mathrm{Hom}_G(X, \mathrm{Hom}(\Gamma, A)) \longrightarrow$$
$$\xrightarrow{(1,\hat{\imath})} \mathrm{Hom}_G(X, \mathrm{Hom}(I, A)) \longrightarrow 0$$

Let us write $\kappa f = c \in \mathrm{Hom}_G(X_0, \mathrm{Hom}(I, A))$, so that $c[\cdot] = f$. Then there exist $b \in \mathrm{Hom}_G(X_0, \mathrm{Hom}(\Gamma, A))$ such that $(1, \hat{\imath})\, b = c$ and $a \in \mathrm{Hom}_G(X_1, \mathrm{Hom}(\mathbf{Z}, A))$ such that $(1, \hat{\varepsilon})\, a = \delta b$—so that $\delta_*(\kappa f)$ is the class of the 1-cocycle a.

Now, let us put $b[\cdot] = g \in \mathrm{Hom}(\Gamma, A)$, so that

$$\hat{\imath} g = \hat{\imath}(b[\cdot]) = c[\cdot] = f$$

Thus, $g \circ i = f$, and f is the restriction of g to I. Since $\Gamma = I \oplus \mathbf{Z} \cdot 1$, we may reverse matters and define g to be the unique extension of f to Γ for which $g(1) = 0$; then define b by $b[\cdot] = g$, so that b and a satisfy all the conditions mentioned above.

Next, we note that $\hat{\varepsilon} \circ a = \delta b \in \mathrm{Hom}_G(X_1, \mathrm{Hom}(\Gamma, A))$ is determined by its action on the 1-cells, and

$$\delta b([\sigma]) = b(\sigma[\cdot] - [\cdot]) = \sigma b[\cdot] - b[\cdot] = g^\sigma - g \in \mathrm{Hom}(\Gamma, A)$$

But $g^\sigma - g$ vanishes on I since

$$g(\tau - 1) = f(\tau - 1) = f^\sigma(\tau - 1) = g^\sigma(\tau - 1)$$

—so (by (#)) there exists, for each $\sigma \in G$, $h_\sigma \in \mathrm{Hom}(\mathbf{Z}, A)$ such that $\hat{\varepsilon} h_\sigma = g^\sigma - g$. It follows that

$$\hat{\varepsilon}(a[\sigma]) = (\delta b)([\sigma]) = \hat{\varepsilon} h_\sigma \qquad \text{all} \quad \sigma \in G$$

and then $a[\sigma] = h_\sigma \in \mathrm{Hom}(\mathbf{Z}, A)$.

It remains to determine which cocycle in $\mathrm{Hom}_G(X_1, A)$ corresponds to $a \in \mathrm{Hom}_G(X_1, \mathrm{Hom}(\mathbf{Z}, A))$. For this we compute, $h_\sigma(1) = h_\sigma(\varepsilon(1)) = (\hat{\varepsilon} h_\sigma)(1) = g^\sigma(1) - g(1) = \sigma g \sigma^{-1}(1) - \sigma g(1)$ (since $g(1) = 0) = \sigma g(\sigma^{-1} - 1) = \sigma f(\sigma^{-1} - 1) = f(1 - \sigma) = -u_f([\sigma])$. In other words, a corresponds to the cocycle $-u_f$, so that $\delta_*(\kappa f) = -\alpha$, and the proof is complete. ▮

In virtue of (4-5-2), given $\alpha \in H^1(G, A) \approx H^1(G, \mathrm{Hom}(\mathbf{Z}, A))$ there exists $f \in \mathrm{Hom}_G(I, A)$ such that $-\delta_*(\kappa f) = \alpha$, and by (4-5-1) there is an associated 1-cocycle $u_f \in \mathrm{Hom}_G(X_1, A)$ which represents α. Conversely, any cocycle $u \in \mathrm{Hom}_G(X_1, A)$ representing α is of form $u = u_f$ for some $f \in \mathrm{Hom}_G(I, A)$. We wish to examine the cup product of α with any element of $H^{-2}(G, \mathbf{Z})$—which element can be expressed (according to (3-5-2)) as $\zeta_\sigma = \delta_{-1}^* \eta(\sigma - 1)$ for some $\sigma \in G$. We shall also make use of the formula given in the proof of (4-4-2), and apply it to the case $B' = I$, $B = \Gamma$, $B'' = C = \mathbf{Z}$.

4-5-3. Proposition. Let $\theta : A \times \mathbf{Z} \longrightarrow A$ be the natural G-pairing, $\theta(a, n) = na$. Then the cup product

$$\cup_\theta : H^1(G, A) \times H^{-2}(G, \mathbf{Z}) \longrightarrow H^{-1}(G, A)$$

is given explicitly by the formula

$$\alpha \cup_\theta \zeta_\sigma = \eta(u[\sigma])$$

where $u = u_f$ is any cocycle representing α, and $f \in \mathrm{Hom}_G(I, A)$.

Proof: For the purposes of this proof, let

$$\theta' : \mathrm{Hom}(I, A) \times I \longrightarrow A \quad \text{and} \quad \theta'' : \mathrm{Hom}(\mathbf{Z}, A) \times \mathbf{Z} \longrightarrow A$$

denote the canonical G-pairings; in particular, $\theta'' \approx \theta$. We have then, for any $\sigma \in G$

$$\begin{aligned}
\alpha \cup_\theta \zeta_\sigma &= -\delta_*(\kappa f) \cup_{\theta''} \zeta_\sigma && \text{(from the identification)} \\
&= \kappa f \cup_{\theta'} \delta_* \zeta_\sigma && \text{(by (4-4-2), with } p = 0) \\
&= \kappa f \cup_{\theta'} \eta(\sigma - 1) && \text{(by (3-5-2))} \\
&= ({}^f\theta')_*(\eta(\sigma - 1)) && \text{(by (4-3-5))} \\
&= f_*(\eta(\sigma - 1)) && ({}^f\theta' = f) \\
&= \eta(f(\sigma - 1)) && \text{(by (2-2-7))} \\
&= \eta(u_f[\sigma]) && \text{(by definition of } u_f)
\end{aligned}$$

This completes the proof. ∎

Suppose that $\theta : \mathbf{Z} \times A \longrightarrow A$ is the natural G-pairing, $\theta(n, a) = na$. Then we have the cup product

$$\cup_\theta : H^{-2}(G, \mathbf{Z}) \times H^q(G, A) \longrightarrow H^{q-2}(G, A) \qquad \text{all } q$$

and the relation

$$\zeta_\sigma \cup_\theta \alpha = (-1)^{-2q} \alpha \cup_{\theta^\top} \zeta_\sigma = \alpha \cup_{\theta^\top} \zeta_0$$

This indicates that in studying cupping by ζ_σ it does not matter very much whether we cup on the left or the right. Of course, $\theta^\top$ is then the G-pairing used in (4-5-3), so that, in particular, for $q = 1$, we have

$$\zeta_\sigma \cup_\theta \alpha = \alpha \cup_{\theta^\top} \zeta_0 = \eta(u_f[\sigma])$$

Given any G-module A, we have then for each $\sigma \in G$ and each integer q a mapping $\alpha \longrightarrow \zeta_\sigma \cup_\theta \alpha$ of $H^q(G, A) \longrightarrow H^{q-2}(G, A)$. Our objective is to give an explicit cochain description of these maps for $q \geqslant 1$. Let us define, for $\sigma \in G$ and $q \geqslant 1$, a homomorphism

$$\sigma\# : \mathrm{Hom}_G (X_q , A) \longrightarrow \mathrm{Hom}_G (X_{q-2} , A)$$

of standard cochain groups. Consider $u \in \mathrm{Hom}_G (X_q , A)$ and define $\sigma^\# u$ on the $(q - 2)$-cells (and then extend linearly to get $\sigma^\# u \in \mathrm{Hom}_G (X_{q-2} , A)$) by the formulas:

$$(\sigma^\# u)(\langle\cdot\rangle) = u[\sigma] \qquad q = 1$$

$$(\sigma^\# u)([\cdot]) = \sum_{\rho\in G} u[\rho, \sigma] \qquad q = 2$$

$$(\sigma^\# u)([\sigma_1 ,\ldots, \sigma_{q-2}] = \sum_{\rho\in G} u[\sigma_1 ,\ldots, \sigma_{q-2} , \rho, \sigma] \qquad q > 2$$

4-5-4. Proposition. The homomorphism $\sigma^\#$ takes cocycles into cocycles and coboundaries into coboundaries. Therefore, it induces a homomorphism of cohomology groups

$$(\sigma^\#)_* : H^q(G, A) \longrightarrow H^{q-2}(G, A) \qquad q \geqslant 1$$

Proof: First let us verify that $\delta \circ \sigma^{\#} = \sigma^{\#} \circ \delta$. For $q = 1$, consider the standard 1-cochain $u \in \mathrm{Hom}_G(X_1, A)$. Then

$$(\sigma^{\#}(\delta u))([\cdot]) = \sum_{\rho} (\delta u)[\rho, \sigma] = \sum_{\rho} u(\rho[\sigma] - [\rho\sigma] + [\rho])$$
$$= \sum_{\rho} (\rho u[\sigma] - u[\rho\sigma] + u[\rho]) = \sum_{\rho} \rho u[\sigma] = S(u[\sigma])$$
$$= S((\sigma^{\#}u)\langle\cdot\rangle) = (\sigma^{\#}u)(S\langle\cdot\rangle) = (\sigma^{\#}u)(\partial[\cdot]) = (\delta(\sigma^{\#}u))([\cdot])$$

For $q = 2$, consider a standard 2-cochain $u \in \mathrm{Hom}_G(X_2, A)$. Then

$$(\sigma^{\#}(\delta u))([\tau]) = \sum_{\rho} (\delta u)[\tau, \rho, \sigma]$$
$$= \sum_{\rho} (\tau u[\rho, \sigma] - u[\tau\rho, \sigma] + u[\tau, \rho\sigma] - u[\tau, \rho])$$
$$= (\tau - 1) \sum_{\rho} u[\rho, \sigma]$$

while

$$(\delta(\sigma^{\#}u))([\tau]) = (\sigma^{\#}u)(\tau[\cdot] - [\cdot]) = (\tau - 1)[(\sigma^{\#}(u))([\cdot])]$$
$$= (\tau - 1) \sum_{\rho} u[\rho, \sigma]$$

The verification that $\delta \circ \sigma^{\#} = \sigma^{\#} \circ \delta$ for $q > 2$ proceeds in the same fashion, and may be left to the reader.

This implies that $\sigma^{\#}$ takes q cocycles into $q - 2$ cocycles for $q \geqslant 1$ and q coboundaries into $q - 2$ coboundaries for $q \geqslant 2$. It remains to check that $\sigma^{\#}$ takes 1 coboundaries into -1 coboundaries. Suppose then that $u \in \mathrm{Hom}_G(X_1, A)$ is a 1 coboundary—so $u = \delta v$ with $v \in \mathrm{Hom}_G(X_0, A)$, and

$$(\sigma^{\#}u)\langle\cdot\rangle = u[\sigma] = \delta v[\sigma] = v(\sigma[\cdot] - [\cdot]) = (\sigma - 1)v[\cdot] \in IA$$

As seen in the proof of (1-5-9), this means that $\sigma^{\#}u$ is a -1 coboundary. This completes the proof. ∎

4-5-5. Proposition. The map $(\sigma^{\#})_*$ commutes with the coboundary arising from exact sequences. In other words, if

$$0 \longrightarrow A \xrightarrow{i} B \xrightarrow{j} C \longrightarrow 0$$

is an exact G-sequence, then the following diagram commutes for $q \geqslant 1$:

$$\begin{array}{ccc} H^q(G, C) & \xrightarrow{\sigma_*^\#} & H^{q-2}(G, C) \\ \delta_* \downarrow & & \downarrow \delta_* \\ H^{q+1}(G, A) & \xrightarrow{\sigma_*^\#} & H^{q-1}(G, A) \end{array}$$

Proof: Roughly speaking, this is a case of induced maps commuting with coboundaries. In more detail, put

$$X^{(1)} = \sum_{q \geqslant 1} \oplus X_q \qquad \text{and} \qquad X^{(-1)} = \sum_{q \geqslant -1} \oplus X_q$$

and consider the commutative diagram

$$\begin{array}{ccccccccc} 0 \longrightarrow & \operatorname{Hom}_G(X^{(1)}, A) & \xrightarrow{(1,i)} & \operatorname{Hom}_G(X^{(1)}, B) & \xrightarrow{(1,j)} & \operatorname{Hom}_G(X^{(1)}, C) & \longrightarrow 0 \\ & \sigma^\# \downarrow & & \sigma^\# \downarrow & & \sigma^\# \downarrow & \\ 0 \longrightarrow & \operatorname{Hom}_G(X^{(-1)}, A) & \xrightarrow{(1,i)} & \operatorname{Hom}_G(X^{(-1)}, B) & \xrightarrow{(1,j)} & \operatorname{Hom}_G(X^{(-1)}, C) & \longrightarrow 0 \end{array}$$

Because (4-5-4) holds, the desired result follows easily (see, for example, (1-1-4), (2-1-9), and (2-2-5)) when a bit of care is exercised for $q = 1$. ∎

4-5-6. Theorem. For any G-module A, let

$$\theta = \theta_A : \mathbf{Z} \times A \longrightarrow A$$

be the natural G-pairing. Then for any $\sigma \in G$ and any $q \geqslant 1$ the maps $\zeta_\sigma \smile_\theta$ (that is, θ-cupping on the left by ζ_σ) and $\sigma_*^\#$ of $H^q(G, A) \longrightarrow H^{q-2}(G, A)$ are identical. In other words, for $\alpha \in H^q(G, A)$, $q \geqslant 1$,

$$\sigma_*^\#(\alpha) = \zeta_\sigma \smile_{\theta_A} \alpha$$

Proof: Consider first the case $q = 1$. As in (4-5-2) and (4-5-3) let $u_f \in \operatorname{Hom}_G(X_1, A)$ be a cocycle representing α—so that $\zeta_\sigma \smile_\theta \alpha = \eta(u_f[\sigma])$. On the other hand, $\sigma_*^\# \alpha =$ the cohomology class of the standard 1-cocycle $\sigma^\# u_f = \eta((\sigma^\# u_f)(\langle\cdot\rangle)) = \eta(u_f[\sigma])$. Therefore, $\sigma_*^\# = \zeta_\sigma \smile_\theta$ in dimension 1.

Now, suppose inductively that $\sigma_*^{\#} = \zeta_\sigma \cup_\theta$ in dimension q. Let $0 \longrightarrow A \longrightarrow B \longrightarrow C \longrightarrow 0$ be an exact G-sequence with middle term G-regular. For any G-module M let $\theta_M : \mathbf{Z} \times M \longrightarrow M$ denote the natural G-pairing. Then the following diagram is commutative and has G-exact rows:

$$\begin{array}{ccccccccc} 0 & \longrightarrow & \mathbf{Z} \otimes A & \longrightarrow & \mathbf{Z} \otimes B & \longrightarrow & \mathbf{Z} \otimes C & \longrightarrow & 0 \\ & & \downarrow{\scriptstyle \theta_A} & & \downarrow{\scriptstyle \theta_B} & & \downarrow{\scriptstyle \theta_C} & & \\ 0 & \longrightarrow & A & \longrightarrow & B & \longrightarrow & C & \longrightarrow & 0 \end{array}$$

Therefore, according to (4-3-5) with $p = -2$, the following diagram commutes:

$$\begin{array}{ccc} H^q(G, C) & \xrightarrow{\zeta_\sigma \cup \theta_C} & H^{q-2}(G, C) \\ \downarrow{\scriptstyle \delta_*} & & \downarrow{\scriptstyle \delta_*} \\ H^{q+1}(G, A) & \xrightarrow{\zeta_\sigma \cup \theta_A} & H^{q-1}(G, A) \end{array}$$

If the horizontal maps are replaced by $\sigma_*^{\#}$, then according to (4-5-5) this diagram commutes. Since both maps δ_* are isomorphisms onto, it follows that $\sigma_*^{\#} = \zeta_\sigma \cup_{\theta_A}$ in dimension $q + 1$, and the proof is complete. ▌

Of course, the same result holds for the pairing $\theta : A \times \mathbf{Z} \longrightarrow A$; that is, $\alpha \cup_\theta \zeta_\sigma = (\sigma^{\#})_* (\alpha)$.

4-5-7. Remark. In virtue of (4-5-6) we can give an explicit formula for the cup product pairing

$$H^{-2}(G, \mathbf{Z}) \times H^2(G, A) \longrightarrow H^0(G, A)$$

For this, consider $\zeta_\sigma \in H^{-2}(G, \mathbf{Z})$ and $\alpha \in H^2(G, A)$, and let $u \in \mathrm{Hom}_G (X_2 , A)$ be any cocycle representing α. Then

$$\zeta_\sigma \cup_\theta \alpha = \kappa \Big(\sum_{\rho \in G} u[\rho, \sigma] \Big) \qquad \theta = \theta_A$$

since $\zeta_\sigma \cup_\theta \alpha = \sigma^{\#}_{*}(\alpha)$ = the cohomology class of the 0-cocycle $\sigma^{\#}u = \kappa((\sigma^{\#}u)([\cdot])) = \kappa(\sum_\rho u[\rho, \sigma])$. Note that it follows automatically from the discussion that $\sum_\rho u[\rho, \sigma] \in A^G$; this fact can also be verified directly from the identity for 2-cocycles.

Thus, for fixed $\alpha \in H^2(G, A)$ and cocycle u representing α, the map

$$\sigma \longrightarrow \zeta_\sigma \cup_\theta \alpha = \kappa\left(\sum_\rho u[\rho, \sigma]\right)$$

is a homomorphism of $G \longrightarrow H^0(G, A)$. We may also drop reference to cohomology groups, and arrive at a homomorphism of

$$G \longrightarrow \frac{A^G}{SA}$$

given by

$$\sigma \longrightarrow \sum_\rho u[\rho, \sigma] \pmod{SA} \qquad (*)$$

Of course, the commutator subgroup G^c is contained in the kernel, and we get a homomorphism of $G/G^c \longrightarrow A^G/(SA)$. The formula $(*)$ was introduced directly by Nakayama (see [51]) and used to give an explicit formula for the reciprocity law isomorphism of local class field theory.

The remainder of this section is devoted to some by-products of the preceding discussion.

4-5-8. Remark. Consider the natural G-pairing

$$\theta = \theta_{\mathbf{Q}/\mathbf{Z}} : \mathbf{Z} \times \mathbf{Q}/\mathbf{Z} \longrightarrow \mathbf{Q}/\mathbf{Z}$$

and the resulting cup product pairing

$$H^{-2}(G, \mathbf{Z}) \times H^1(G, \mathbf{Q}/\mathbf{Z}) \xrightarrow{\cup_\theta} H^{-1}(G, \mathbf{Q}/\mathbf{Z})$$

Now, $H^{-2}(G, \mathbf{Z})$ may be identified with G/G^c (see (3-5-2)), and $H^1(G, \mathbf{Q}/\mathbf{Z})$ may be identified with $(\widehat{G/G^c})$, the character group of G/G^c (and also of G) in $\mathbf{Q}/\mathbf{Z}$, since

$$H^1\left(G, \frac{\mathbf{Q}}{\mathbf{Z}}\right) = \mathscr{Z}^1\left(G, \frac{\mathbf{Q}}{\mathbf{Z}}\right) \approx \operatorname{Hom}\left(G, \frac{\mathbf{Q}}{\mathbf{Z}}\right) = \hat{G} = \widehat{\left(\frac{G}{G^c}\right)}$$
$$= \operatorname{Hom}\left(\frac{G}{G^c}, \frac{\mathbf{Q}}{\mathbf{Z}}\right)$$

In addition, if G has order n, then

$$\eta = \eta_{\mathbf{Q}/\mathbf{Z}} : \left(\frac{1}{n}\mathbf{Z}\right)/\mathbf{Z} = \left(\frac{\mathbf{Q}}{\mathbf{Z}}\right)_S \longrightarrow H^{-1}\left(G, \frac{\mathbf{Q}}{\mathbf{Z}}\right)$$

is an isomorphism onto, and we may view it as an identification.

On the other hand, we have the natural pairing of

$$\frac{G}{G^c} \times \widehat{\left(\frac{G}{G^c}\right)} \longrightarrow \left(\frac{1}{n}\mathbf{Z}\right)/\mathbf{Z}$$

given by $(\sigma G^c, \chi) = \chi(\sigma G^c) = \chi(\sigma)$. (There is no harm in using χ to denote a character of G/G^c and also the corresponding character of G.) This pairing is compatible with the previous cup product pairing—that is, for $\chi \in (\widehat{G/G^c}) \approx H^1(G, \mathbf{Q}/\mathbf{Z})$,

$$\zeta_\sigma \cup_\theta \chi = \eta(\chi(\sigma))$$

To see this, one simply applies (4-5-3), with both α and the cocycle $u = u_f$ representing it replaced by χ.

Now, let us give an explicit form for the isomorphisms that express the periodicity of the cohomology groups of a cyclic group. For this, we have need of the following simple result.

4-5-9. Exercise. Let G have order n, and consider the exact G-sequence $0 \longrightarrow \mathbf{Z} \xrightarrow{i} \mathbf{Q} \xrightarrow{j} \mathbf{Q}/\mathbf{Z} \longrightarrow 0$. An element

$$c \in \left(\frac{1}{n}\mathbf{Z}\right)/\mathbf{Z} = \left(\frac{\mathbf{Q}}{\mathbf{Z}}\right)_S$$

may also be viewed as an element of

$$\left(\frac{1}{n}\mathbf{Z}\right)/\mathbf{Z} \approx \frac{\mathbf{Z}}{n\mathbf{Z}} \xrightarrow{\bar{\kappa}} H^0(G, \mathbf{Z})$$

and the map $\delta_* : H^{-1}(G, \mathbf{Q}/\mathbf{Z}) \longrightarrow H^0(G, \mathbf{Z})$ (which is an isomorphism onto) is given by

$$\delta_*(\eta c) = \bar{\kappa}(c)$$

Suppose that G is a cyclic group of order n, and fix a generator σ. With σ we can associate a canonical generator χ_σ of the character group $\hat{G} = \text{Hom}\,(G, \mathbf{Q}/\mathbf{Z})$ (which is also a cyclic group of order n) by putting

$$\chi_\sigma(\sigma^r) = \frac{r}{n} \quad (\text{mod } \mathbf{Z})$$

Of course, we may also view χ_σ as an element of $H^1(G, \mathbf{Q}/\mathbf{Z})$.

4-5-10. Theorem. Let G be a cyclic group of order n with generator σ; then for every G-module A and every integer q we have an isomorphism

$$H^q(G, A) \approx H^{q-2}(G, A)$$

In fact, if $\theta = \theta_A : \mathbf{Z} \times A \longrightarrow A$ is the natural G-pairing then, for every q, the maps

$$\zeta_\sigma \cup_\theta : H^q(G, A) \longrightarrow H^{q-2}(G, A)$$

and

$$\delta_*\chi_\sigma \cup_\theta : H^{q-2}(G, A) \longrightarrow H^q(G, A)$$

are inverses of each other.

Proof: For $\alpha \in H^q(G, A)$ we have

$$\begin{aligned}
\delta_*\chi_\sigma \cup_\theta (\zeta_\sigma \cup_\theta \alpha) &= (\delta_*\chi_\sigma \cup_{\theta_{\mathbf{Z}}} \zeta_\sigma) \cup_\theta \alpha \\
&= \delta_*(\chi_\sigma \cup_{\theta^{\mathsf{T}}_{\mathbf{Q}/\mathbf{Z}}} \zeta_\sigma) \cup_\theta \alpha \\
&= \delta_*(\zeta_\sigma \cup_{\theta_{\mathbf{Q}/\mathbf{Z}}} \chi_\sigma) \cup_\theta \alpha \\
&= \delta_*(\eta(\chi_\sigma(\sigma))) \cup_\theta \alpha \\
&= \bar{\kappa}(\chi_\sigma(\sigma)) \cup_\theta \alpha \\
&= \kappa(1) \cup_\theta \alpha \\
&= ({}^1\theta)_*(\alpha) \\
&= \alpha
\end{aligned}$$

Furthermore, for $\alpha \in H^{q-2}(G, A)$ we have

$$\begin{aligned}\zeta_\sigma \smile_\theta (\delta_* \chi_\sigma \smile_\theta \alpha) &= (\zeta_\sigma \smile_{\theta \mathbf{Z}} \delta_* \chi_\sigma) \smile_\theta \alpha \\ &= (\delta_* \chi_\sigma \smile_{\theta \mathbf{Z}} \zeta_\sigma) \smile_\theta \alpha = \delta_* \chi_\sigma \smile_\theta (\zeta_\sigma \smile_\theta \alpha) = \alpha\end{aligned}$$

and the proof is complete. ∎

4-6. PROBLEMS AND SUPPLEMENTS

4-6-1. Show that if $0 \longrightarrow A' \xrightarrow{i} A$ is an exact G-sequence, and B is an arbitrary G-module, then the sequence

$$0 \longrightarrow A' \otimes B \xrightarrow{i \otimes 1} A \otimes B$$

need not be exact. On the other hand, if A' is a direct $\mathbf{Z}$-summand —that is, $A \approx A' \oplus A''$, direct sum of abelian groups—then

$$A \otimes B \approx (A' \otimes B) \oplus (A'' \otimes B) \qquad \text{(as groups)}$$

and the G-sequence $0 \longrightarrow A' \otimes B \xrightarrow{i \otimes 1} A \otimes B$ is exact.

4-6-2. Given two left G-modules A and B we may form their tensor product as $\mathbf{Z}$-modules, $A \otimes B$. This may be defined concretely as follows. Let F be the free abelian group generated by all the pairs (a, b) $a \in A$, $b \in B$, and let R be the subgroup of F generated by all the elements of form $(a_1 + a_2, b) - (a_1, b) - (a_2, b)$, $(a, b_1 + b_2) - (a, b_1) - (a, b_2)$. Then $A \otimes B$ is the quotient group F/R, and the image of (a, b) in $A \otimes B$ is denoted by $a \otimes b$. The rules

$$(a_1 + a_2) \otimes b = a_1 \otimes b + a_2 \otimes b$$

$$a \otimes (b_1 + b_2) = a \otimes b_1 + a \otimes b_2$$

then hold.

Now, A may be viewed as a right G-module by putting $a\sigma = \sigma^{-1}a$ $a \in A$, $\sigma \in G$ and we may form the tensor product of the right Γ-module A and the left Γ-module B, $A \otimes_\Gamma B$, as follows. Let R' be the subgroup of F generated by all elements of form

$(a_1 + a_2, b) - (a_1, b) - (a_2, b)$, $(a, b_1 + b_2) - (a, b_1) - (a, b_2)$, $(a\gamma, b) - (a, \gamma b)$, $\gamma \in \Gamma$. Then $A \otimes_\Gamma B$ is the quotient group F/R' and the image of (a, b) in $A \otimes_\Gamma B$ is denoted by $a \otimes_\Gamma b$. The rules

$$(a_1 + a_2) \otimes_\Gamma b = a_1 \otimes_\Gamma b + a_2 \otimes_\Gamma b$$

$$a \otimes_\Gamma (b_1 + b_2) = a \otimes_\Gamma b_1 + a \otimes_\Gamma b_2$$

$$a\gamma \otimes_\Gamma b = a \otimes_\Gamma \gamma b$$

then hold.

Show that we have an isomorphism of groups

$$A \otimes_\Gamma B \approx \frac{A \otimes B}{I(A \otimes B)}$$

where (as in Section 4-1) $A \otimes B$ is a G-module with

$$\sigma(a \otimes b) = \sigma a \otimes \sigma b$$

4-6-3. For any G-module C, let us write $C_G = C/(IC)$. Then the trace map $S : C \longrightarrow C$ induces a homomorphism

$$S^* : C_G \longrightarrow C^G$$

Moreover, if C is G-regular then S^* is an isomorphism onto.

4-6-4. (i) Suppose that X and A are G-modules, and as usual, let $\hat{X} = \text{Hom}(X, \mathbf{Z})$. Then the map

$$\mu : X \otimes A \longrightarrow \text{Hom}(\hat{X}, A)$$

such that

$$\mu(x \otimes a)f = f(x)a \qquad x \in X, \quad a \in A, \quad f \in \hat{X}$$

is a G-homomorphism. (Of course, if X and A are simply $\mathbf{Z}$-modules then μ is a homomorphism of groups.)

(ii) Suppose further that X has a finite Γ-basis; then X has a finite $\mathbf{Z}$-basis, and so does $\hat{X}$. In fact (see (1-4-4)), if

$\{x_i \mid i = 1,\ldots, r\}$ is a $\mathbf{Z}$-basis of X, then $\{\hat{x}_i \mid i = 1,\ldots, r\}$ is a $\mathbf{Z}$-basis of $\hat{X}$. Then the map

$$\nu : \operatorname{Hom}(\hat{X}, A) \longrightarrow X \otimes A$$

given by

$$\nu(\varphi) = \sum_{i=1}^{r} x_i \otimes \varphi(\hat{x}_i) \qquad \varphi \in \operatorname{Hom}(\hat{X}, A)$$

is a homomorphism which is the inverse of μ. In particular, μ is a G-isomorphism.

(iii) Combining this fact with (4-6-2) and (4-6-3) we have group isomorphisms

$$X \otimes_{\Gamma} A \approx \frac{X \otimes A}{I(X \otimes A)} \approx (X \otimes A)^G \approx \operatorname{Hom}_G(\hat{X}, A)$$

Furthermore, since $\hat{\hat{X}}$ may be identified with X, we have an isomorphism

$$\hat{X} \otimes_{\Gamma} A \approx \operatorname{Hom}_G(X, A)$$

which is given explicitly by λ, where

$$\lambda(f \otimes a)(x) = \sum_{\sigma \in G} f(\sigma^{-1}x)(\sigma a) \qquad f \in \hat{X}, \quad x \in X, \quad a \in A$$

(iv) Compare the dimension shifting sequences

$$0 \longrightarrow I \otimes A \longrightarrow \Gamma \otimes A \longrightarrow A \longrightarrow 0$$
$$0 \longrightarrow A \longrightarrow \Gamma \otimes A \longrightarrow J \otimes A \longrightarrow 0$$

of (4-1-6) with the dimension shifting sequences

$$0 \longrightarrow A \longrightarrow \operatorname{Hom}(\Gamma, A) \longrightarrow \operatorname{Hom}(I, A) \longrightarrow 0$$
$$0 \longrightarrow \operatorname{Hom}(J, A) \longrightarrow \operatorname{Hom}(\Gamma, A) \longrightarrow A \longrightarrow 0$$

of (3-3-9). Note that we have G-isomorphisms $\hat{\Gamma} \approx \Gamma$, $\hat{I} \approx J$, $\hat{J} \approx I$ (see (3-3-8)).

4-6-5. If the finite group G is cyclic, then the G-modules I and J are isomorphic, and it follows that for any G-module A and any integer n, $H^n(G, A) \approx H^{n+2}(G, A)$.

4-6-6. Suppose that H is a normal subgroup of G and A is a G-module. Let X and Y denote the standard G and G/H complexes, respectively. If $\pi : G \longrightarrow G/H$ is the canonical map then, in order to compute

$$\inf = (\pi, i)_* : H^n(G/H, A^H) \longrightarrow H^n(G, A) \qquad n \geqslant 1$$

we make use of a G-homomorphism $\Lambda : X \longrightarrow Y$ (in dimensions $\geqslant 0$) which may be given explicitly (see (2-1-5), (2-1-8), (2-3-6), and Sec. (2-5)). In particular, we have the commutative diagram

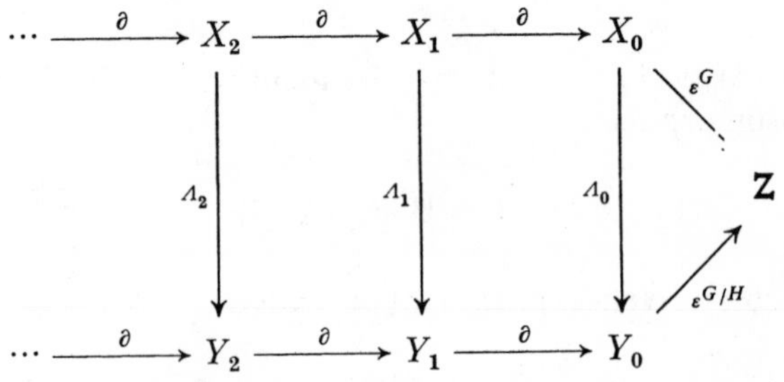

Dualizing this diagram leads to the commutative diagram

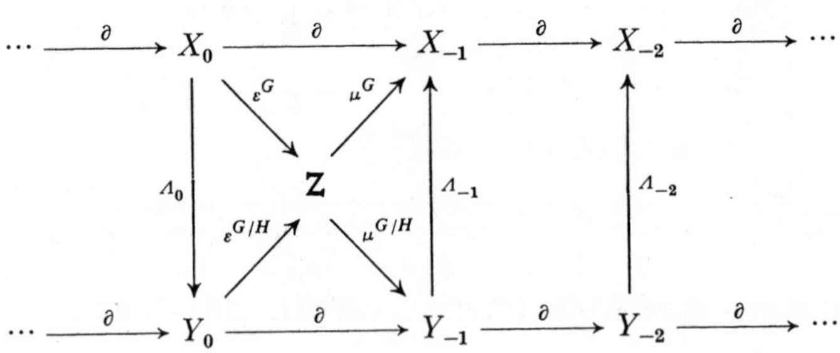

where $X_{-r} = \hat{X}_{r-1}$, $Y_{-r} = \hat{Y}_{r-1}$, $\Lambda_{-r} = (\Lambda_{r-1}, 1)$ for $r \geqslant 1$. The maps Λ_{-r} are G-homomorphisms, they are monomorphisms (since they are duals of epimorphisms) and may be found explicitly.

Now consider the commutative diagram

$$\begin{array}{ccccccc}
\cdots \xrightarrow{\delta} & \mathrm{Hom}_G(X_{-2}, A) & \xrightarrow{\delta} & \mathrm{Hom}_G(X_{-1}, A) & \xrightarrow{\delta} & \mathrm{Hom}_G(X_0, A) \\
& \downarrow{\scriptstyle (\Lambda_{-2},1)} & & \downarrow{\scriptstyle (\Lambda_{-1},1)} & & \uparrow{\scriptstyle (\Lambda_0,1)} \\
\cdots \xrightarrow{\delta} & \mathrm{Hom}_G(Y_{-2}, A) & \xrightarrow{\delta} & \mathrm{Hom}_G(Y_{-1}, A) & \xrightarrow{\delta} & \mathrm{Hom}_G(Y_0, A)
\end{array}$$

By making use of the explicit form of the Λ_{-r}, we see that the image of $(\Lambda_{-r}, 1)$ is contained in $\mathrm{Hom}_G(Y_{-r}, A^H)$, and arrive at the commutative diagram

$$\begin{array}{ccccccc}
\cdots \xrightarrow{\delta} & \mathrm{Hom}_G(X_{-2}, A) & \xrightarrow{\delta} & \mathrm{Hom}_G(X_{-1}, A) & \xrightarrow{\delta} & \mathrm{Hom}_G(X_0, A) \\
& \downarrow{\scriptstyle (\Lambda_{-2},1)} & & \downarrow{\scriptstyle (\Lambda_{-1},1)} & & \uparrow{\scriptstyle (\Lambda_0,1)} \\
\cdots \xrightarrow{\delta} & \mathrm{Hom}_{G/H}(Y_{-2}, A^H) & \xrightarrow{\delta} & \mathrm{Hom}_{G/H}(Y_{-1}, A^H) & \xrightarrow{\delta} & \mathrm{Hom}_G(Y_0, A)
\end{array}$$

It follows that there is a cohomological map, called the **deflation**

$$\mathrm{defl} : H^{-r}(G, A) \longrightarrow H^{-r}(G/H, A^H) \qquad r \geqslant 1$$

By coherence, the deflation is carried over to arbitrary G and (G/H)-complexes. Furthermore, let us recall that in (3-4-4) the deflation map was defined in dimension 0.

Among the properties of deflation the following may be verified:

(1) In dimension -1, deflation corresponds to the H-trace S_H.

(2) Deflation commutes with induced homomorphisms in all dimensions $\leqslant 0$—that is, for all $r \geqslant 0$.

(3) Deflation commutes with restriction and corestriction for all $r \geqslant 0$.

(4) Deflation commutes with coboundaries arising from exact sequences; more precisely, if $0 \longrightarrow A \longrightarrow B \longrightarrow C \longrightarrow 0$ is an exact G-sequence such that

$$0 \longrightarrow A^H \longrightarrow B^H \longrightarrow C^H \longrightarrow 0$$

is an exact (G/H)-sequence, then the following diagram commutes:

$$\begin{array}{ccc}
H^{-r}(G, C) & \xrightarrow{\delta_*} & H^{-r+1}(G, A) \\
{\scriptstyle \mathrm{defl}}\downarrow & & \downarrow{\scriptstyle \mathrm{defl}} \\
H^{-r}(G/H, C^H) & \xrightarrow{\delta_*} & H^{-r+1}(G/H, A^H)
\end{array} \qquad r \geqslant 1$$

(5) With the same hypotheses as in (4), the following diagram commutes:

$$\begin{array}{ccc} H^0(G, C) & \xrightarrow{\delta_*} & H^1(G, A) \\ {\scriptstyle \text{defl}}\downarrow & & \uparrow{\scriptstyle \text{inf}} \\ H^0(G/H, C^H) & \xrightarrow{\delta_*} & H^1(G/H, A^H) \end{array}$$

(6) Deflation is transitive.

(7) defl $\circ$ cor $= 0$ for all $r \geqslant 0$.

(8) If $H^{-s}(H, A) = (0)$ for $s = 0, 1, 2, \ldots, r - 1$, then

$$H^{-r}(H, A) \xrightarrow{\text{cor}} H^{-r}(G, A) \xrightarrow{\text{defl}} H^{-r}(G/H, A^H) \longrightarrow 0$$

is exact.

(9) If $\theta : A \times B \longrightarrow C$ is a G-pairing then all meaningful commutativities in the following diagram are valid

$$\begin{array}{ccc} H^p(G, A) \times H^q(G, B) & \xrightarrow{\cup_\theta} & H^{p+q}(G, C) \\ {\scriptstyle \text{defl}}\downarrow\uparrow{\scriptstyle \text{inf}} \quad {\scriptstyle \text{defl}}\downarrow\uparrow{\scriptstyle \text{inf}} & & {\scriptstyle \text{defl}}\downarrow\uparrow{\scriptstyle \text{inf}} \\ H^p(G/H, A^H) \times H^q(G/H, B^H) & \xrightarrow{\cup_\theta} & H^{p+q}(G/H, C^H) \end{array}$$

Finally, the reader may compare

$$\text{defl} : H^{-2}(G, \mathbf{Z}) \longrightarrow H^{-2}(G/H, \mathbf{Z})$$

with the map induced by $\pi : G \longrightarrow G/H$.

4-6-7. Suppose that H is a subgroup of G, and A is a G-module. If $\rho \in H$ and $\alpha \in H^n(G, A)$, then

$$\text{cor}\,(\zeta_\rho \cup_\theta \text{res}\,\alpha) = \zeta_\rho \cup_\theta \alpha$$

where $\theta : \mathbf{Z} \times A \longrightarrow A$ is the natural G-pairing.

4-6-8. Suppose that H is a normal subgroup of G with

$\#(H) = m$, and let $\sigma \longrightarrow \bar{\sigma}$ denote the canonical map of $G \longrightarrow G/H$. If $\sigma \in G$ and $\alpha \in H^n(G/H, A^H)$, $n > 2$, then

$$\zeta_\sigma \smile_\theta \inf \alpha = m \inf (\zeta_{\bar{\sigma}} \smile_\theta \alpha)$$

where $\theta : \mathbf{Z} \times A \longrightarrow A$ is the natural G-pairing and the induced pairing $\theta : \mathbf{Z} \times A^H \longrightarrow A^H$ is the natural (G/H)-pairing. What happens if $n = 2$?

4-6-9. Suppose that $\theta : A \times B \longrightarrow C$ is a G-pairing; then for $a \in A^G$, $b \in B_S$ we have

$$\kappa a \smile_\theta \eta b = \eta(\theta(a, b))$$

4-6-10. Develop, from first principles, the theory of cup products

$$H^{p_1}(G, A_1) \times \cdots \times H^{p_n}(G, A_n) \longrightarrow H^{p_1 + \cdots + p_n}(G, A_1 \otimes \cdots \otimes A_n) \quad n \geqslant 2$$

and the consequent theory of θ-cup products for G-multilinear maps

$$\theta : A_1 \times \cdots \times A_n \longrightarrow C$$

4-6-11. Consider $\alpha \in H^{-1}(G, A)$, $\beta \in H^1(G, B)$ and the cup product $H^{-1}(G, A) \times H^1(G, B) \longrightarrow H^0(G, A \otimes B)$. If $a \in A_S$ is such that $\eta a = \alpha$ and $f \in \mathrm{Hom}_G (X_1, B)$ is a standard 1-cocycle representing β, then $c = - \sum_{\sigma \in G} (\sigma a \otimes f[\sigma]) \in (A \otimes B)^G$ and $\alpha \smile \beta = \kappa c$.

4-6-12. Suppose that A and B are G-modules and let $\Phi \in \mathfrak{C}^r(G, A)$ and $\Psi \in \mathfrak{C}^r(G, B)$ be homogeneous cochains of degrees $r \geqslant 1$ and $s \geqslant 1$, respectively (see (1-5-13)). Define $\Phi \cup \Psi \in \mathfrak{C}^{r+s}(G, A \otimes B)$ by

$$(\Phi \cup \Psi)(\sigma_0, \sigma_1, \ldots, \sigma_{r+s}) = \Phi(\sigma_0, \ldots, \sigma_r) \otimes \Psi(\sigma_r, \ldots, \sigma_{r+s})$$

This cup product is additive in each variable and satisfies the coboundary relation

$$\delta(\Phi \cup \Psi) = \delta\Phi \cup \Psi + (-1)^r \Phi \cup \delta\Psi$$

Thus, it determines, after the usual identifications (see (1-5-13)), a bilinear map of

$$H^r(G, A) \times H^s(G, B) \longrightarrow H^{r+s}(G, A \otimes B) \qquad r, s \geqslant 1$$

If $f \in \mathscr{C}^r(G, A)$ and $g \in \mathscr{C}^s(G, B)$ are the non-homogeneous (that is, standard) cochains corresponding to Φ and Ψ, respectively, then $f \cup g \in \mathscr{C}^{r+s}(G, A \otimes B)$ corresponding to $\Phi \cup \Psi$ satisfies

$$(f \cup g)(\sigma_1, \sigma_2, ..., \sigma_{r+s}) = f(\sigma_1, ..., \sigma_r) \otimes \sigma_1\sigma_2 \cdots \sigma_r\, g(\sigma_{r+1}, ..., \sigma_{r+s})$$

How is this cup product related to the one we have defined (in Section 4-1) via the $\varphi_{p\,q}$'s? Can it be extended to all dimensions r and s?

4-6-13. Let $f : A \longrightarrow A'$ be a G-homomorphism of G-modules, then for $\zeta \in H^p(G, \mathbf{Z})$ and $\alpha \in H^q(G, A)$

$$f_*(\zeta \cup_{\theta_A} \alpha) = \zeta \cup_{\theta_A} f_*(\alpha)$$

—in other words, the following diagram commutes:

$$\begin{array}{ccccc} H^p(G, \mathbf{Z}) & \times & H^q(G, A) & \xrightarrow{\cup_{\theta_A}} & H^{p+q}(G, A) \\ \downarrow 1 & & \downarrow f_* & & \downarrow f_* \\ H^p(G, \mathbf{Z}) & \times & H^q(G, A') & \xrightarrow{\cup_{\theta_{A'}}} & H^{p+q}(G, A') \end{array}$$

We also have $f_*(\alpha \cup_{\theta_A^\mathsf{T}} \zeta) = f_*(\alpha) \cup_{\theta_A^\mathsf{T}} \zeta$.

4-6-14. Suppose that G is a G-module and $\theta : A \times \mathbf{Z} \longrightarrow A$ is the natural G-pairing. In view of the identification

$$\hat{G} = \mathrm{Hom}\,(G, \mathbf{Q}/\mathbf{Z}) \approx H^1(G, \mathbf{Q}/\mathbf{Z}) \overset{\delta_*}{\rightarrowtail\!\!\!\rightarrow} H^2(G, \mathbf{Z})$$

we may define a pairing of

$$A^G \times \hat{G} \longrightarrow H^2(G, A)$$

by putting

$$\langle a, \chi\rangle = \kappa a \cup_\theta \delta_*\chi \qquad a \in A^G, \quad \chi \in \hat{G}$$

If H is a subgroup of G, then (with notation as in (3-5-4))

$$\operatorname{res}\langle a, \chi\rangle = \langle a, \operatorname{res}\chi\rangle \qquad a \in A^G, \quad \chi \in \hat{G}$$

and if H is normal in G, then

$$\inf\langle a, \chi'\rangle = \langle a, \inf\chi'\rangle \qquad a \in A^G, \quad \chi' \in \widehat{(G/H)}$$

What about cor?

Given $\chi \in \hat{G}$, let $\bar{\chi}$ be a representative of χ in $\mathbf{Q}$; by this we mean that $\bar{\chi} : G \longrightarrow \mathbf{Q}$ is a function with $\bar{\chi}(\sigma)$ (mod $\mathbf{Z}$) equal to $\chi(\sigma)$ for every $\sigma \in G$. Then $\delta_*\chi \in H^2(G, \mathbf{Z})$ is represented by the standard 2-cocycle $u = \delta\bar{\chi}$ for which

$$u[\sigma, \tau] = \bar{\chi}(\sigma) + \bar{\chi}(\tau) - \bar{\chi}(\sigma\tau)$$

Furthermore, $\kappa a \smile_\theta \delta_*\chi$ is represented by the standard 2-cocycle v for which

$$v[\sigma, \tau] = (u[\sigma, \tau])a$$

4-6-15. If the G-module A is a ring which satisfies the condition

$$\sigma(ab) = (\sigma a)(\sigma b) \qquad \sigma \in G, \quad a, b \in A$$

we say that A is a **G-ring.** In this situation, ring multiplication in A (that is, the map $\theta : A \times A \longrightarrow A$ given by $\theta(a, b) = ab$) is a G-pairing. The abelian group $H(G, A) = \sum_{-\infty}^{+\infty} \oplus H^r(G, A)$ becomes a ring when multiplication is defined in terms of cupping with respect to θ; it is called the **cohomology ring of A.**

If A is commutative, $H(G, A)$ is skew-commutative. This means, in particular, that for $\alpha \in H^p(G, A)\, \beta \in H^q(G, A)$

$$\alpha \smile_\theta \beta = (-1)^{pq} \beta \smile_\theta \alpha$$

If A has an identity 1, then $1 \in A^G$ and $\kappa(1) \in H^0(G, A)$ is an identity of $H(G, A)$.

Suppose further that M is a G-module that is also a (left) A-module and such that the action of G is compatible with the module operation of A—this means that

$$\sigma(am) = (\sigma a)(\sigma m) \qquad \sigma \in G, \quad a \in A, \quad m \in M$$

Thus, the map $\theta' : A \times M \longrightarrow M$ given by $\theta'(a, m) = am$ is a G-pairing. The abelian group $H(G, M) = \sum_{-\infty}^{+\infty} \oplus H^r(G, M)$ becomes an $H(G, A)$-module under the operation defined via cupping with respect to θ'. If A has an identity 1, this module is unitary.

An important example of a cohomology ring is $H(G, \mathbf{Z})$. This ring is known to be finitely generated—see [25]. Note that for any G-module M, $H(G, M)$ becomes a unitary $H(G, \mathbf{Z})$-module.

4-6-16. Suppose that G is a finite group of order $n > 1$. If $H^q(G, \mathbf{Z})$ is a cyclic group of order n then the integer q is said to be a **period** for G, and an element $\xi \in H^q(G, \mathbf{Z})$ which generates this cyclic group is called a **maximal generator**. Consider the cohomology ring with identity (see (4-6-15)) $H(G, \mathbf{Z}) = \sum_r \oplus H^r(G, \mathbf{Z})$ with multiplication (which we denote by $\smile$) arising from the natural G-pairing of $\mathbf{Z} \times \mathbf{Z} \longrightarrow \mathbf{Z}$.

(i) For $\xi \in H^q(G, \mathbf{Z})$ the following conditions are equivalent:

(a) ξ is a maximal generator.

(b) ξ has order n.

(c) There exists $\xi' \in H^{-q}(G, \mathbf{Z})$ such that $\xi' \smile \xi = 1$.

(d) The map $\alpha \longrightarrow \alpha \smile_\theta \xi$ is an isomorphism of $H^r(G, A)$ onto $H^{r+q}(G, A)$ for all $r \in \mathbf{Z}$ and all G-modules A, where $\theta : A \times \mathbf{Z} \longrightarrow A$ is the natural G-pairing.

(ii) If $\xi \in H^q(G, \mathbf{Z})$ is a maximal generator then ξ' is unique, so we denote it by ξ^{-1}, and $\xi^{-1} \in H^{-q}(G, \mathbf{Z})$ is a maximal generator. If $\zeta \in H^p(G, \mathbf{Z})$ is also a maximal generator, then so is $\xi \smile \zeta \in H^{p+q}(G, \mathbf{Z})$. The periods of G form a subgroup of $\mathbf{Z}$, and all the periods are even.

If q is a period of G then it is a period of every subgroup H of G; in fact, if $\xi \in H^q(G, \mathbf{Z})$ is a maximal generator, then so is $\mathrm{res}_{G \to H}\, \xi \in H^q(H, \mathbf{Z})$.

(iii) Suppose that H is a p-Sylow subgroup of G, and that $\xi \in H^q(H, \mathbf{Z})$ is a maximal generator. Let n be an integer such that $r^n \equiv 1 \pmod{\#(H)}$ for every integer r prime to p. Then $\xi^n \in H^{qn}(H, \mathbf{Z})$ is stable (see (3-7-13)) and $\mathrm{cor}_{H \to G}\, \xi^n$ is of order $\#(H)$.

(iv) G has a period $>0 \Leftrightarrow$ every Sylow group G_p has a period >0. Furthermore, a p-group has a period $>0 \Leftrightarrow$ it is cyclic or a generalized quaternion group.

(A generalized quaternion group is one with two generators σ, τ with relations

$$\sigma^t = \tau^2 = (\sigma\tau)^2$$

where t is a power of 2. Such a group then has order $4t$, and every element can be expressed uniquely in the form

$$\sigma^r\tau^s \qquad\qquad 0 \leqslant r < 2t \quad s = 0, 1$$

We know (see [83, p. 118] or [64, p. 252]) that a p-group contains exactly one subgroup of order $p \Leftrightarrow$ it is cyclic or a generalized quaternion group. By constructing a special periodic complex, one shows that a generalized quaternion group has periodic cohomology groups. On the other hand, if G is a p-group that has a period >0, then it has a cyclic subgroup H_1 of order p contained in the center; if H_2 is another cyclic subgroup of G of order p, then $H_1 + H_2$ is a group of order p^2 which has no period >0 since $H^q(H_1 + H_2, \mathbf{Z})$ is not cyclic of order p^2—so G has only one cyclic subgroup of order p.)

4-6-17. Let B be a multiplicative group, finite or infinite, and consider a free resolution of $\mathbf{Z}$ over

$$\Gamma = \mathbf{Z}[G] = \left\{ \sum_{\sigma \in G} n_\sigma \sigma \mid \text{almost all } n_\sigma = 0 \right\}$$

By this is meant an exact G-sequence

$$\cdots \longrightarrow X_n \xrightarrow{\partial_n} X_{n-1} \longrightarrow \cdots \xrightarrow{\partial_1} X_0 \xrightarrow{\varepsilon} \mathbf{Z} \longrightarrow 0$$

where each X_n is G-free. (Actually, it suffices to deal with a projective resolution—that is, one in which each X_n is G-projective.) Thus, a free resolution is simply the positive part of a G-complex (see Section 1-3) in which the requirement that each X_n have a finite G-basis is dropped.

Let A be a G-module and consider the differential graded groups

$$0 \longrightarrow \mathrm{Hom}_G(X_0, A) \xrightarrow{\delta_1} \cdots \longrightarrow \mathrm{Hom}_G(X_{n-1}, A) \xrightarrow{\delta_n} \mathrm{Hom}_G(X_n, A) \longrightarrow \cdots \qquad (*)$$

$$\cdots \longrightarrow X_n \otimes_\Gamma A \xrightarrow{\partial_n \otimes 1} X_{n-1} \otimes_\Gamma A \longrightarrow \cdots \longrightarrow X_0 \otimes_\Gamma A \longrightarrow 0 \qquad (**)$$

where $\delta_n = (\partial_n, 1)$. The derived groups of $(*)$,

$$\mathscr{H}^n(G, A) = \ker \delta_{n+1}/\mathrm{im}\, \delta_n$$

are called the cohomology groups of G in A. The derived groups of $(**)$, $\mathscr{H}_n(G, A) = \ker(\partial_n \otimes 1)/\mathrm{im}(\partial_{n+1} \otimes 1)$, are called the homology groups of G in A. Both the homology and cohomology groups are independent of the choice of resolution (the proof makes use of homotopies). Moreover, $\mathscr{H}^0(G, A) \approx A^G$, and $\mathscr{H}_0(G, A) = (X_0 \otimes_\Gamma A)/\mathrm{im}(\partial_1 \otimes 1) \approx A/(IA) = A_G$. Given the short exact G-sequence $0 \longrightarrow A \longrightarrow B \longrightarrow C \longrightarrow 0$ there arise the exact homology and cohomology sequences

$$\cdots \longrightarrow \mathscr{H}^n(G, B) \longrightarrow \mathscr{H}^n(G, C) \xrightarrow{\delta_*} \mathscr{H}^{n+1}(G, A) \longrightarrow \mathscr{H}^{n+1}(G, B) \longrightarrow \cdots$$

$$n \geqslant 0$$

$$\cdots \longrightarrow \mathscr{H}_{n+1}(G, B) \longrightarrow \mathscr{H}_{n+1}(G, C) \xrightarrow{d_*} \mathscr{H}_n(G, A) \longrightarrow \mathscr{H}_n(G, B) \longrightarrow \cdots$$

If G is finite these sequences may be joined to give an exact sequence which is infinite in both directions. This may be done as follows. We have the homomorphism

$$S^* : \mathscr{H}_0(G, A) \longrightarrow \mathscr{H}^0(G, A)$$

induced by the trace (see (4-6-3)). Put

$$\hat{\mathscr{H}}_0(G, A) = \ker S^* \qquad \hat{\mathscr{H}}^0(G, A) = \mathrm{coker}\, S^*$$

Thus,

$$\hat{\mathscr{H}}_0(G, A) \approx \frac{A_S}{IA} \qquad \text{and} \qquad \hat{\mathscr{H}}^0(G, A) \approx \frac{A^G}{SA}$$

Now, we have a commutative diagram with exact rows

$$\begin{array}{ccccccccc} \mathscr{H}_1(G, C) & \longrightarrow & \mathscr{H}_0(G, A) & \longrightarrow & \mathscr{H}_0(G, B) & \longrightarrow & \mathscr{H}_0(G, C) & \longrightarrow & 0 \\ \downarrow & & \downarrow S_A^* & & \downarrow S_B^* & & \downarrow S_C^* & & \downarrow \\ 0 & \longrightarrow & \mathscr{H}^0(G, A) & \longrightarrow & \mathscr{H}^0(G, B) & \longrightarrow & \mathscr{H}^0(G, C) & \longrightarrow & \mathscr{H}^1(G, A) \end{array}$$

From this there arises a homomorphism

$$\delta : \ker S_C^* \longrightarrow \operatorname{coker} S_A^*$$

and we have the exact sequence

$$\cdots \longrightarrow \mathscr{H}_1(G, C) \longrightarrow \hat{\mathscr{H}}_0(G, A) \longrightarrow \hat{\mathscr{H}}_0(G, B) \longrightarrow \hat{\mathscr{H}}_0(G, C) \xrightarrow{\delta} \hat{\mathscr{H}}^0(G, A) \longrightarrow \hat{\mathscr{H}}^0(G, B) \longrightarrow \hat{\mathscr{H}}^0(G, C) \longrightarrow \mathscr{H}^1(G, A) \longrightarrow \cdots$$

Now put

$$\begin{aligned} H^n(G, A) &= \mathscr{H}^n(G, A) && n \geqslant 1 \\ H^0(G, A) &= \hat{\mathscr{H}}^0(G, A) \approx A^G/SA \\ H^{-1}(G, A) &= \hat{\mathscr{H}}_0(G, A) \approx A_S/IA \\ H^{-n}(G, A) &= \mathscr{H}_{n-1}(G, A) && n \geqslant 2 \end{aligned}$$

These are our customary cohomology groups of G in A. In fact, this is clear in dimensions $\geqslant -1$. As for the remaining dimensions, using the resolution given by the positive part of the standard complex, we recall (see (4-6-4)) that $X_n \otimes_\Gamma A$ is isomorphic to $\operatorname{Hom}_G(\hat{X}_n, A)$. Moreover, under this isomorphism the boundary and coboundary correspond—that is, the following diagram commutes:

$$\begin{array}{ccc} X_n \otimes_\Gamma G & \longleftrightarrow & \operatorname{Hom}_G(\hat{X}_n, A) \\ \partial\otimes 1 \downarrow & & \downarrow \delta=(\partial,1) \\ X_{n-1} \otimes_\Gamma A & \longleftrightarrow & \operatorname{Hom}_G(\hat{X}_{n-1}, A) \end{array}$$

It follows that for $n \geqslant 2$, $H^{-n}(G, A)$ is our usual cohomology group.

Historically, the procedure we have outlined for defining the cohomology groups in all dimensions via cohomology and homology and then connecting these by the Tate linking (see (3-7-7), (2-2-8), and (2-2-9)) comes first. The use of a full G-complex is a matter of technical convenience. Note that since $\hat{\hat{X}}_n \approx X_n$ and $\hat{X}_n \otimes_\Gamma A \approx \mathrm{Hom}_G(X_n, A)$, we could get all the cohomology groups by tensoring the full G-complex rather than taking Hom.

We leave it to the reader to find the explicit formulas (using the standard complex) for the various cohomological maps when the cohomology groups are given in terms of the tensor product.

CHAPTER

V

Group Extensions

In this chapter we study some of the basic properties of group extensions and their connection with cohomology. The main objective is the group theoretical version of the principal ideal theorem.

5-1. THE EXTENSION PROBLEM AND H^2

Consider the exact sequence of multiplicative groups

$$1 \longrightarrow A \xrightarrow{\ i\ } U \xrightarrow{\ j\ } G \longrightarrow 1 \tag{5.1}$$

where G may be finite or infinite. There is no loss of generality in assuming that i is the inclusion map; thus, A is a normal subgroup of U and $U/A \approx G$.

For each $\sigma \in G$, let us choose a representative $u_\sigma \in U$—so that $j(u_\sigma) = \sigma$. We shall refer to $\{u_\sigma \mid \sigma \in G\}$ as a **complete system of representatives** for G in U, or more briefly, as a **section** of G in U. (Of course, it is really the mapping $\sigma \longrightarrow u_\sigma$ of $G \longrightarrow U$ which should be called a section.) Since $j(u_\sigma u_\tau) = \sigma\tau = j(u_{\sigma\tau})$ for all $\sigma, \tau \in G$, there exist elements $a_{\sigma,\tau} \in A$ such that

$$u_\sigma u_\tau = a_{\sigma,\tau} u_{\sigma\tau} \qquad \forall \sigma, \tau \in G \tag{5.2}$$

Clearly, $\{a_{\sigma,\tau} \mid \sigma, \tau \in G\}$ may be viewed as a standard 2-cochain of G in A.

For purposes of simplicity and, even more, because it suffices for our needs, we shall always assume that **A is abelian.** (A good deal of the theory can be carried over when A is not abelian—see, for example, [31, Chapter 15] or [45, Chapter 12]). Now, U acts on A by inner automorphisms and, because A is abelian, the elements of A act trivially on A—so there is induced a "natural" action of G on A. In more detail, for $a \in A$, $\sigma \in G$ define an exponential action of σ on a by

$$a^\sigma = u_\sigma a u_\sigma^{-1} \tag{5.3}$$

Then A is indeed a G-module; in fact,

$$(ab)^\sigma = u_\sigma ab u_\sigma^{-1} = u_\sigma a u_\sigma^{-1} u_\sigma b u_\sigma^{-1} = a^\sigma b^\sigma,$$

$a^1 = u_1 a u_1^{-1} = a$ since $u_1 \in A$, and

$$(a^\tau)^\sigma = u_\sigma(u_\tau a u_\tau^{-1})u_\sigma^{-1} = a_{\sigma,\tau}(u_{\sigma\tau} a u_{\sigma\tau}^{-1})a_{\sigma,\tau}^{-1} = u_{\sigma\tau} a u_{\sigma\tau}^{-1} = a^{\sigma\tau}$$

Moreover, this action of G on A is independent of the choice of representatives $\{u_\sigma\}$; in fact, if $\{v_\sigma \mid \sigma \in G\}$ is another section of G in U, then for each $\sigma \in G$ there exists $c_\sigma \in A$ such that $v_\sigma = c_\sigma u_\sigma$, and then $v_\sigma a v_\sigma^{-1} = c_\sigma u_\sigma a u_\sigma^{-1} c_\sigma^{-1} = u_\sigma a u_\sigma^{-1}$.

Once the section $\{u_\sigma \mid \sigma \in G\}$ is fixed, every element $u \in U$ has a unique expression of form

$$u = a u_\sigma \qquad a \in A, \quad \sigma \in G$$

which is given by $\sigma = j(u)$ and $a = u u_\sigma^{-1}$. Because Eq. (5.3) can be written as

$$a^\sigma u_\sigma = u_\sigma a \qquad a \in A, \quad \sigma \in G \tag{5.4}$$

it follows that multiplication in U can be described in terms of the multiplications in A and in G, the action of G on A, and the 2-cochain $\{a_{\sigma,\tau}\}$. Thus, if $u, v \in U$ are of form $u = a u_\sigma$, $v = b u_\tau$, then

$$uv = a b^\sigma u_\sigma u_\tau = a b^\sigma a_{\sigma,\tau} u_{\sigma\tau} \tag{5.5}$$

Furthermore, associativity in U leads to the equality (for all $\sigma, \tau, \rho \in G$) of

$$(u_\sigma u_\tau)u_\rho = a_{\sigma,\tau}u_{\sigma\tau}u_\rho = a_{\sigma,\tau}a_{\sigma\tau,\rho}u_{\sigma\tau\rho}$$

and

$$u_\sigma(u_\tau u_\rho) = u_\sigma a_{\tau,\rho}u_{\tau\rho} = a^\sigma_{\tau,\rho}u_\sigma u_{\tau\rho} = a^\sigma_{\tau,\rho}a_{\sigma,\tau\rho}u_{\sigma\tau\rho}$$

Therefore, for all $\sigma, \tau, \rho \in G$,

$$a_{\sigma,\tau}a_{\sigma\tau,\rho} = a^\sigma_{\tau,\rho}a_{\sigma,\tau\rho} \qquad \text{or} \qquad a^\sigma_{\tau,\rho}a^{-1}_{\sigma\tau,\rho}a_{\sigma,\tau\rho}a^{-1}_{\sigma,\tau} = 1 \tag{5.6}$$

This formula is precisely (see (1-5-12)) the multiplicative form of the coboundary formula for standard cochains of G in A in dimension 2; consequently, the 2-cochain $\{a_{\sigma,\tau}\}$ is a standard 2-cocycle of G in A (or, in classical terminology, $\{a_{\sigma,\tau}\}$ is a **factor set**), and we may express this by $(\delta a)_{\sigma,\tau,\rho} = 1$. Of course, if $\{v_\sigma = c_\sigma u_\sigma\}$ is any other section and we write $v_\sigma v_\tau = b_{\sigma,\tau}v_{\sigma\tau}$, then $\{b_{\sigma,\tau}\}$ is also a standard 2-cocycle of G in A. From

$$v_\sigma v_\tau = (c_\sigma u_\sigma)(c_\tau u_\tau) = c_\sigma c^\sigma_\tau c^{-1}_{\sigma\tau}\, a_{\sigma,\tau}v_{\sigma\tau}$$

it follows that $b_{\sigma,\tau}a^{-1}_{\sigma,\tau} = c_\sigma c^\sigma_\tau c^{-1}_{\sigma\tau}$, and the right side is the coboundary $(\delta c)_{\sigma,\tau} = c_\sigma c^\sigma_\tau c^{-1}_{\sigma\tau}$ of the standard 1-cochain $\{c_\sigma\}$ of G in A. This means that the cocycles $\{a_{\sigma,\tau}\}$ and $\{b_{\sigma,\tau}\}$ belong to the same cohomology class in $H^2(G, A)$. It should be noted that the action of G on A used for cohomology is derived from the sequence (5.1) and is expressed by the formula (5.3).

Suppose now that G is a multiplicative group, finite or infinite, and that A (also written multiplicatively) is a G-module with the action of G given exponentially. By a **solution of the extension problem for the pair (G, A)** we shall mean an exact sequence of the form (5.1) (which shall also be denoted simply as a triple $\{U, i, j\}$) such that the action of G on A determined by (5.1) and expressed by (5.3) coincides with the given action of G on A. We shall then say that $\{U, i, j\}$ is an **extension** of A by G. Thus, the upshot of the foregoing discussion is that if $\{U, i, j\}$ is an extension of A by G, then it determines an element of $H^2(G, A)$.

Our next objective is to show that, conversely, an element $\alpha \in H^2(G, A)$ determines a solution $\{U, i, j\}$ of the extension problem, and such that the associated cohomology class in $H^2(G, A)$

is precisely α. For this, let $\{a_{\sigma,\tau}\}$ be a standard 2-cocycle of G in A which belongs to the class α. In particular, Eq. (5.6) holds. By substituting $\sigma = \tau = 1$ or $\tau = \rho = 1$ or $\tau = \sigma^{-1} = \rho^{-1}$, respectively, one arrives at the formulas (for all $\sigma \in G$)

$$\begin{aligned} a_{1,\sigma} &= a_{1,1} \\ a_{\sigma,1} &= a_{1,1}^{\sigma} \\ a_{\sigma,\sigma^{-1}}a_{1,1} &= a_{\sigma^{-1},\sigma}^{\sigma}a_{1,1}^{\sigma} \end{aligned} \tag{5.7}$$

Now, for each $\sigma \in G$, choose a formal symbol u_σ and let

$$U = \{(a, u_\sigma) \mid a \in A, \ \sigma \in G\} \tag{5.8}$$

Define multiplication in U according to the rule

$$(a, u_\sigma)(b, u_\tau) = (ab^\sigma a_{\sigma,\tau}, u_{\sigma\tau}) \qquad a, b \in A \quad \sigma, \tau \in G \tag{5.9}$$

From the coboundary relation $(\delta a)_{\sigma,\tau,\rho} = 1$ and commutativity of A, it follows that multiplication in U is associative; in fact,

$$((a, u_\sigma)(b, u_\tau))(c, u_\rho) = (ab^\sigma a_{\sigma,\tau}, u_{\sigma\tau})(c, u_\rho) = (ab^\sigma a_{\sigma,\tau}c^{\sigma\tau}a_{\sigma\tau,\rho}, u_{\sigma\tau\rho})$$

and this is equal to

$$(a, u_\sigma)((b, u_\tau)(c, u_\rho)) = (a, u_\sigma)(bc^\tau a_{\tau,\rho}u_{\tau\rho}) = (a(bc^\tau a_{\tau,\rho})^\sigma a_{\sigma,\tau\rho}, u_{\sigma\tau\rho})$$

To exhibit a left identity in U, one seeks an element (x, u_ρ) such that $(x, u_\rho)(a, u_\sigma) = (a, u_\sigma)$ for all $a \in A$, $\sigma \in G$—that is, $(xa^\rho a_{\rho,\sigma}, u_{\rho\sigma}) = (a, u_\sigma)$. Consequently, $\rho = 1$, and, making use of the first part of (5.7), it is clear that $(a_{1,1}^{-1}, u_1)$ is a left identity. Finally, given $(a, u_\sigma) \in U$, we see that

$$(a_{1,1}^{-1}a^\sigma a_{\sigma^{-1},\sigma}^{-1}, u_{\sigma^{-1}})(a, u_\sigma) = (a_{1,1}^{-1}, u_\sigma)$$

so that $(a_{1,1}^{-1}a^\sigma a_{\sigma^{-1},\sigma}^{-1}, u_{\sigma^{-1}})$ is a left inverse for (a, u_σ). This shows that U is a group. The maps $i : A \longrightarrow U$ and $j : U \longrightarrow G$ are now defined by

$$i(a) = (aa_{1,1}^{-1}, u_1) \qquad j(a, u_\sigma) = \sigma \tag{5.10}$$

It is easy to check that i is a monomorphism, that j is an epimorphism, and that $\operatorname{im} i = \{(a, u_1) | a \in A\} = \ker j$. This means that

$$1 \longrightarrow A \xrightarrow{i} U \xrightarrow{j} G \longrightarrow 1$$

is an exact sequence.

We must also verify that the 2-cocycle of G in A determined by this exact sequence coincides with the one from which the construction started. Consider then the section $\{(1, u_\sigma) | \sigma \in G\}$ of G in U; we must check that

$$(1, u_\sigma)(1, u_\tau) = i(a_{\sigma,\tau})(1, u_{\sigma\tau})$$

But this is straightforward, both sides being equal to $(a_{\sigma,\tau}, u_{\sigma\tau})$. Finally, it remains to verify that the action of G on A determined by the exact sequence coincides with the original action—that is,

$$i(a^\sigma) = (1, u_\sigma)(aa_{1,1}^{-1}, u_1)(1, u_\sigma)^{-1}$$

Now, the left side is $(a^\sigma a_{1,1}^{-1}, u_1)$ while the right side is

$$((aa_{1,1}^{-1})^\sigma a_{\sigma,1}, u_\sigma)(a_{1,1}^{-1} 1^\sigma a_{\sigma^{-1},\sigma}^{-1}, u_{\sigma^{-1}})$$

$$= (a^\sigma (a_{1,1}^\sigma)^{-1} a_{\sigma,1} (a_{1,1}^\sigma)^{-1} (a_{\sigma^{-1},\sigma}^\sigma)^{-1} a_{\sigma,\sigma^{-1}}, u_1)$$

$$= (a^\sigma a_{1,1}^{-1}, u_1) \qquad \text{(in virtue of (5.7))}$$

The foregoing discussion shows (among other things) that a 2-cocycle $\{a_{\sigma,\tau}\}$ belonging to the cohomology class $\alpha \in H^2(G, A)$ leads to an extension $\{U, i, j\}$ of A by G. Since any other cocycle $\{b_{\sigma,\tau}\}$ belonging to α also leads to a solution of the extension problem, it is desirable to compare solutions of the same extension problem, and we are led to the following definitions.

Suppose that A is a G-module with $\{U, i, j\}$ an extension of A by G and that A' is a G'-module with $\{U', i', j'\}$ an extension of A' by G'. By a **homomorphism** (or **translation**) of $\{U, i, j\}$ into $\{U', i', j'\}$ we shall mean a triple (f, φ, λ) of homomorphisms

$f : A \longrightarrow A'$, $\lambda : G \longrightarrow G'$, $\varphi : U \longrightarrow U'$ such that the following diagram commutes:

$$\begin{array}{ccccccccc} 1 & \longrightarrow & A & \xrightarrow{i} & U & \xrightarrow{j} & G & \longrightarrow & 1 \\ & & {\scriptstyle f}\downarrow & & {\scriptstyle \varphi}\downarrow & & {\scriptstyle \lambda}\downarrow & & \\ 1 & \longrightarrow & A' & \xrightarrow{i'} & U' & \xrightarrow{j'} & G' & \longrightarrow & 1 \end{array} \tag{5.11}$$

Of course, since i and i' are inclusions, f is the restriction of φ to A. When f, φ, λ are all isomorphisms onto, we shall say that the translation (f, φ, λ) is an **isomorphism**. In the special case where $A = A'$, $G = G'$, $f = 1$, $\lambda = 1$, we say that the extensions $\{U, i, j\}$ and $\{U', i', j'\}$ of A by G are **equivalent** if there exists a homomorphism $(f, \varphi, \lambda) = (1, \varphi, 1) : \{U, i, j\} \longrightarrow \{U', i', j'\}$. This definition can be restated as the requirement that there exist a homomorphism $\varphi : U \longrightarrow U'$ such that

$$\begin{array}{ccccccc} & & & U & & & \\ & & {\scriptstyle i}\nearrow & & \searrow{\scriptstyle j} & & \\ 1 \longrightarrow & A & & {\scriptstyle \varphi}\downarrow & & G & \longrightarrow 1 \\ & & {\scriptstyle i'}\searrow & & \nearrow{\scriptstyle j'} & & \\ & & & U' & & & \end{array} \tag{5.12}$$

is a commutative diagram. It should be noted that, as is easily checked, such a φ is automatically an isomorphism onto, so we are dealing with an equivalence relation and φ may be referred to as an **equivalence** of $\{U, i, j\}$ and $\{U', i', j'\}$. If $\{U', i', j'\} = \{U, i, j\}$, we say that φ is an **automorphism** of $\{U, i, j\}$.

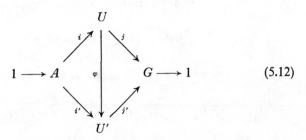

It is easy to see that equivalent extensions of A by G determine the same cohomology class $\alpha \in H^2(G, A)$; in fact, matters may be arranged so that they determine the same cocycle. To see this, suppose that $\{U, i, j\}$ and $\{U', i', j'\}$ are equivalent extensions with equivalence $\varphi : U \longrightarrow U'$, and let $\{u_\sigma\}$ be a section of G in U. Then $\{a_{\sigma,\tau}\}$, which is defined via $u_\sigma u_\tau = a_{\sigma,\tau} u_{\sigma\tau}$, is a 2-cocycle belonging to α. If we put $u'_\sigma = \varphi(u_\sigma)$, then $j'(u'_\sigma) = \sigma$. Therefore, $\{u'_\sigma \mid \sigma \in G\}$ is a section of G in U' and $u'_\sigma u'_\tau = a_{\sigma,\tau} u'_{\sigma\tau}$—which says that $\{a_{\sigma,\tau}\}$ is a cocycle associated with $\{U', i', j'\}$. It is also true that cocycles belonging to the same class $\alpha \in H^2(G, A)$ determine equivalent extensions of A by G; this will be a consequence of the more general discussion which follows.

Suppose that $\{U, i, j\}$ is an extension of A by G, that $\{U', i', j'\}$ is an extension of A' by G', and that $f: A \longrightarrow A'$, $\lambda: G \longrightarrow G'$ are homomorphisms. We would like to decide if there exists a homomorphism $\varphi: U \longrightarrow U'$ such that (f, φ, λ) is a homomorphism of group extensions (see the diagram (5.11)), and if so to count them. To study this question, choose actions $\{u_\sigma \mid \sigma \in G\} \subset U$ and $\{u'_{\sigma'} \mid \sigma' \in G'\} \subset U'$, and let $\{a_{\sigma,\tau}\}$ and $\{a'_{\sigma',\tau'}\}$ be the corresponding 2-cocycles. Suppose that we have such a φ. Then for any $u = au_\sigma \in U$ we have

$$\varphi(u) = f(a)\,\varphi(u_\sigma) \tag{5.13}$$

so that φ is determined completely as soon as the values $\varphi(u_\sigma) \in U'$ are prescribed. Because $j'(\varphi u_\sigma) = \lambda(ju_\sigma) = \lambda(\sigma) = j'(u'_{\lambda(\sigma)})$, there exists for each $\sigma \in G$ an element $c'_\sigma \in A'$ such that

$$\varphi(u_\sigma) = c'_\sigma u'_{\lambda(\sigma)} \tag{5.14}$$

In the usual way (see (2-1-1)), we may define $(a')^\sigma = (a')^{\lambda(\sigma)}$ and thus view A' as a G-module. When this is done, $\{c'_\sigma \mid \sigma \in G\}$ is a standard 1-cochain of G in A' which, in virtue of (5.13) and (5.14), serves to describe φ.

Let us observe further that $f: A \longrightarrow A'$ is a G-homomorphism —that is:

$$f(a^\sigma) = f(a)^\sigma \qquad a \in A, \quad \sigma \in G \tag{5.15}$$

In more detail, we have

$$\begin{aligned} f(a^\sigma) = \varphi(a^\sigma) = \varphi(u_\sigma a u_\sigma^{-1}) = \varphi(u_\sigma) f(a)\,\varphi(u_\sigma)^{-1} &= c'_\sigma u'_{\lambda(\sigma)} f(a)\, u'^{-1}_{\lambda(\sigma)} c'^{-1}_\sigma \\ &= u'_{\lambda(\sigma)} f(a)\, u'^{-1}_{\lambda(\sigma)} = (f(a))^{\lambda(\sigma)} = (f(a))^\sigma \end{aligned}$$

This implies, in particular, that $(1, f): (G, A) \longrightarrow (G, A')$ is a homomorphism of pairs (see Section 2-1). Of course,

$$(\lambda, 1): (G', A') \longrightarrow (G\ A')$$

is also a homomorphism of pairs. We assert that if $\{a_{\sigma,\tau}\}$ belongs to $\alpha \in H^2(G, A)$ and $\{a'_{\sigma',\tau'}\}$ belongs to $\alpha' \in H^2(G', A')$, then

$$f_*(\alpha) = (1, f)_*\alpha = (\lambda, 1)_* \quad \alpha' \in H^2(G, A') \tag{5.16}$$

To see this, it suffices (see the beginning of Section 2-5) to show that the 2-cocycles $\{f(a_{\sigma,\tau})\}$ and $\{a'_{\lambda(\sigma),\lambda(\tau)}\}$ of G in A' differ by the coboundary of the 1-cochain $\{c'_\sigma\}$. More precisely, we have

$$f(a_{\sigma,\tau}) = a'_{\lambda(\sigma),\lambda(\tau)}(c'^{\sigma}_{\tau} c'^{-1}_{\sigma\tau} c'_\sigma) \qquad \forall \sigma, \tau \in G \tag{5.17}$$

because $\varphi(u_\sigma u_\tau) = \varphi(a_{\sigma,\tau} u_{\sigma\tau}) = f(a_{\sigma,\tau})\, c'_{\tau\sigma} u'_{\lambda(\sigma\tau)}$ while

$$\varphi(u_\sigma u_\tau) = \varphi(u_\sigma)\,\varphi(u_\tau) = c'_\sigma u'_{\lambda(\sigma)} c'_\tau u'_{\lambda(\tau)} = c'_\sigma (c'_\tau)^{\lambda(\sigma)} a'_{\lambda(\sigma),\lambda(\tau)} u'_{\lambda(\sigma\tau)}$$

This proves half of the following result.

5-1-1. Proposition. Let $\{U, i, j\}$ be an extension of A by G which determines $\alpha \in H^2(G, A)$, and let $\{U', i', j'\}$ be an extension of A' by G' which determines $\alpha' \in H^2(G', A')$. Suppose that $f : A \longrightarrow A'$ and $\lambda : G \longrightarrow G'$ are homomorphisms and that A' is also viewed as a G-module by putting $(a')^\sigma = (a')^{\lambda(\sigma)}$. Then there exists a homomorphism $\varphi : U \longrightarrow U'$ such that

$$\begin{array}{ccccccccc} 1 & \longrightarrow & A & \xrightarrow{i} & U & \xrightarrow{j} & G & \longrightarrow & 1 \\ & & {\scriptstyle f}\downarrow & & {\scriptstyle \varphi}\downarrow & & {\scriptstyle \lambda}\downarrow & & \\ 1 & \longrightarrow & A' & \xrightarrow{i'} & U' & \xrightarrow{j'} & G' & \longrightarrow & 1 \end{array}$$

is a commutative diagram (that is, there exists a translation $(f, \varphi, \lambda) : \{U, i, j\} \longrightarrow \{U', i', j'\}$) if and only if

(i) f is a G-homomorphism.

(ii) $f_*(\alpha) = (1, f)_*\, \alpha = (\lambda, 1)_*\, \alpha'$.

Proof: Suppose that (i) and (ii) hold, and choose sections $\{u_\sigma\} \subset U$ and $\{u'_{\sigma'}\} \subset U'$. If we write

$$u_\sigma u_\tau = a_{\sigma,\tau} u_{\sigma\tau}\,, \qquad u'_{\sigma'} u'_{\tau'} = a'_{\sigma',\tau'} u'_{\sigma'\tau'}$$

then the cocycles $\{a_{\sigma,\tau}\}$ and $\{a'_{\sigma',\tau'}\}$ belong to α and α', respectively. Condition (i) says that Eq. (5.15) holds, while (ii) implies that there exists a 1-cochain $\{c'_\sigma\}$ of G in A' such that (5.17) holds. Define $\varphi : U \longrightarrow U'$ by

$$\varphi(au_\sigma) = f(a)c'_\sigma u'_{\lambda(\sigma)} \tag{5.18}$$

Then

$$\varphi(au_\sigma bu_\tau) = f(a)\,f(b)^\sigma f(a_{\sigma,\tau})\,c'_{\sigma\tau}u'_{\lambda(\sigma\tau)}$$

and

$$\varphi(au_\sigma)\,\varphi(bu_\tau) = f(a)\,f(b)^{\lambda(\sigma)}c'_\sigma c_\tau'^{\lambda(\sigma)}a'_{\lambda(\sigma),\lambda(\tau)}u'_{\lambda(\sigma\tau)}$$

so that φ is a homomorphism. Because $u_1 \in A$, we have $1 = \varphi(1) = \varphi(u_1^{-1}u_1) = f(u_1)^{-1}\,c_1u'_1$, and therefore $f(u_1) = c'_1u'_1$. Then $i' \circ f = \varphi \circ i$, since for $a \in A$,

$$\varphi(a) = \varphi(au_1^{-1}u_1) = f(au_1^{-1})c'_1u'_1 = f(a)$$

Finally, $\lambda \circ j = j' \circ \varphi$ since $(\lambda \circ j)(au_\sigma) = \lambda(\sigma)$ and

$$(j' \circ \varphi)(au_\sigma) = j'(f(a)c'_\sigma u'_{\lambda(\sigma)}) = \lambda(\sigma)$$

This completes the proof. ∎

5-1-2. Theorem. Suppose that the abelian group A is a G-module (G is finite or infinite). Then there is a natural 1–1 correspondence between the equivalence classes of extensions of A by G and the elements of $H^2(G, A)$.

Proof: It remains only to show that if $\{a_{\sigma,\tau}\}$ and $\{b_{\sigma,\tau}\}$ are cocycles belonging to $\alpha \in H^2(G, A)$ and $\{U, i, j\}$, $\{U', i', j'\}$ are the extensions they determine according to our construction, then these extensions are equivalent. To do this, simply apply (5-1-1) with $A' = A$, $G' = G$, $f = 1$, $\lambda = 1$, $\alpha' = \alpha$. Since conditions (i) and (ii) are trivially satisfied, the existence of $\varphi : U \longrightarrow U'$ is assured, and the theorem is proved. ∎

5-1-3. Exercise. In virtue of (5-1-2), the equivalence classes of extensions of A by G can be made into an abelian group by carrying over the multiplication from $H^2(G, A)$. This multiplication of extensions was first defined in a direct fashion in [8]; it is consequently known as **Baer multiplication.** The details may be organized as follows:

Suppose that $1 \longrightarrow A \xrightarrow{i} U \xrightarrow{j} G \longrightarrow 1$ is a group extension corresponding to $\alpha \in H^2(G, A)$, so that for a section $\{u_\sigma\}$ the cocycle $\{a_{\sigma,\tau}\}$ given by $u_\sigma u_\tau = a_{\sigma,\tau} u_{\sigma\tau}$ belongs to α. Suppose, further, that $1 \longrightarrow A \xrightarrow{i'} U' \xrightarrow{j'} G \longrightarrow 1$ is a group extension corresponding to $\alpha' \in H^2(G, A)$, so that the cocycle $\{a'_{\sigma,\tau}\}$ determined by the section $\{u'_\sigma\}$ belongs to α'. Form the direct sum

$$U \oplus U' = \{(u, u') \mid u \in U, u' \in U'\}$$

and consider the subgroup $\langle U, U'\rangle = \{(u, u') \mid j(u) = j'(u')\}$. This leads immediately to an exact sequence

$$1 \longrightarrow A \oplus A \longrightarrow \langle U, U'\rangle \longrightarrow G \longrightarrow 1$$

Now consider the subgroup $D = \{(a, a^{-1}) \mid a \in A\}$ of $A \oplus A$. When $A \oplus A$ is viewed as a subgroup of $\langle U, U'\rangle$, D becomes a normal subgroup of $\langle U, U'\rangle$. If we put $U'' = \langle U, U'\rangle / D$, then A is isomorphic with $(A \oplus A)/D$, and with suitable definitions of i'' and j''

$$1 \longrightarrow A \xrightarrow{i''} U'' \xrightarrow{j''} G \longrightarrow 1$$

is a group extension of A by G. Moreover, the cocycle $\{a''_{\sigma,\tau}\}$ associated with this extension (for the section $\{u''_\sigma\} = \{(u_\sigma, u'_\sigma)D\}$) is precisely $\{a_{\sigma,\tau} a'_{\sigma,\tau}\}$. This enables us to define multiplication in the set of equivalence classes of extensions of A by G, and it becomes a group isomorphic to $H^2(G, A)$. ∎

Now, let us place ourselves once again in the situation of (5-1-1) (where conditions (i) and (ii) hold, and the same notation applies) and investigate the number of φ's that exist (for fixed f and λ). Let $\mathscr{S}$ denote the set of all homomorphisms $\varphi : U \longrightarrow U'$ such that $(f, \varphi, \lambda) : \{U, i, j\} \longrightarrow \{U', i', j'\}$ is a translation. According to (5-1-1), $\mathscr{S} \neq \phi$, and we may **fix a $\varphi \in \mathscr{S}$.** Then, by (5.14), there is associated with φ a 1-cochain $c' = \{c'_\sigma\} \in \mathscr{C}^1(G, A')$ given by $c'_\sigma = \varphi(u_\sigma)\, u'^{-1}_{\lambda(\sigma)}$. For any $\psi \in \mathscr{S}$, let us denote its associated 1-cochain by $b' = \{b'_\sigma\} \in \mathscr{C}^1(G, A')$, where $b'_\sigma = \psi(u_\sigma)\, u'^{-1}_{\lambda(\sigma)}$. Since

Eq. (5.17) applies for both φ and ψ, it follows that c' and b' have the same coboundary (namely, $f(a_{\sigma,\tau})\, a'^{-1}_{\lambda(\sigma),\lambda(\tau)}$). Consequently, their quotient $d' = b'c'^{-1}$, where

$$d'_\sigma = b'_\sigma c'^{-1}_\sigma = \psi(u_\sigma)\, \varphi(u_\sigma)^{-1} \tag{5.19}$$

is a cocycle—that is, $d' = \{d'_\sigma\} \in \mathscr{Z}^1(G, A')$. Furthermore, d' is independent of the choice of section of G in U; in fact, if $\{v_\sigma = a_\sigma u_\sigma\}$, $a_\sigma \in A$, is any other section then the cocycle determined by this procedure is

$$\psi(v_\sigma)\, \varphi(v_\sigma)^{-1} = f(a)\, \psi(u_\sigma)\, \varphi(u_\sigma^{-1})\, f(a)^{-1} = f(a)\, d_\sigma f(a)^{-1} = d'_\sigma$$

With each $\psi \in \mathscr{S}$ we have associated a $d' \in \mathscr{Z}^1(G, A')$. Conversely, given $d' \in \mathscr{Z}^1(G, A')$, the steps in the discussion are reversible, so there is defined a $\psi \in \mathscr{S}$ which in turn determines d'. It follows that there is a 1–1 correspondence

$$\psi \longleftrightarrow d'$$

between $\mathscr{S}$ and $\mathscr{Z}^1(G, A')$. This enables us to make $\mathscr{S}$ into a group isomorphic with $\mathscr{Z}^1(G, A')$, but, of course, this isomorphism depends on the choice of $\varphi \in \mathscr{S}$. On the other hand, we may define a "natural" action of $\mathscr{Z}^1(G, A')$ on $\mathscr{S}$; namely, if $e' \in \mathscr{Z}^1(G, A')$ and $\psi \in \mathscr{S}$ with $\psi \longleftrightarrow d'$, then $e'd' \in \mathscr{Z}^1(G, A)$, and we define $e'\psi$ to be the element of $\mathscr{S}$ associated with $e'd'$. It follows that

$$(e'\psi)(u_\sigma) = e'_\sigma \psi(u_\sigma) \qquad \sigma \in G \tag{5.20}$$

Since this relation suffices to determine $e'\psi$, we see that the action of $\mathscr{Z}^1(G, A')$ on $\mathscr{S}$ is independent of the choice of φ. It is clear that the action of $\mathscr{Z}^1(G, A')$ on $\mathscr{S}$ is transitive and without fixed points, and that $e''(e'\psi) = (e''e')\psi$ for $e'', e' \in \mathscr{Z}^1(G, A)$.

Next, let us see which elements of $\mathscr{S}$ correspond to coboundaries. Suppose then that $\psi \longleftrightarrow d$, $d \in \mathscr{B}^1(G, A')$. There exists $a' \in A'$ such that $d'_\sigma = a'^{\sigma-1}$ for all $\sigma \in G$, so that

$$\psi(au_\sigma) = f(a)\, d'_\sigma c'_\sigma u'_{\lambda(\sigma)} = f(a)\, c'_\sigma a'^{-1} a'^{\sigma} u'_{\lambda(\sigma)} = a'^{-1}(f(a)\, c'_\sigma u'_{\lambda(\sigma)})\, a'$$

If, for $a' \in A'$ we let $I_{a'}$ denote the inner automorphism of U' given by $I_{a'}u' = a'^{-1}u'a'$, then $\psi = I_{a'} \circ \varphi$. This means that φ and ψ are **equivalent** ($\varphi \approx \psi$) under the equivalence relation defined by: $\psi_1 \approx \psi_2$ if and only if there exists $a' \in A'$ such that $\psi_2 = I_{a'} \circ \psi_1$.

On the other hand, if $\varphi \approx \psi$ (where φ, ψ are the ones appearing in (5.19)) so that $\psi = I_{a'} \circ \varphi$, then

$$\psi(u_\sigma) = a'^{-1}c'_\sigma u'_{\lambda(\sigma)}a' = a'^{\sigma-1}c'_\sigma u'_{\lambda(\sigma)}$$

But $\psi(u_\sigma) = b'_\sigma u'_{\lambda(\sigma)}$; hence $\psi \longleftrightarrow d'$ where $d'_\sigma = a'^{\sigma-1}$—that is d' is a coboundary. We have shown that $\psi \approx \varphi \Longleftrightarrow \psi$ corresponds to a $d' \in \mathscr{B}^1(G, A')$. (We leave it to the reader to state the necessary and sufficient condition for $\psi_1 \approx \psi_2$.)

It follows that $H^1(G, A')$ acts on $\bar{\mathscr{S}}$, the set of equivalence classes of elements of $\mathscr{S}$, and we have proved the following result:

5-1-4. Proposition. Let the hypotheses be as in (5-1-1), and suppose that conditions (i) and (ii) are satisfied. Then $H^1(G, A')$ and $\bar{\mathscr{S}}$ have the same number of elements; in fact, $H^1(G, A')$ operates transitively and without fixed points on $\bar{\mathscr{S}}$. In particular, if $H^1(G, A') = \{1\}$ then all the elements of $\mathscr{S}$ are equivalent—in other words, any two elements of $\mathscr{S}$ differ (with respect to composition of maps) by an inner automorphism of U' by an element of A'.

An easy consequence of (5-1-4) is the following:

5-1-5. Exercise. Let $\{U, i, j\}$ be a group extension of A by G. Then the automorphisms of $\{U, i, j\}$ form a group $\mathscr{S}$ which is canonically isomorphic to $\mathscr{Z}^1(G, A)$. The inner automorphisms $I_a : u \longrightarrow a^{-1}ua$ of $\{U, i, j\}$ by an element of A form a subgroup $\mathscr{S}'$ isomorphic to $\mathscr{B}^1(G, A)$. Therefore, the equivalence classes of automorphisms of $\{U, i, j\}$ form a group $\bar{\mathscr{S}} = \mathscr{S}/\mathscr{S}'$ isomorphic to $H^1(G, A)$. In particular, if $H^1(G, A) = \{1\}$ then every automorphism of $\{U, i, j\}$ is an inner automorphism determined by an element of A.

5-2. COMMUTATOR SUBGROUPS IN GROUP EXTENSIONS

5-2-1. Remark. It is convenient to recall some facts and notation from (1-6-7). For an arbitrary multiplicative group U, we write $\hat{U} = \text{Hom}\,(U, \mathbf{Q}/\mathbf{Z})$. This is an additive abelian group (since $\mathbf{Q}/\mathbf{Z}$ is additive) called the **group of characters** of U in $\mathbf{Q}/\mathbf{Z}$.

If A is an abelian group and B is a subgroup of A, then any character $f: B \longrightarrow \mathbf{Q}/\mathbf{Z}$ can be extended to a character $g: A \longrightarrow \mathbf{Q}/\mathbf{Z}$. (In other words, if

$$(1) \longrightarrow B \xrightarrow{i} A \xrightarrow{j} C \longrightarrow (1)$$

is an exact sequence of abelian groups, then

$$0 \longrightarrow \hat{C} \xrightarrow{(j,1)} \hat{A} \xrightarrow{(i,1)} \hat{B} \longrightarrow 0$$

is exact.) In fact, if $a \in A$, $a \notin B$, then g may be chosen so that $g(a) \neq 0$. If $A \subset U$, with A abelian and U not abelian, then a character of A need not be extendible to a character of U; a condition for extendibility is discussed in (5-2-2).

Let $B^{\perp} = \{g \in \hat{A} \mid g(B) = 0\} = \ker(i, 1)$, and call it the **annihilator** of B in $\hat{A}$. If B_1, B_2 are subgroups of A, then $(B_1 + B_2)^{\perp} = B_1^{\perp} \cap B_2^{\perp}$ and $B_1 < B_2 \Longrightarrow B_2^{\perp} < B_1^{\perp}$. Consequently, for any subgroups B_1, B_2, $B_1^{\perp} = B_2^{\perp} \Longleftrightarrow B_1 = B_2$.

5-2-2. Lemma. Consider the group extension

$$1 \longrightarrow A \longrightarrow U \longrightarrow G \longrightarrow 1$$

and a character $f \in \hat{A}$, and let U^c denote the commutator subgroup of U. Then

$$f \text{ can be extended to a character } g \in \hat{U} \Longleftrightarrow f(U^c \cap A) = 0$$

Proof: In other words, we must show that if $(U^c \cap A)^{\perp}$ denotes the annihilator of $U^c \cap A$ in $\hat{A}$, then

$$(U^c \cap A)^{\perp} = \{f \in \hat{A} \mid f \text{ can be extended to } g \in \hat{U}\} \tag{5.21}$$

Suppose first that $f \in \hat{A}$ can be extended to $g \in \operatorname{Hom}(U, \mathbf{Q}/\mathbf{Z})$. Then $U^c \subset \ker g$ because $\mathbf{Q}/\mathbf{Z}$ is abelian—so $f(U^c \cap A) = 0$. For the converse, consider $f \in (U^c \cap A)^{\perp} \subset \operatorname{Hom}(A, \mathbf{Q}/\mathbf{Z})$. Then f determines a character $\bar{f}$ of $A/(U^c \cap A)$ given by $\bar{f}[a(U^c \cap A)] = f(a)$. Since U^c is a normal subgroup of U, $A/(U^c \cap A) \approx (U^cA)/U^c$ and we have a character $\bar{g}$ of $(U^cA)/U^c$ given by $\bar{g}(aU^c) = \bar{f}[a(U^c \cap A)]$. Now $(U^cA)/U^c$ is a subgroup of the abelian group U/U^c, so $\bar{g}$ can be extended to a character g' of U/U^c; and then g' determines a character g of U given by $g(u) = g'(uU^c)$. This $g \in \hat{U}$ is the desired extension of f. ▌

5-2-3. Proposition. Let G be a finite group and suppose that

$$1 \longrightarrow A \xrightarrow{\ i\ } U \xrightarrow{\ j\ } G \longrightarrow 1$$

is a group extension associated with $\alpha \in H^2(G, A)$. Then

$$U^c \cap A = \eta_A^{-1}\{\alpha \cup_{\theta_A^T} H^{-3}(G, \mathbf{Z})\} = \eta_A^{-1}\{H^{-3}(G, \mathbf{Z}) \cup_{\theta_A} \alpha\}$$

where $\theta_A : \mathbf{Z} \times A \longrightarrow A$ is the canonical G-pairing and

$$\eta_A : A_N \twoheadrightarrow H^{-1}(G, A)$$

is the usual map in dimension -1.

Proof: Suppose that $f \in \hat{A} = \operatorname{Hom}(A, \mathbf{Q}/\mathbf{Z})$. In virtue of (5-2-1), it suffices to show that

$$f \in (U^c \cap A)^\perp \iff f[\eta_A^{-1}(H^{-3}(G, \mathbf{Z}) \cup_{\theta_A} \alpha)] = 0 \qquad (5.22)$$

To accomplish this, we note first that by (5-2-2), $f \in (U^c \cap A)^\perp \iff f$ can be extended to $\varphi \in \hat{U}$. Therefore, let us consider the diagram

$$\begin{array}{ccccccccc}
1 & \longrightarrow & A & \xrightarrow{\ i\ } & U & \xrightarrow{\ j\ } & G & \longrightarrow & 1 \\
 & & {\scriptstyle f}\downarrow & & {\scriptstyle \varphi}\vdots & & \downarrow{\scriptstyle \lambda=0} & & \\
0 & \longrightarrow & \mathbf{Q}/\mathbf{Z} & \xrightarrow{\ 1\ } & \mathbf{Q}/\mathbf{Z} & \longrightarrow & (0) & \longrightarrow & 0
\end{array}$$

where, in the bottom row (which is additive), $\mathbf{Q}/\mathbf{Z}$ is viewed as an extension of $\mathbf{Q}/\mathbf{Z}$ by the (0) group. Now, apply (5-1-1) (note that the action of G on $\mathbf{Q}/\mathbf{Z}$ defined via $\lambda = 0$ is precisely the customary trivial action, and that $\alpha' = 0 \in H^2((0), \mathbf{Q}/\mathbf{Z}) = (0)$) which says that

$$f \text{ can be extended to } \varphi \in \hat{U} \iff \begin{cases} \text{(i)} & f \text{ is a } G\text{-homomorphism} \\ \text{(ii)} & (1, f)_*\alpha = (\lambda, 1)_*\alpha' \\ & \qquad = 0 \in H^2(G, \mathbf{Q}/\mathbf{Z}) \end{cases}$$

Condition (i) says that $f(a^\sigma) = \sigma f(a) = f(a)$ for all $a \in A$, $\sigma \in G$—so it is equivalent to the condition that f vanishes on IA. To rephrase condition (ii), we apply the duality theorem (4-4-6) with $p = -2$, $B = \mathbf{Z}$, $\hat{B} = \operatorname{Hom}(\mathbf{Z}, \mathbf{Q}/\mathbf{Z}) \approx \mathbf{Q}/\mathbf{Z}$. Thus, $f_*(\alpha) = 0 \in H^2(G, \mathbf{Q}/\mathbf{Z}) \iff$ the character of $H^{-3}(G, \mathbf{Z})$ in $\mathbf{Q}/\mathbf{Z}$ given by $\zeta \longrightarrow \eta_{\mathbf{Q}/\mathbf{Z}}^{-1}(f_*(\alpha) \cup_{\theta_{\mathbf{Q}/\mathbf{Z}}^{\top}} \zeta)$ is the trivial character

$$\iff \eta_{\mathbf{Q}/\mathbf{Z}}^{-1}(f_*(\alpha) \cup_{\theta_{\mathbf{Q}/\mathbf{Z}}^{\top}} \zeta) = 0 \text{ for all } \zeta \in H^{-3}(G, \mathbf{Z})$$

$$\iff \eta_{\mathbf{Q}/\mathbf{Z}}^{-1} f_*(\alpha \cup_{\theta_A^{\top}} \zeta) = 0 \text{ for all } \zeta \in H^{-3}(G, \mathbf{Z}) \text{ see (4-6-13))}$$

In other words,

$$f \in (U^c \cap A)^\perp \iff \begin{cases} (1) & f(IA) = 0 \\ (2) & \eta_{\mathbf{Q}/\mathbf{Z}}^{-1} f_*(\alpha \cup_{\theta_A^{\top}} H^{-3}(G, \mathbf{Z})) = (0) \end{cases} \tag{5.23}$$

Now, let us consider the diagram

$$\begin{array}{ccc} A_N & \xrightarrow{\ f\ } & (\mathbf{Q}/\mathbf{Z})_S \\ \eta_A \downarrow & & \downarrow \eta_{\mathbf{Q}/\mathbf{Z}} \\ H^{-1}(G, A) & \xrightarrow{\ f_*\ } & H^{-1}(G, \mathbf{Q}/\mathbf{Z}) \end{array} \tag{5.24}$$

If (1) holds, then f is a G-homomorphism and by (2-2-7) the diagram commutes. Then, since $\eta_{\mathbf{Q}/\mathbf{Z}}$ is an isomorphism onto, $f \circ \eta_A^{-1} = \eta_{\mathbf{Q}/\mathbf{Z}}^{-1} \circ f_*$, and (2) implies that

$$f\eta_A^{-1}(H^{-3}(G, \mathbf{Z}) \cup_{\theta_A} \alpha) = 0 \tag{5.25}$$

since for $\zeta \in H^{-3}(G, \mathbf{Z})$, $\zeta \cup_{\theta_A} \alpha = \alpha \cup_{\theta_A^{\top}} \zeta$. Conversely, suppose that (5.25) holds. Then f is a G-homomorphism because

$\eta_A^{-1}(0) = IA$, so that (5.24) commutes and $\eta_{\mathbf{Q}/\mathbf{Z}}^{-1} \circ f_* = f \circ \eta_A^{-1}$. By applying (5.23) we conclude that $f \in (U^c \cap A)^\perp$. Thus (5.22) holds, and the proof is complete. ∎

5-2-4. Remark. It is an incidental by-product of the preceding proof that

$$IA \subset U^c \cap A \subset A_N \tag{5.26}$$

These inclusions may also be proved in a more direct fashion. For $a \in A$, $\sigma \in G$ we have $a^{\sigma-1} = u_\sigma a u_\sigma^{-1} a^{-1} \in U^c \cap A$, and the first inclusion holds. To show that $U^c \cap A \subset A_N$, one makes use of the group theoretical transfer (see (3-5-4))

$$V = V_{U\to A} : U \longrightarrow \frac{A}{A^c} = A$$

The section $\{u_\sigma\}$ provides a complete system of representatives for the cosets of A in U, $U = \bigcup_{\sigma\in G} Au_\sigma$. Fix $u \in U$, and for $\sigma \in G$ let us write $j(u_\sigma u) = \sigma'$; thus σ' is the unique element of G such that $u_\sigma u u_{\sigma'}^{-1} \in A$. It then follows from the transfer formula (see (3-5-4)) that

$$V(u) = \prod_{\sigma\in G} u_\sigma u u_{\sigma'}^{-1} \qquad \sigma' = j(u_\sigma u) \tag{5.27}$$

In particular, for $u = a \in A$ we have $\sigma' = j(u_\sigma u) = \sigma$, and

$$V(a) = \prod_{\sigma\in G} u_\sigma a u_\sigma^{-1} = \prod_\sigma a^\sigma = N(a)$$

Since the homomorphism V maps $U^c \longrightarrow 1$ it follows that $U^c \cap A \subset A_N$.

5-2-5. Proposition. Let G be a finite group and suppose that

$$1 \longrightarrow A \xrightarrow{\ i\ } U \xrightarrow{\ j\ } G \longrightarrow 1$$

is a group extension and that $\{u_\sigma\}$ is a section of G in U with associated 2-cocycle $\{a_{\sigma,\tau}\}$. Then the transfer $V = V_{U\to A} : U \longrightarrow A$ is given by the formulas

(i) $V(a) = N(a) \in A^G \qquad a \in A.$

(ii) $V(u_\tau) = \prod_{\sigma\in G} a_{\sigma,\tau} \in A^G \qquad \tau \in G.$

(iii) $V(au_\tau) = N(a) \prod_\sigma a_{\sigma,\tau} \in A^G.$

Proof: Part (i) has been proved in (5-2-4). As for part (ii), if $\sigma \in G$, then (according to (5.27)) $\sigma' = j(u_\sigma u_\tau) = \sigma\tau$, so that $V(u_\tau) = \prod_\sigma u_\sigma u_\tau u_{\sigma\tau}^{-1} = \prod_\sigma a_{\sigma,\tau}$. Moreover, for $\rho \in G$ we have

$$\Big(\prod_\sigma a_{\sigma,\tau}\Big)^\rho = \prod_\sigma (a_{\rho,\sigma}a_{\rho\sigma,\tau}a_{\rho,\sigma\tau}^{-1}) = \prod_\sigma a_{\rho\sigma,\tau} = \prod_\sigma a_{\sigma,\tau}$$

Consequently, the homomorphism V is given by (iii) and takes values in A^G. ∎

It should be observed that, in virtue of (ii), V leads to a mapping $\tau \longrightarrow \prod_\sigma a_{\sigma\,\tau}$ of $G \longrightarrow A^G$. This mapping need not be a homomorphism, but when the right side is reduced modulo the group of norms NA we have precisely the Nakayama map (see (4-5-7)).

5-2-6. Proposition. Let the hypotheses be those of (5-2-4) and (5-2-5). Then the following diagram is commutative and has exact rows:

$$\begin{array}{ccccccccc}
1 \longrightarrow & \dfrac{A}{U^c \cap A} & \approx \dfrac{U^cA}{U^c} & \xrightarrow{\bar{\imath}} & \dfrac{U}{U^c} & \xrightarrow{\bar{\jmath}} & \dfrac{G}{G^c} \approx H^{-2}(G, \mathbf{Z}) & \longrightarrow 0 \\
& \downarrow{\scriptstyle \bar{N}} & & & \downarrow{\scriptstyle \bar{V}} & & \downarrow{\scriptstyle \alpha_{-2}} & \\
1 \longrightarrow & NA & & \xrightarrow{i} & A^G & \xrightarrow{\kappa} & H^0(G, A) & \longrightarrow 1
\end{array}$$

Here, $\bar{\imath}, \bar{\jmath}, \bar{N}$ are induced from i, j, N in the obvious fashion, $\bar{V}$ is the reduced transfer (see (3-5-4), (3-5-5), (3-5-6)), and for each integer n, $\alpha_n : H^n(G, \mathbf{Z}) \longrightarrow H^{n+2}(G, A)$ is the map given by

$$\alpha_n : \beta \longrightarrow \beta \cup_{\theta_A} \alpha \qquad \beta \in H^n(G, \mathbf{Z}) \quad (5.28)$$

Proof: The bottom row is clearly exact. As for the top row, $\bar{\imath}$ is given by $\bar{\imath}[a(U^c \cap A)] = aU^c$, so it is a monomorphism with image $(AU^c)/U^c$; and $\bar{j}$ is given by $\bar{j}(uU^c) = j(u)\,G^c$, so it is an epimorphism. Since j maps U onto G with kernel A and $j(U^c) = G^c$, it follows that $\ker \bar{j} = (AU^c)/U^c$. Therefore, the top row is exact.

The square on the left commutes because, according to (5-2-5), $\bar{V}\bar{i}[a(U^c \cap A)] = \bar{V}(aU^c) = V(a) = N(a) = \bar{N}[a(U^c \cap A)]$. As for the square on the right, consider $u = au_\tau \in U$. Then, making use of (4-5-7), we have

$$\kappa\bar{V}(uU^c) = \kappa\left[N(a)\prod_\sigma a_{\sigma,\tau}\right] = \kappa\left(\prod_\sigma a_{\sigma,\tau}\right) = \zeta_\tau \smile_{\theta_A} \alpha$$

On the other hand, $\bar{j}(uU^c) = j(au_\tau)\,G^c = \tau G^c$ is identified with $\zeta_\tau \in H^{-2}(G, \mathbf{Z})$, and by definition $\alpha_{-2}(\zeta_\tau) = \zeta_\tau \smile_{\theta_A} \alpha$. ▌

5-2-7. Corollary. Let us place ourselves in the situation of (5-2-6) and consider $\bar{V} : U/U^c \longrightarrow A^G$. Then

(i) $\bar{V}$ is an epimorphism $\Longleftrightarrow$ α_{-2} is an epimorphism.

(ii) $\bar{V}$ is a monomorphism $\Longleftrightarrow$ α_{-2} and α_{-3} are monomorphisms.

Proof: Viewing all groups additively, we complete the diagram of (5-2-6) as follows:

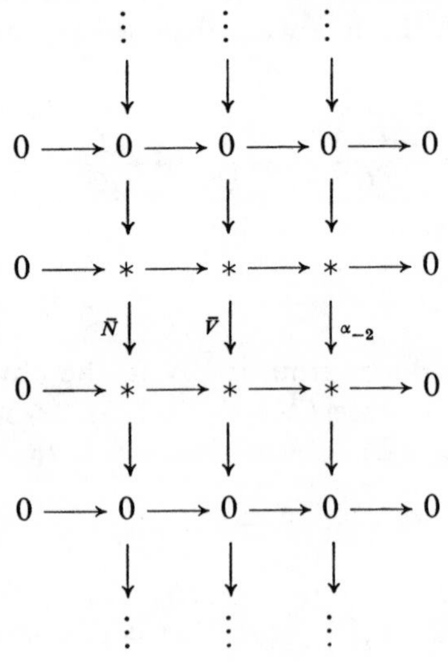

The columns may be viewed as differential graded groups, so by (1-1-4) we have the exact sequence of derived groups:

$$\cdots \longrightarrow (0) \longrightarrow \ker \bar{N} = \frac{A_N}{U^c \cap A} \longrightarrow \ker \bar{V} \longrightarrow \ker \alpha_{-2} \longrightarrow \frac{NA}{\operatorname{im} \bar{N}} = (0) \longrightarrow \frac{A^G}{\operatorname{im} \bar{V}} \longrightarrow \frac{H^0(G, A)}{\operatorname{im} \alpha_{-2}} \longrightarrow (0) \longrightarrow \cdots$$

Now, $\bar{V}$ is an epimorphism $\Longleftrightarrow$ $\operatorname{im} \alpha_{-2} = H^0(G, A) \Longleftrightarrow \alpha_{-2}$ is an epimorphism, thus proving (i). Furthermore, $\bar{V}$ is a monomorphism $\Longleftrightarrow \ker \alpha_{-2} = (0)$ and $A_N/(U^c \cap A) = (0)$. From (5-2-3) it follows that $\operatorname{im} \alpha_{-3} \approx (U^c \cap A)/(IA)$, so that

$$\frac{H^{-1}(G, A)}{\operatorname{im} \alpha_{-3}} \approx \frac{A_N}{U^c \cap A}$$

and (ii) is immediate. ▮

5-3. FACTOR EXTENSIONS

Let A be a multiplicative G-module (where the group G is finite or infinite) and let $\{U, i, j\}$ be a group extension of A by G associated with $\alpha \in H^2(G, A)$; in other words,

$$(1) \longrightarrow A \xrightarrow{\ i\ } U \xrightarrow{\ j\ } G \longrightarrow (1)$$

is exact, and if $\{u_\sigma \mid \sigma \in G\}$ is a section of G in U with $u_\sigma u_\tau = a_{\sigma,\tau} u_{\sigma\tau}$, then $\{a_{\sigma,\tau}\}$ is a standard 2-cocycle belonging to α.

Suppose that H is a subgroup of G. If we put $W = j^{-1}(H)$, then the sequence

$$(1) \longrightarrow A \xrightarrow{\ i\ } W \xrightarrow{\ j\ } H \longrightarrow (1) \tag{5.29}$$

is exact. Even more, $\{W, i, j\}$ is an extension of A by H and may be referred to as a **subextension** of $\{U, i, j\}$—in fact, $\{u_\rho \mid \rho \in H\}$ is a section of H in W and for $a \in A$, $\rho \in H$, $a^\rho = u_\rho a u_\rho^{-1}$. The cohomology class in $H^2(H, A)$ associated with the subextension $\{W, i, j\}$ is precisely $\operatorname{res}_{G\to H} \alpha$; this is immediate because

$\{a_{\sigma,\tau} \mid \sigma, \tau \in H\}$ is the restriction of the cocycle $\{a_{\sigma,\tau} \mid \sigma, \tau \in G\}$ and is also the cocycle determined by the section $\{u_\sigma \mid \sigma \in H\}$.

Now, let us turn to the situation where H is a normal subgroup of G. Here $W = j^{-1}(H) = \{au_\rho \mid a \in A, \rho \in H\} = \{u_\rho a \mid a \in A, \rho \in H\}$ is a normal subgroup of U, and $U/W \approx G/H$. If we let $\pi : G \longrightarrow G/H$ denote the natural map (and i = inclusion, as usual), then we have the exact sequence

$$(1) \longrightarrow W \xrightarrow{i} U \xrightarrow{\pi \circ j} \frac{G}{H} \longrightarrow (1) \tag{5.30}$$

Unfortunately, this cannot be considered within our framework for group extensions—W is not abelian and there is no way to define an action of G/H on W by conjugation. To overcome this, let us consider

$$(1) \longrightarrow \frac{W}{W^c} \xrightarrow{i'} \frac{U}{W^c} \xrightarrow{j'} \frac{G}{H} \longrightarrow (1) \tag{5.31}$$

where $i'(wW^c) = i(w)\, W^c = wW^c$ and $j'(uW^c) = j(u)\, H = \pi(j(u))$. First of all, it should be noted that W^c is normal in U—for a generator of W^c may be taken of form $w_1 w_2 w_1^{-1} w_2^{-1}$ where $w_1, w_2 \in W$ and then for $u \in U$,

$$uw_1w_2w_1^{-1}w_2^{-1}u^{-1} = (uw_1u^{-1})(uw_2u^{-1})(uw_1^{-1}u^{-1})(uw_2^{-1}u^{-1})$$

which is the commutator of the elements uw_1u^{-1} and uw_2u^{-1} of W. From the definitions of i' and j', it is immediate that the sequence (5.31) is exact. Because $A \subset W$ there is no natural way to derive an action of G/H on W/W^c from the action of G on A; however, we shall view W/W^c as a G/H module according to the standard procedure arising from the exact sequence (5.31)—that is, through conjugation by representatives of G/H in U/W^c.

In this way, an element $\alpha \in H^2(G, A)$ leads to an element $\alpha' \in H^2(G/H, W/W^c)$ associated with the factor extension $\{U/W^c, i', j'\}$. This is not quite satisfactory cohomologically, because A is the basic module. For this reason, let us consider the group theoretical transfer (see (3-5-6)) $V_{W \to A} : W \longrightarrow A/A^c = A$. Note that for this we must have $(W : A) < \infty$; in other words, H must be a finite group. From (5-2-5), applied to the subextension

(5.30), it follows that $V_{W\to A}$ maps W into A^H, so that the reduced transfer is a homomorphism

$$\bar{V} = \bar{V}_{W\to A} : \frac{W}{W^c} \longrightarrow A^H$$

Moreover, $\bar{V}$ is a (G/H)-homomorphism (as we shall next verify) and we have the induced homomorphism of cohomology

$$V_* = (\bar{V}_{W\to A})_* : H^2\left(\frac{G}{H}, \frac{W}{W^c}\right) \longrightarrow H^2\left(\frac{G}{H}, A^H\right)$$

To show that $\bar{V}$ is a (G/H)-homomorphism, note (see (5.27)) that $\bar{V}(wW^c) = V(w) = \prod_{\rho\in H} u_\rho w u_{\rho'}^{-1} \in A^H$ where $\rho' = j(u_\rho w)$, and that for $a \in A^H$, $\sigma \in G$ we have $a^{(\sigma H)} = a^\sigma = u_\sigma a u_\sigma^{-1}$. Since, for any $\sigma \in G$, $\sigma H \sigma^{-1} = H$ and $j(u_\sigma u_\rho u_\sigma^{-1}) = \sigma\rho\sigma^{-1}$, we may also use $\{u_\sigma u_\rho u_\sigma^{-1} \mid \rho \in H\}$ as a section of H in W for the subextension (5.29). Consequently,

$$\begin{aligned}\{\bar{V}(wW^c)\}^{(\sigma H)} &= u_\sigma \left(\prod_{\rho\in H} u_\rho w u_{\rho'}^{-1}\right) u_\sigma^{-1} \\ &= \prod_{\rho\in H} (u_\sigma u_\rho u_\sigma^{-1})(u_\sigma w u_\sigma^{-1})(u_\sigma u_{\rho'} u_\sigma^{-1})^{-1}\end{aligned}$$

and because $j(u_\sigma u_\rho u_\sigma^{-1} u_\sigma w u_\sigma^{-1}) = \sigma\rho'\sigma^{-1}$, this expression is simply $V_{W\to A}(u_\sigma w u_\sigma^{-1})$ computed with respect to the section $\{u_\sigma u_\rho u_\sigma^{-1} \mid \rho \in H\}$; but $V(u_\sigma w u_\sigma^{-1}) = \bar{V}\{(wW^c)^{(\sigma H)}\}$, thus proving the assertion.

5-3-1. Exercise. The foregoing discussion shows that if A is a G-module and H is a finite normal subgroup of G, then there exists a mapping

$$v_{G\to G/H} : H^2(G, A) \longrightarrow H^2\left(\frac{G}{H}, A^H\right)$$

defined in the following manner. Give $\alpha \in H^2(G, A)$, let $\{U, i, j\}$ be an extension of A by G associated with α; putting $W = j^{-1}(H)$ there arises the factor extension (5.31) which is associated with $\alpha' \in H^2(G/H, W/W^c)$; then $v_{G\to G/H}(\alpha) = V_*(\alpha') = (\bar{V}_{W\to A})_*(\alpha')$. Show that $v_{G\to G/H}(\alpha)$ is independent of the choice of the extension $\{U, i, j\}$ belonging to the equivalence class which corresponds to α.

5-3-2. Proposition. Let H be a normal subgroup of the finite group G; then the map

$$v_{G\to G/H} : H^2(G, A) \longrightarrow H^2\left(\frac{G}{H}, A^H\right)$$

is a homomorphism. Moreover, if we let $\bar{r} \in G$ denote the representative of the coset $r = \bar{r}H$, then $v_{G\to G/H}$ is induced by the map of cocycles $\{a_{\sigma,\tau}\} \longrightarrow \{(va)_{r,s}\}$ where

$$(va)_{r,s} = [N_H(a_{\bar{r},\bar{s}}a_{\lambda,\overline{rs}}^{-1})]\left[\prod_{\rho\in H} a_{\rho,\lambda}\right]$$

and $\lambda = \lambda_{r,s} = \bar{r}\bar{s}\overline{rs}^{-1} \in H$.

Proof: The fact that $v_{G\to G/H}$ is a homomorphism will follow immediately from the formula for $(va)_{r,s}$ since both N_H and $\prod_{\rho\in H}$ are multiplicative.

Starting from the section $\{u_\sigma \mid \sigma \in G\}$ of G in U (which determines the cocycle $\{a_{\sigma,\tau}\}$ via $u_\sigma u_\tau = a_{\sigma,\tau}u_{\sigma\tau}$) we know that $\{u_\rho \mid \rho \in H\}$ is a section of H in W. If for $r \in G/H$ we put $u_r = u_{\bar{r}}$, then $\{u_r \mid r \in G/H\}$ is a full set of representatives for G/H in U (see (5.30)), and $\{u_r W^c \mid r \in G/H\}$ is a section of G/H in U/W^c (see (5.31)). It is clear that $u_{\bar{r}}u_{\bar{s}} = a_{\bar{r},\bar{s}}u_{\bar{r}\bar{s}}$, and writing $\lambda = \bar{r}\bar{s}\overline{rs}^{-1}$ (so that λ is a function of r and s in G/H, and $\lambda \in H$) we have $u_\lambda u_{\overline{rs}} = a_{\lambda,\overline{rs}}u_{\bar{r}\bar{s}}$.

Suppose the factor extension (5.31) is described by the 2-cocycle $\{w_{r,s}W^c\}$—this means that

$$(u_r W^c)(u_s W^c) = w_{r,s}u_{rs}W^c$$

and that if $\{a_{\sigma,\tau}\}$ belongs to the class $\alpha \in H^2(G, A)$, then the cocycle $\{w_{r,s}W^c\}$ belongs to the class $\alpha' \in H^2(G/H, A^H)$. Thus $v_{G\to G/H}(\alpha)$ is represented by

$$\begin{aligned}\bar{V}_{W\to A}(w_{r,s}W^c) &= \bar{V}(a_{\bar{r},\bar{s}}a_{\lambda,\overline{rs}}^{-1}u_\lambda W^c)\\ &= \bar{V}(a_{\bar{r},\bar{s}}a_{\lambda,\overline{rs}}^{-1}W^c)\,\bar{V}(u_\lambda W^c)\\ &= N_H(a_{\bar{r},\bar{s}}a_{\lambda,\overline{rs}}^{-1})\prod_{\rho\in H} a_{\rho,\lambda} \in A^H \qquad \text{(by (5-2-5))}\end{aligned}$$

This completes the proof. ∎

5-3-3. Proposition. Let the hypotheses be as in (5-3-2); then for $\alpha \in H^2(G, A)$

$$\inf_{(G/H)\to G} v_{G\to G/H}\alpha = \alpha^{(H:1)}$$

Proof: For $\sigma \in G$, denote the representative of the coset to which σ belongs by $\bar{\sigma}$—so $\bar{\sigma} = \overline{\sigma H}$. It will also be convenient to let h denote a generic element of H. When all cohomology groups are computed with respect to the standard complex, $\inf v\alpha$ is represented by the 2-cocycle $\inf va$ for which

$$(\inf va)_{\sigma,\tau} = (va)_{\sigma H,\tau H} = [N_H(a_{\bar{\sigma},\bar{\tau}} a^{-1}_{\bar{\sigma}\bar{\tau}\overline{\sigma\tau}^{-1},\overline{\sigma\tau}})] \left[\prod_{h\in H} a_{h,\bar{\sigma}\bar{\tau}\overline{\sigma\tau}^{-1}}\right]$$

$$= \prod_{h\in H} (a^h_{\bar{\sigma},\bar{\tau}} a^{-h}_{\bar{\sigma}\bar{\tau}\overline{\sigma\tau}^{-1},\overline{\sigma\tau}} a_{h,\bar{\sigma}\bar{\tau}\overline{\sigma\tau}^{-1}})$$

Since $\{a_{\sigma,\tau}\}$ is a 2-cocycle we have the coboundary relation

$$(\delta a)_{\sigma,\tau,\rho} = a^{\rho}_{\sigma,\tau} a^{-1}_{\rho\sigma,\tau} a_{\rho,\sigma\tau} a^{-1}_{\rho,\sigma} = 1 \qquad \sigma, \tau, \rho \in G$$

In this formula we substitute first $\sigma \longleftrightarrow \bar{\sigma}$, $\tau \longleftrightarrow \bar{\tau}$, $\rho \longleftrightarrow h$ and then $\sigma \longleftrightarrow \bar{\sigma}\bar{\tau}\overline{\sigma\tau}^{-1}$, $\tau \longleftrightarrow \overline{\sigma\tau}$, $\rho \longleftrightarrow h$, which enables us to write

$$(\inf va)_{\sigma,\tau} = \prod_{h\in H} (a_{h\bar{\sigma},\bar{\tau}} a^{-1}_{h,\bar{\sigma}\bar{\tau}} a_{h,\bar{\sigma}})(a_{h\bar{\sigma}\bar{\tau}\overline{\sigma\tau}^{-1},\overline{\sigma\tau}} a_{h,\bar{\sigma}\bar{\tau}} a^{-1}_{h,\bar{\sigma}\bar{\tau}\overline{\sigma\tau}^{-1}})(a_{h,\bar{\sigma}\bar{\tau}\overline{\sigma\tau}^{-1}})$$

$$\prod_{h\in H} (a_{\sigma h,\bar{\tau}} a_{h,\bar{\sigma}} a^{-1}_{h,\overline{\sigma\tau}})$$

because $H\bar{\sigma} = \sigma H$, and $H\bar{\sigma}\bar{\tau}\overline{\sigma\tau}^{-1} = H$ since $\bar{\sigma}\bar{\tau}\overline{\sigma\tau}^{-1} \in H$. To simplify this expression consider the 1-cochain of G in A $\{c_\sigma = \prod_{h\in H} a_{h,\bar{\sigma}}\}$; its coboundary is

$$(\delta c)_{\sigma,\tau} = c^{\sigma}_{\tau} c^{-1}_{\sigma\tau} c_{\sigma} = \prod_{h\in H} a^{\sigma}_{h,\bar{\tau}} a^{-1}_{h,\overline{\sigma\tau}} a_{h,\bar{\sigma}}$$

and dividing by this coboundary gives

$$(\inf va)_{\sigma,\tau} \sim \prod_{h\in H} (a_{\sigma h,\bar{\tau}} a^{-\sigma}_{h,\bar{\tau}})$$

Since $(\delta a)^{-1}_{h,\bar{\tau},\sigma} = 1$ the right side is equal to

$$\prod_{h\in H} a_{\sigma,h\bar{\tau}} a^{-1}_{\sigma,h} = \prod_{h\in H} (a_{\sigma,\tau h} a^{-1}_{\sigma,h})$$

and since $(\delta a)_{\tau,h,\sigma} = 1$ this is equal to

$$\prod_{h\in H} (a_{\tau,h}^{-\sigma} a_{\sigma\tau,h} a_{\sigma,\tau})(a_{\sigma,h}^{-1}) = \left\{ \prod_{h\in H} (a_{\tau,h}^{\sigma} a_{\sigma\tau,h}^{-1} a_{\sigma,h}) \right\}^{-1} a_{\sigma,\tau}^{(H;1)} = (\delta d)_{\sigma,\tau}^{-1} a_{\sigma,\tau}^{(H;1)}$$

where d is the 1-cochain $\{d_\sigma = \prod_{h\in H} a_{\sigma,h}\}$. Therefore,

$$(\inf va)_{\sigma,\tau} \sim a_{\sigma,\tau}^{(H;1)}$$

and the proof is complete. ∎

5-4. THE PRINCIPAL IDEAL THEOREM

Suppose that A is a G-module, and consider $\alpha \in H^2(G, A)$. Let $1 \longrightarrow A \longrightarrow U \longrightarrow G \longrightarrow 1$ be a group extension associated with α. Thus, if $\{u_\sigma\}$ is a section of G in U and we write $u_\sigma u_\tau = a_{\sigma,\tau} u_{\sigma\tau}$, then $\{a_{\sigma,\tau}\}$ is a standard 2-cocycle belonging to α. Of course, in choosing the section $\{u_\sigma\}$ we may always take $u_1 = 1 \in U$; and when this is done

$$a_{\sigma,1} = a_{1\,\sigma} = 1 \qquad \text{all } \sigma \in G \quad (5.32)$$

In this section, when dealing with a cocycle $\{a_{\sigma,\tau}\}$ we shall always assume that (5.32) is satisfied.

If B is a G-module containing A such that for the inclusion map $i : A \longrightarrow B$ we have $i_*(\alpha) = 1 \in H^2(G, B)$, the B is said to be a **splitting module** for α. In other words, a cocycle belonging to α becomes a coboundary when viewed in the larger module B.

5-4-1. Proposition. A splitting module exists for each $\alpha \in H^2(G, A)$.

Proof: We construct a G-module B containing A (which we view additively) for which there exists a standard 1-cochain $\{x_\sigma\}$ such that $a_{\sigma,\tau} = (\delta x)_{\sigma,\tau} = \sigma x_\tau - x_{\sigma\tau} + x_\sigma$ for all $\sigma, \tau \in G$. To do this, take a formal symbol x_τ for each $\tau \neq 1 \in G$. It is also convenient to put $x_1 = 0$. Now, put

$$B = A \oplus \sum_{\tau \neq 1} \oplus \mathbf{Z} x_\tau$$

—so B is an abelian group. Define the action of $\sigma \in G$ on B by keeping the given action of σ on A and putting

$$\sigma x_\tau = x_{\sigma\tau} - x_\sigma + a_{\sigma,\tau} \qquad \tau \neq 1 \quad (5.33)$$

Of course, this suffices for the definition of the homomorphism $\sigma : B \longrightarrow B$. Note that Eq. (5.33) is also true for $\tau = 1$. It is clear that $1 \in G$ acts as the identity on B. Finally,

$$\rho(\sigma x_\tau) = \rho(x_{\sigma\tau} - x_\sigma + a_{\sigma,\tau}) = x_{\rho\sigma\tau} - x_\rho + a_{\rho,\sigma\tau} - x_{\rho\sigma} + x_\rho - a_{\rho,\sigma} + \rho a_{\sigma,\tau}$$
$$= x_{(\rho\sigma)\tau} - x_{\rho\sigma} + a_{\rho\sigma,\tau} = (\rho\sigma)x_\tau$$

This completes the proof. ∎

5-4-2. Exercise. Suppose that A is an additive G-module, and fix $n \geqslant 1$. For each $i = 0, 1,\ldots, n$ let us say that the standard n-cochain $\in \mathscr{C}^n = \mathscr{C}^n(G, A)$ is ***i*-normalized** (or ***i*-normal**) if $f[\sigma_1, \ldots, \sigma_n] = 0$ whenever one (or more) of $\sigma_1, \ldots, \sigma_i$ is 1. In particular, the 0-normal cochains are simply the cochains of $\mathscr{C}^n$. The n-normal cochains are referred to simply as **normal cochains**.

(i) The normal cochains form a subgroup $\bar{\mathscr{C}}^n$ of $\mathscr{C}^n$; we also put $\bar{\mathscr{C}}^0 = \mathscr{C}^0$. The cocycles which are normal form a subgroup $\bar{\mathscr{Z}}^n$ of $\mathscr{Z}^n$. In virtue of the coboundary formula, the coboundary of an i-normal cochain is i-normal. In particular, $\bar{\mathscr{B}}^n = \delta\bar{\mathscr{C}}^{n-1}$ is a subgroup of $\bar{\mathscr{Z}}^n$, and we may form the normalized cohomology groups $\bar{H}^n(G, A) = \bar{\mathscr{Z}}^n/\bar{\mathscr{B}}^n$. At the beginning of this section (see formula (5.32)) it was observed that every element of $H^2(G, A)$ may be represented by a normal cocycle. We shall see in (ii) that, even more, $\bar{H}^n(G, A) \approx H^n(G, A)$ for all $n \geqslant 1$.

(ii) Given any cochain $f \in \mathscr{C}^n$ let us construct, inductively, cochains $f_0, f_1, \ldots, f_n \in \mathscr{C}^n$ and $g_1, \ldots, g_n \in \mathscr{C}^{n-1}$ in the following manner:

$$f_0 = f \qquad f_i = f_{i-1} - \delta g_i \qquad i = 1,\ldots, n$$

$$g_i[\sigma_1, \ldots, \sigma_{n-1}] = (-1)^{i-1} f_{i-1}[\sigma_1, \ldots, \sigma_{i-1}, 1, \sigma_i, \ldots, \sigma_{n-1}]$$

It is clear that $\delta f = \delta f_0 = \delta f_i$. Furthermore, if δf is normal, then f_i is i-normal for $i = 0, 1,\ldots, n$. (To prove this, observe that the case $i = 0$ is trivial, and then proceed inductively. By the induction hypothesis, f_i is i-normal; hence, so are g_{i+1}, δg_{i+1} and f_{i+1}. Thus, it remains to verify that

$$f_{i+1}[\sigma_1, \ldots, \sigma_i, 1, \sigma_{i+2}, \ldots, \sigma_n] = 0$$

For $i \geqslant 1$, one may verify that

$$f_{i+1}[\sigma_1, \ldots, \sigma_i, 1, \sigma_{i+2}, \ldots, \sigma_n] = (-1)^i \, \delta f_i[\sigma_1, \ldots, \sigma_i, 1, 1, \sigma_{i+2}, \ldots, \sigma_n]$$

which is 0 since δf is $(i+1)$-normal. A similar argument takes care of the case $i = 0$.) It follows that every normal cochain which is a coboundary is the coboundary of a normal cochain—so that $\overline{\mathscr{B}}^n = \delta\overline{\mathscr{C}}^{n-1} = \overline{\mathscr{C}}^n \cap \delta\mathscr{C}^{n-1} = \overline{\mathscr{C}}^n \cap \mathscr{B}^n = \overline{\mathscr{Z}}^n \cap \mathscr{B}^n$ and also that every cocycle is cohomologous to a normal cocycle. Consequently, $\overline{H}^n$ is isomorphic to H^n for $n \geqslant 1$.

(iii) For $\alpha \in H^n(G, A)$ define the notion of a splitting module for α in the same way as was done for $n = 2$. Then for every α (and $n \geqslant 1$), there exists a splitting module.

We shall not make use of (5-4-2) in the sequel.

5-4-3. Proposition. Let B be the splitting module for $\alpha \in H^2(G, A)$ constructed in (5-4-1); then the factor module B/A is G-isomorphic to I; more precisely, the following G-sequence is exact

$$(0) \longrightarrow A \xrightarrow{\ i\ } B \xrightarrow{\ j\ } I \longrightarrow (0)$$

where i is inclusion and j is given by $j(A) = 0$, $j(x_\tau) = \tau - 1$ for $\tau \neq 1$.

Proof: Since $\{\tau - 1 \mid \tau \neq 1\}$ is a $\mathbf{Z}$-basis for I, it is clear that j is a $\mathbf{Z}$-homomorphism of B onto I with kernel A. Note that $jx_\tau = \tau - 1$ even for $\tau = 1$. Moreover, j is a G-homomorphism, because

$$j(\sigma x_\tau) = j(x_{\sigma\tau} - x_\sigma + a_{\sigma,\tau}) = (\sigma\tau - 1) - (\sigma - 1) = \sigma(\tau - 1) = \sigma(jx)_\tau). \quad \blacksquare$$

5-4-4. Lemma. Let

$$\gamma = \sum_{\sigma\in G} m_\sigma \sigma \in \mathbf{Z}[G] \qquad \text{and} \qquad S = \sum_{\sigma\in G} \sigma \in \mathbf{Z}[G]$$

then

$$\gamma B \subset A \Longleftrightarrow \gamma = mS \quad \text{for some} \quad m \in \mathbf{Z}$$

Proof: In virtue of (5-4-3) we have

$$\gamma B \subset A \Longleftrightarrow \gamma I = (0) \Longleftrightarrow \gamma(\tau - 1) = 0 \text{ for all } \tau \in G$$
$$\Longleftrightarrow m_\sigma \text{ are equal.} \quad \blacksquare$$

We observe that (5-4-4) is the first place in the discussion where G is required to be finite, and in this case the trace provides a G-homomorphism $S : B \longrightarrow A$ which induces a G-homomorphism $\bar{S} : B/(IB) \longrightarrow A$.

In the following, we shall view A as a multiplicative subgroup of U, $U/A \approx G$, and as an additive subgroup of B.

5-4-5. Lemma. We have an isomorphism of abelian groups

$$\frac{B}{IB} \approx \frac{U}{U^c}$$

under the correspondence

$$a + x_\sigma + IB \longleftrightarrow au_\sigma U^c$$

Proof: We define a map $\log : U \longrightarrow B/(IB)$ by

$$\log(au_\sigma) = a + x_\sigma + IB$$

then log is a homomorphism because the difference of

$$\log(au_\sigma bu_\tau) = \log(ab^\sigma a_{\sigma,\tau} u_{\sigma\tau}) = a + \sigma b + a_{\sigma,\tau} + x_{\sigma\tau} + IB$$

and

$$\log(au_\sigma) + \log(bu_\tau) = a + x_\sigma + b + x_\tau + IB \qquad (\text{where } a, b \in A)$$

is

$$(\sigma b - b) + (x_{\sigma\tau} - x_\sigma + a_{\sigma,\tau}) - x_\tau + IB = (\sigma - 1)b + (\sigma - 1)x_\tau + IB = IB.$$

Since $B/(IB)$ is abelian, there is an induced homomorphism

$$\overline{\log} : \frac{U}{U^c} \longrightarrow \frac{B}{IB}$$

On the other hand, we may define a homomorphism $\exp : B \longrightarrow U/U^c$ by $\exp a = aU^c$ for $a \in A$ and $\exp x_\tau = u_\tau U^c$ for $\tau \neq 1$, and then extending to all of B. Note that for $\tau = 1$, $x_\tau = 0$, and $u_\tau = 1$, so that $\exp x_\tau = u_\tau U^c$ holds in this case, too. Moreover, exp vanishes on IB because

$$\exp (\sigma - 1)a = a^{\sigma-1}U^c = u_\sigma a u_\sigma^{-1} a^{-1} U^c = U^c$$

and

$$\exp (\sigma - 1)x_\tau = \exp (x_{\sigma\tau} - x_\sigma + a_{\sigma,\tau} - x_\tau) = u_{\sigma\tau} u_\sigma^{-1} a_{\sigma,\tau} u_\tau^{-1} U^c = U^c$$

Thus, we have an induced homomorphism $\overline{\exp} : B/(IB) \longrightarrow U/U^c$ which is clearly the inverse of $\overline{\log}$. ∎

5-4-6. Lemma. The following diagram commutes:

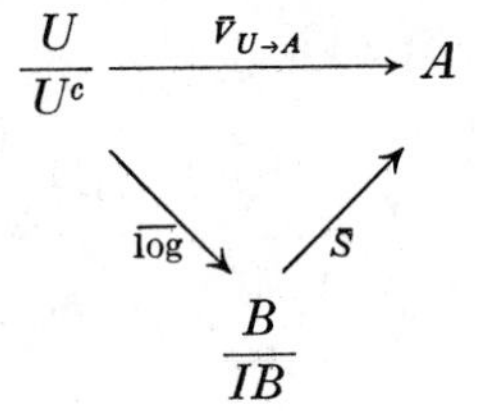

Proof: For $a \in A$, $\bar{V}_{U\to A}(aU^c) = N(a)$ while

$$\bar{S}\,\overline{\log}\,(aU^c) = \bar{S}(a + IB) = S(a)$$

and because we are viewing A both multiplicatively and additively, these are equal. Furthermore, $\bar{V}(u_\tau U^c) = \prod_{\sigma\in G} a_{\sigma,\tau}$ while

$$\begin{aligned}\bar{S}\,\overline{\log}\,(u_\tau U^c) &= \bar{S}(x_\tau + IB) = S(x_\tau)\\ &= \sum_{\sigma\in G} \sigma x_\tau = \sum_{\sigma\in G} (x_{\sigma\tau} - x_\sigma + a_{\sigma,\tau}) = \sum_{\sigma\in G} a_{\sigma,\tau}\end{aligned}$$

thus completing the proof. ∎

We are now in a position to prove the group theoretic formulation of the theorem, first stated by Hilbert (see [2], [27], [39], [49]) that every ideal of an algebraic number field becomes principal in its absolute class field.

5-4-7. Theorem. Let U be a multiplicative group such that U^c is finitely generated and $(U : U^c) < \infty$. Suppose that A is an abelian subgroup of U which contains U^c. If $(A : U^c) = e$, then

$$\bar{V}^e_{U\to A} : \frac{U}{U^c} \longrightarrow A$$

is the trivial map.

Proof: For $a \in A$, $u \in U$ we have $uau^{-1}a^{-1} \in U^c \subset A$; so $uau^{-1} \in A$ and A is normal in U. The factor group $G = U/A$ is then a finite abelian group, and upon writing $\#(G) = (U : A) = n$ we have $(U : U^c) = ne$. In the exact sequence

$$(1) \longrightarrow A \longrightarrow U \longrightarrow G \longrightarrow (1)$$

A may be made into a G-module in the usual way (that is, by conjugation via a section of G in U). Let $\alpha \in H^2(G, A)$ be the cohomology class associated with this group extension, and let B be the splitting module for α as constructed in (5-4-1). Since $\bar{V}^e_{U\to A}$ means apply $\bar{V}_{U\to A}$ and then raise to the eth power, it follows from (5-4-5) and (5-4-6) that it suffices to verify that

$$eS : B \longrightarrow A$$

is the trivial map. (Here, again, A is viewed additively as a subgroup of B and multiplicatively as a subgroup of U.)

Since $B/A \approx I$ is $\mathbf{Z}$-free on $(n - 1)$ generators, A/U^c is a finite abelian group and U^c is finitely generated, it follows that B is a finitely generated $\mathbf{Z}$-module. In addition, $B/(IB)$ is finite abelian with $(B : IB) = (U : U^c) = ne$. Let $b_1, \ldots, b_m \in B$ be elements which are representatives of a basis of $B/(IB)$, and if e_i, $i = 1, \ldots, m$, is the order of b_i (mod IB), then $ne = \prod_1^m e_i$. Now IB, being a subgroup of the finitely generated abelian group B, is itself finitely generated—say by $b_{m+1}, \ldots, b_s \in B$. If we put $e_{m+1} = \cdots = e_s = 1$, then the elements $b_1, \ldots, b_s$ generate B, $e_i b_i \in IB$ for $i = 1, \ldots, s$, and $ne = \prod_1^s e_i$.

Since $B = \sum_{j=1}^s \mathbf{Z} b_j$ and $IB = \sum_{j=1}^s I b_j$, there exist $\lambda_{ij} \in I$ such that $e_i b_i = \sum_{j=1}^s \lambda_{ij} b_j$, $i, j = 1, \ldots, s$. In other words,

$$\sum_{j=1}^{s} (e_i \delta_{ij} - \lambda_{ij}) b_j = 0 \qquad i = 1, \ldots, s$$

Because $\Gamma = \mathbf{Z}[G]$ is a commutative ring, we may write

$$\Delta = \det\,(e_i\delta_{ij} - \lambda_{ij}) \in \mathbf{Z}[G]$$

and Cramer's rule says that

$$\Delta b_j = 0 \qquad\qquad j = 1,\ldots, s$$

Therefore, $\Delta B = (0)$, and by (5-4-4) there exists $t \in \mathbf{Z}$ such that $\Delta = tS$.

It suffices then to show that $t = e$, and this follows from a simple application of the map $\varepsilon : \mathbf{Z}[G] \longrightarrow \mathbf{Z}$. In fact, $\varepsilon(\Delta) = t\varepsilon(S) = tn$ and since $\varepsilon(\lambda_{ij}) = 0$,

$$\varepsilon(\Delta) = \varepsilon\{\det\,(e_i\delta_{ij} - \lambda_{ij})\} = \det\,(e_i\delta_{ij}) = \prod_1^s e_i = ne$$

Thus, $t = e$ and the proof is complete. ▮

5-4-8. Principal Ideal Theorem. Let U be a multiplicative group for which U/U^c is finite and $U^c/(U^c)^c$ is finitely generated; then

$$\bar{V}_{U\to U^c} : \frac{U}{U^c} \longrightarrow \frac{U^c}{(U^c)^c}$$

is the trivial map.

Proof: Let $W = U/(U^c)^c$; then $W^c = U^c/(U^c)^c$, $(W^c)^c = (1)$, $W/W^c \approx U/U^c$, $W^c/(W^c)^c \approx U^c/(U^c)^c$, and W satisfies the hypotheses of the theorem. The following diagram commutes:

$$\begin{array}{ccc} \dfrac{U}{U^c} & \xrightarrow{\bar{V}_{U\to U^c}} & \dfrac{U^c}{(U^c)^c} \\ \downarrow & & \downarrow \\ \dfrac{W}{W^c} & \xrightarrow{\bar{V}_{W\to W^c}} & \dfrac{W^c}{(W^c)^c} \end{array}$$

Therefore, it suffices to prove our result for W—in other words, we may assume at the start that $(U^c)^c = (1)$, and prove the result for such U only.

In this situation, U^c is abelian, and (5-4-7) may be applied with $A = U^c$. Since $(A : U^c) = e = 1$, the proof is complete. ∎

5-4-9. Example. Let us exhibit an example, due to Witt [80], which shows that in the principal ideal theorem the hypothesis that $U^c/(U^c)^c$ is finitely generated cannot be discarded.

Consider the quaternions Q over the reals, $\mathbf{R}$; thus, Q is a 4-dimensional space over $\mathbf{R}$ with basis $\{1, i, j, k\}$, and multiplication in Q is determined by the relations $i^2 = j^2 = k^2 = -1$, $ij = k = -ji$, $jk = i = -kj$, $ki = j = -ik$. Of course Q is a division ring which contains $\mathbf{C} = \mathbf{R} \cdot 1 \oplus \mathbf{R} \cdot i$. In the multiplicative group of Q consider

$$U = \{j^m e^{\pi i r} \mid m \in \mathbf{Z}, \quad r \in \mathbf{Q}\}$$

Because $e^{\pi i(r+1)} = -e^{\pi i r}$ and $j^2 = -1$, we need only take $m = 0$ or 1. Now, U is a group. Closure under multiplication follows from $e^{\pi i r} j = (\cos \pi r + i \sin \pi r)\, j = j \cos \pi r - ji \sin \pi r = je^{\pi i(-r)}$. Associativity and the existence of an identity are clear. The inverse of $e^{\pi i r}$ is $e^{\pi i(-r)}$, and the inverse of $je^{\pi i r}$ is $-je^{\pi i r} = je^{\pi i(r+1)}$.

Let $G = \{+1, -1\}$ be the cyclic group of order 2, and map U onto G by $j^m e^{\pi i r} \longrightarrow (-1)^m$. This is a homomorphism with kernel $A = \{e^{\pi i r} \mid r \in \mathbf{Q}\}$. Thus, $U/A \approx G$ is abelian and $U^c \subset A$. Furthermore, the commutator in U of j and $e^{\pi i r}$ is

$$(j)(e^{\pi i r})(-j)(e^{-\pi i r}) = e^{+2\pi i(-r)}$$

—and since $\{e^{2\pi i(-r)} \mid r \in \mathbf{Q}\} = \{e^{2\pi i r} \mid r \in \mathbf{Q}\}$, it follows that $A = U^c$. Therefore, U/U^c is finite, and $U^c/(U^c)^c = U^c$ (since U^c is abelian) is not finitely generated.

It remains to show that $\overline{V}_{U \to U^c}$ is not the trivial map. In the extension $(1) \longrightarrow U^c = A \longrightarrow U \longrightarrow G \longrightarrow (1)$, take as representatives of G in U, $u_1 = 1$ and $u_{-1} = j$. Consequently, $a_{1,1} = a_{1,-1} = a_{-1,1} = 1$ and $a_{-1,-1} = j^2 = -1$; and then using (5-2-5) $\overline{V}_{U \to U^c}(u_{-1}U^c) = \prod_{\sigma \in G} a_{\sigma,-1} = -1$, so that $\overline{V}_{U \to U^c}$ is not trivial and the conclusion of the principal ideal theorem does not hold.

CHAPTER

VI

Abstract Class Field Theory

This chapter is concerned with the purely cohomological aspects of class field theory (both local and global), and its contents, which are due to Artin and Tate, are rather standard (see, for example, [6], [43], [46]). The central feature is the somewhat formal notion of a class formation. Its function is to provide a set of axioms from which one is able to prove, by cohomological techniques, the basic theorems of class field theory—especially, the reciprocity law with its many consequences. It follows then that, from this point of view, the verification of the axioms for a class formation constitutes the arithmetic part of class field theory. Needless to say, this arithmetic part, which is a deep and complicated story in itself, is beyond the scope of this book.

6-1. FORMATIONS

6-1-1. Definition. A **formation** is a composite object $\{G, \{G_F\}, A\}$ where G is a group (usually infinite), A is a G-module, $\{G_F\}_{F\in\Sigma}$ is a nonempty collection of subgroups G_F of G which are indexed by some set Σ, and such that the following conditions hold:

(1) $A = \bigcup_{F\in\Sigma} A^{G_F}$ where, as usual, A^{G_F} denotes the set of

elements of A invariant under the action of every element of G_F. In other words, this condition says that every element of A is left fixed by some member of the family $\{G_F\}$. For simplicity, we denote A^{G_F} by A_F. With a view to the applications, the indices $F \in \Sigma$ are called **fields**, A is called the **formation module**, and A_F is called the ***F*-level** of the formation.

(2) G is a topological group (this includes Hausdorff) such that

(a) $\{G_F\}$ is a fundamental system of neighborhoods of the identity in G.

(b) Every open subgroup of G is a member of $\{G_F\}$.

(c) Every G_F is of finite index in G.

The group G is called the **Galois group** of the formation.

6-1-2. Remark. Condition (2) says that the open subgroups of G form a fundamental system of neighborhoods of the identity, and that they are all of finite index in G. Thus, G being totally disconnected but not quite compact, is very much like a Galois group in the usual sense. Actually, the topological properties of G do not play a role until much later, and it is convenient to formulate condition (2) in the following completely equivalent, but entirely algebraic, fashion:

(2) G is a group such that

(a′) The intersection of two members of $\{G_F\}$ is an element of $\{G_F\}$. (Of course, the intersection of two subgroups of finite index is of finite index.) Since we *assume* that for $F, F' \in \Sigma$, $G_F = G_{F'} \Longleftrightarrow F = F'$, this condition says that given G_{F_1}, G_{F_2} in $\{G_F\}$ there exists a unique element of Σ, called the **composite** of the fields F_1 and F_2, denoted by F_1F_2, such that $G_{F_1} \cap G_{F_2} = G_{F_1F_2}$.

(b′) Any conjugate of an element of $\{G_F\}$ is again an element of $\{G_F\}$. In other words, for any $G_F \in \{G_F\}$ and $\sigma \in G$, $G_F^\sigma = \sigma G_F \sigma^{-1}$ is an element of $\{G_F\}$—so there exists an $F^\sigma \in \Sigma$ (F^σ is said to be a **conjugate field** of F) such that $G_{F^\sigma} = G_F^\sigma$. (We may also write $F^\sigma = \sigma F$.)

(c′) $\bigcap_{F\in\Sigma} G_F = \{1\}$.

(d′) Every element of $\{G_F\}$ is of finite index in G.

(e′) Any subgroup of G which contains an element of $\{G_F\}$ is itself an element of $\{G_F\}$.

It is well-known (see, for example, [59, p. 55] or [10, p. 6]) that conditions (a′), (b′), (c′) together say that G can be made into a topological group with $\{G_F\}$ as a fundamental system of neighborhoods of the identity. It is, therefore, immediate that the two versions of (2) are equivalent.

6-1-3. Example. The canonical example of a formation, which provides motivation and suggests terminology, is that of ordinary Galois theory. Let k be any field and let Ω be a Galois extension (this means algebraic, normal, and separable) of k. In particular, Ω is often taken as a separable closure of k. Put $G = \mathscr{G}(\Omega/k)$, the Galois group of the extension. Let further $\Sigma = \{F, E, K, L, ...\}$ denote the set of all finite extensions of k which are contained in Ω. For every $F \in \Sigma$, let $G_F = \mathscr{G}(\Omega/F)$. As usual, G may be made into a topological group (which is compact and totally disconnected) by taking $\{G_F\}$, $F \in \Sigma$, as a fundamental system of neighborhoods of the identity. Each G_F is of finite index in G; in fact, $(G : G_F) = [F : k]$. The fundamental theorem of Galois theory for infinite extensions says, in particular, that there is a 1–1 correspondence between the set of all intermediate fields between k and Ω and the set of all closed subgroups of G; consequently, every open subgroup of G is a G_F for some $F \in \Sigma$. Let $A = \Omega^*$ be the multiplicative group of Ω; this becomes a G-module in a natural way. By Galois theory, $A_F = A^{G_F}$ is precisely the multiplicative group of F—that is, $A_F = F^*$. It is now immediate that $\{G, \{G_F\}, A\}$ is a formation.

6-1-4. Remarks. In accord with the canonical example, we introduce the following notations and definitions for an arbitrary formation.

If F, $E \in \Sigma$ we write $E \supset F$ whenever $G_E \subset G_F$, and then say that F is a **subfield** of E, or that E is an **extension** field of F. Note that, in this situation, $A_F \subset A_E$ since $A^{G_F} \subset A^{G_E}$. A pair of fields $F \subset E$ determines a **layer** of the formation; denote it by E/F. A_F is called the **ground level** and A_E the **top level** of the layer.

Define the **degree** $[E : F]$ of the layer E/F by $[E : F] = (G_F : G_E)$; note that it is finite.

The layer K/F is said to be **normal** when G_K is a normal subgroup of G_F (we shall try to reserve the notations $K/F, L/F, L/E$ for normal layers and E/F for ordinary ones); the factor group G_F/G_K, which we denote by $G_{K/F}$ or $G(K/F)$ and call the **Galois group** of the normal layer, is finite and acts in a natural way on the top layer $A_K = A^{G_K}$. Namely, for $\sigma \in G_F$, let $\bar\sigma$ be the corresponding element of $G_{K/F}$ and for $a \in A_K$ put $\bar\sigma a = \sigma a$. This is well-defined and $\bar\sigma a \in A_K$, so that A_K is a $G_{K/F}$-module. Moreover, $\bar\sigma a = a \, \forall \bar\sigma \in G_{K/F} \Longrightarrow \sigma a = a \, \forall \sigma \in G_F \Longrightarrow a \in A_F$, so that $A_K^{G_{K/F}} = A_F$—or in other words, the ground level of a normal layer consists precisely of all elements of the top level which are invariant under the action of the Galois group of the layer. (A symbolic proof of this statement looks like $(A_K)^{G_{K/F}} = (A^{G_K})^{G_F/G_K} = A^{G_F} = A_F$.) The normal layer K/F is said to be abelian, cyclic, solvable, etc., when these properties hold for $G_{K/F}$.

In virtue of all this, the **cohomology groups** of a normal layer are defined as

$$H^r\left(\frac{K}{F}\right) = H^r(G_{K/F}, A_K) = H^r\left(\frac{G_F}{G_K}, A^{G_K}\right) \qquad r \in \mathbf{Z}$$

and the order of $H^r(K/F)$ will be denoted by $h_r(K/F)$. Our standard cohomological maps occur in this context in the following forms: If $F \subset E \subset K$ with K/F normal and $\sigma \in G$, then

$$\operatorname{res}_{G_F/G_K \to G_E/G_K} : H^r\left(\frac{G_F}{G_K}, A^{G_K}\right) \longrightarrow H^r\left(\frac{G_E}{G_K}, A^{G_K}\right) \qquad r \in \mathbf{Z}$$

$$\operatorname{cor}_{G_E/G_K \to G_F/G_K} : H^r\left(\frac{G_E}{G_K}, A^{G_K}\right) \longrightarrow H^r\left(\frac{G_F}{G_K}, A^{G_K}\right) \qquad r \in \mathbf{Z}$$

$$\sigma_* : H^r\left(\frac{G_F}{G_K}, A^{G_K}\right) \longrightarrow H^r\left(\left(\frac{G_F}{G_K}\right)^{\sigma}, \sigma(A^{G_K})\right) \qquad r \in \mathbf{Z}$$

(Note that $(G_F/G_K)^\sigma = G_{F^\sigma}/G_{K^\sigma}$, $\sigma(A^{G_K}) = A^{G_K^\sigma} = A^{G_{K^\sigma}}$, and that σ_* here is not exactly the conjugation map discussed in Section 2-3 because $G_{K/F}$ and G_{K^σ/F^σ} are not subgroups of the same group.

Instead, σ_* is gotten from the homomorphism of pairs (see Section 2-3)

$$(\sigma^{-1}, \sigma) : (G_{K/F}, A_K) \longrightarrow (G_{K^\sigma/F^\sigma}, \sigma(A_K))$$

—and all the usual properties of conjugation hold.) If $F \subset K \subset L$, with L/F and K/F both normal, then

$$\inf_{G_F/G_K \to G_F/G_L} : H^r\left(\frac{G_F}{G_K}, A^{G_K}\right) \longrightarrow H^r\left(\frac{G_F}{G_L}, A^{G_L}\right) \qquad r \geqslant 1$$

It is clearly notationally convenient and helpful to write these maps as:

$$\operatorname{res}_{K/F\to K/E} : H^r\left(\frac{K}{F}\right) \longrightarrow H^r\left(\frac{K}{E}\right) \qquad r \in \mathbf{Z}$$

$$\operatorname{cor}_{K/E\to K/F} : H^r\left(\frac{K}{E}\right) \longrightarrow H^r\left(\frac{K}{F}\right) \qquad r \in \mathbf{Z}$$

$$\sigma_* : H^r\left(\frac{K}{F}\right) \longrightarrow H^r\left(\frac{K^\sigma}{F^\sigma}\right) \qquad r \in \mathbf{Z}$$

$$\inf_{K/F\to L/F} : H^r\left(\frac{K}{F}\right) \longrightarrow H^r\left(\frac{L}{F}\right) \qquad r \geqslant 1$$

The discussion in this section has been based, in large part, upon carrying over the statements of ordinary Galois theory to formations; however, this is not always valid. For example, the analog of the fundamental theorem of Galois theory does not apply to formations (if it were really needed, we would have included it among the axioms)—that is, the correspondence $G_E \longleftrightarrow A_E$ need not be 1–1. In fact, it is perfectly possible that G acts trivially on the formation module A, so that all the levels $A_E = A$. On the other hand, among the results which do carry over, the following is quite useful.

6-1-5. Proposition. (i) Given a finite number of layers E_i/F $i = 1, 2, \ldots, s$ over the same ground field, there exists a normal layer K/F containing them all (that is, $F \subset E_i \subset K$).

(ii) Every subgroup of the Galois group $G_{K/F}$ of a normal layer is of form $G_{K/E}$ for some intermediate field E.

Proof: (i) For each i, G_{E_i} is of finite index in G_F; so as σ runs over G_F the number of distinct conjugates $G_{E_i}^{\sigma} = G_{E_i^{\sigma}}$ is finite. Consequently, $\bigcap_{i,\sigma} G_{E_i^{\sigma}} = G_K$ for some $K \in \Sigma$, and it is clear that G_K is normal in G_F and $E_i \subset K$.

(ii) Every subgroup of $G_{K/F} = G_F/G_K$ is of form H/G_K where H is a group between G_K and G_F. Thus, $H = G_E$ for some $E \in \Sigma$; so $F \subset E \subset K$ and the subgroup is $G_E/G_K = G_{K/E}$. ▮

6-1-6. Remark. Given any two fields F_1 and F_2 in a formation we have defined (see (6-1-2)) their compositum F_1F_2 by $G_{F_1} \cap G_{F_2} = G_{F_1F_2}$. Since $G_{F_1F_2}$ is the biggest subgroup of G contained in both G_{F_1} and G_{F_2}, it follows that F_1F_2 is the smallest field containing both F_1 and F_2. We may further define the intersection $F_1 \cap F_2$ by $G_{F_1 \cap F_2} = [G_{F_1} \cup G_{F_2}]$ (the subgroup of G generated by G_{F_1} and G_{F_2}). Since $[G_{F_1} \cup G_{F_2}]$ is the smallest subgroup of G containing G_{F_1} and G_{F_2}, it follows that $F_1 \cap F_2$ is the largest field contained in both F_1 and F_2.

Because M/F abelian means that $G_{M/F} = G_F/G_M$ is abelian, we see that M/F is abelian $\Longleftrightarrow$ $G_M \supset G_F^c$. From this criterion it follows that if M_1/F and M_2/F are abelian then so is $(M_1M_2)/F$. Also, any extension E/F contains a unique maximal abelian subextension M/F. To see this, note that the smallest subgroup H of G_F which contains G_E and for which G_F/H is abelian is precisely $H = [G_E \cup G_F^c]$. If we define M by $G_M = [G_E \cup G_F^c]$, M/F is the desired extension.

We leave it to the reader to check that if M/F is abelian and E/F is an arbitrary extension, then $(ME)/E$ is abelian.

6-1-7. Exercise. Let E/F be an arbitrary layer of degree n, say. If $G_F = \bigcup_{i=1}^{n} \sigma_i G_E$ is a coset decomposition, and for $a \in A_E$ we put $N_{E/F}a = \prod_1^n \sigma_i a = \prod_1^n a^{\sigma_i}$, then $N_{E/F} : A_E \longrightarrow A_F$ is a homomorphism, called the **norm** map, which is independent of the choice of coset representatives, and satisfies the usual properties of the norm in field theory. Among these we have:

(1) If K/F is normal then $N_{K/F}a = \prod_{\rho \in G_{K/F}}^{n} \rho a$, $a \in A_K$—that is, $N_{K/F}$ is the "customary" norm.

(2) If $F \subset E \subset E'$, then $N_{E'/F} = N_{E/F} \circ N_{E'/E}$.

(3) If $a \in A_F$, then $N_{E/F}a = a^{[E:F]}$.

(4) If $F \subset E \subset E'$, then $N_{E'/F}A_{E'} \subset N_{E/F}A_E$.

(5) If $F \subset E \subset K$ with K/F normal, then $N_{K/F}A_K \subset N_{K/E}A_K$.

(Note that some of the properties of the norm have already been treated in (2-4-1).)

6-2. FIELD FORMATIONS

A formation is said to be a **field formation** if it satisfies

Axiom I. $H^1(K/F) = (1)$ for every normal layer K/F.

According to (1-5-4), our canonical example of a formation is a field formation. However, for certain formations (namely, those of global class field theory) Axiom I is not accessible directly but rather through an ostensibly weaker formulation.

6-2-1. Proposition. In a formation, Axiom I is equivalent to

Axiom I′. $H^1(K/F) = (1)$ for every cyclic layer K/F of prime degree.

Proof: This is a special case (with $r = 1$, $\nu = 0$) of the following lemma which shall be used later in the case $r = 2$, $\nu = 1$. ▌

6-2-2. Lemma. Suppose that for the formation $\{G, \{G_F\}, A\}$ there exists a positive integer r such that for all $F \subset K \subset L$ with K/F, L/F both normal, the sequence

$$0 \longrightarrow H^r\left(\frac{K}{F}\right) \xrightarrow{\text{inf}} H^r\left(\frac{L}{F}\right) \xrightarrow{\text{res}} H^r\left(\frac{L}{K}\right)$$

is exact. Suppose, further, that for a fixed integer $\nu \geqslant 0$, $h_r(K/F) \mid [K:F]^\nu$ for all cyclic layers K/F of prime degree; then $h_r(K/F) \mid [K:F]^\nu$ for all normal layers K/F.

Proof: Note first that since the inflation–restriction sequence is always exact in dimension 1 (see (3-4-2)), (6-2-1) does indeed follow from the case $r = 1$, $\nu = 0$ of the lemma. As for the proof of the lemma, we show first that the desired divisibility relation holds for $[K:F] = p^n$, p prime. (Of course, a statement of divisibility includes the finiteness of $h_r(K/F)$.) For $n = 1$, this is asserted by the hypothesis of the lemma. Suppose then, inductively, that divisibility holds for all $m \leqslant n$, we must show that it holds for $n + 1$—that is for p^{n+1}. The group $G_{K/F}$ is then of order p^{n+1} and (by standard properties of p-groups) contains a proper normal subgroup—which has form $G_{K/K'}$ where $F \subset K' \subset K$. Therefore, $p \leqslant [K:K'] \leqslant p^n$ and $p \leqslant [K':F] \leqslant p^n$, so that by the induction hypothesis $h_r(K/K') | [K:K']^\nu$ and $h_r(K'/F) | [K':F]$. Now, it follows from the exact sequence

$$0 \longrightarrow H^r\left(\frac{K'}{F}\right) \xrightarrow{\text{inf}} H^r\left(\frac{K}{F}\right) \xrightarrow{\text{res}} H^r\left(\frac{K}{K'}\right)$$

that $H^r(K/F)$ is a finite group with

$$h_r\left(\frac{K}{F}\right) \Big| \, h_r\left(\frac{K'}{F}\right) h_r\left(\frac{K}{K'}\right) \Big| \, [K':F]^\nu [K:K']^\nu = [K:F]^\nu$$

and this part of the proof is complete.

It remains to consider the case $[K:F] = n$, n not a power of a prime. For each prime $p \mid n$ there exists (by (6-1-5)) a field E_p, with $F \subset E_p \subset K$, such that G_{K/E_p} is a p-Sylow subgroup of $G_{K/F}$. By (3-1-15), the sequence $0 \longrightarrow H^r(K/F)_p \xrightarrow{\text{res}} H^r(K/E_p)$ is exact for each such p. The right side is finite (in fact, by the preceding, $h_r(K/E_p) | [K:E_p]^\nu$), hence so is the left, and $\#\{H^r(K/F)_p\} | \, h_r(K/E_p)$. Since the abelian torsion group $H^r(K/F)$ is the direct product of its p-primary parts, we have

$$h_r\left(\frac{K}{F}\right) = \prod_{p|n} \# \left\{H^r\left(\frac{K}{F}\right)_p\right\} \Big| \prod_{p|n} h_r\left(\frac{K}{E_p}\right) \Big| \prod_{p|n} [K:E_p]^\nu = [K:F]^\nu$$

This completes the proof. ▮

6-2-3. Exercise. In a formation, suppose that for some

integer $\nu \geqslant 0$, $h_0(K/F) \mid [K:F]^\nu$ for all cyclic K/F of prime degree; then $h_0(K/F) \mid [K:F]^\nu$ for all normal layers K/F.

6-2-4. Remark. In a formation, if $F \subset K \subset L \subset M$ with K/F, L/F, and M/F all normal, then by (2-3-6) the following diagram is commutative:

$$\begin{array}{ccc} H^r\left(\dfrac{K}{F}\right) & \xrightarrow{\ \inf_{K/F\to M/F}\ } & H^r\left(\dfrac{M}{F}\right) \\ & {\scriptstyle \inf_{K/F\to L/F}} \searrow \qquad \nearrow {\scriptstyle \inf_{L/F\to M/F}} & \\ & H^r\left(\dfrac{L}{F}\right) & \end{array} \qquad r \geqslant 1$$

If we let $\mathscr{F}$ denote the set of all fields K containing F for which K/F is a normal layer, then $\mathscr{F}$ is partially ordered by inclusion, and for any $K_1, K_2 \in \mathscr{F}$ there exists a field $K \in \mathscr{F}$ with $K_1 \subset K$, $K_2 \subset K$. This means that $\mathscr{F}$ is a **directed set**, and we may form the **injective** or **direct limit** of the groups $H^r(K/F)$ as K runs over $\mathscr{F}$—denote it by

$$H^r\left(\frac{*}{F}\right) = \varinjlim_{K \in \mathscr{F}} H^r\left(\frac{K}{F}\right) \qquad r \geqslant 1$$

(Throughout the discussion here, we assume that the reader is acquainted with the basic properties of direct limits as found, for example, in [24, Chapter VIII].) Now, $H^r(*/F)$ is an abelian group, and for each normal layer K/F there is defined a homomorphism

$$\inf_{K/F\to */F} : H^r\left(\frac{K}{F}\right) \longrightarrow H^r\left(\frac{*}{F}\right)$$

which we call **symbolic inflation**. Furthermore, if $F \subset K \subset L$ with K/F, L/F normal, then

$$\inf_{K/F\to */F} = \inf_{L/F\to */F} \circ \inf_{K/F\to L/F}$$

Suppose next that E/F is any layer in our formation, and consider

$F \subset E \subset K \subset L \subset M$ with M/F, L/F, K/F all normal. According to results from Chapter II, the following diagram commutes

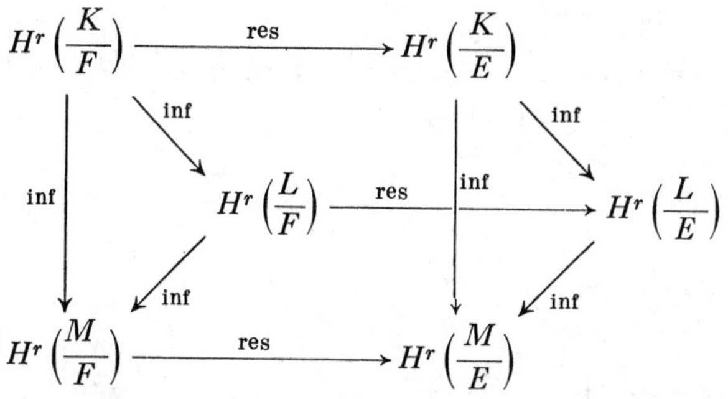

Let $\mathscr{S} = \{K \in \mathscr{F} \mid K \supset E\}$; then $\mathscr{S}$ is a **cofinal** subset of $\mathscr{F}$ (which means that for any element of $\mathscr{F}$ there exists an element of $\mathscr{S}$ containing it) and the direct limit of the $H^r(K/F)$ over $\mathscr{S}$ is the same (isomorphic is more precise) as the direct limit over $\mathscr{F}$. In other words, $H^r(*/F) = \lim_{K \in \mathscr{S}} H^r(K/F)$. Moreover, $\mathscr{S}$ is a cofinal subset of $\mathscr{E}$, the set of all fields K containing E for which K/E is a normal layer. It follows that there exists a **symbolic restriction** homomorphism

$$\text{res}_{F\to E} : H^r\left(\frac{*}{F}\right) \longrightarrow H^r\left(\frac{*}{E}\right) \qquad r \geqslant 1$$

with the expected properties. In the same way, there exists a **symbolic corestriction** map

$$\text{cor}_{E\to F} : H^r\left(\frac{*}{E}\right) \longrightarrow H^r\left(\frac{*}{F}\right) \qquad r \geqslant 1$$

and we know that (when A is additive)

$$\text{cor}_{E\to F} \circ \text{res}_{F\to E} = \text{multiplication by} \quad [E:F]$$

because this relation holds in the layers (see (2-4-9)). Finally, an application of this procedure leads to the existence of a **symbolic conjugation** map

$$\sigma_*^F = \sigma_* : H^r\left(\frac{*}{F}\right) \longrightarrow H^r\left(\frac{*}{F^\sigma}\right) \qquad \sigma \in G, \quad r \geqslant 1$$

which commutes with symbolic restriction and corestriction.

With these general considerations behind us, let us turn to the specific case of primary interest—namely, the case $r = 2$ in a field formation.

6-2-5. Proposition. In a field formation, if $F \subset K \subset L$ with K/F, L/F normal, then the sequence

$$0 \longrightarrow H^2\left(\frac{K}{F}\right) \xrightarrow{\inf_{K/F\to L/F}} H^2\left(\frac{L}{F}\right) \xrightarrow{\operatorname{res}_{L/F\to L/K}} H^2\left(\frac{L}{K}\right)$$

is exact, and so is

$$0 \longrightarrow H^2\left(\frac{K}{F}\right) \xrightarrow{\inf_{K/F\to */F}} H^2\left(\frac{*}{F}\right) \xrightarrow{\operatorname{res}_{F\to K}} H^2\left(\frac{*}{K}\right)$$

Proof: The exactness of the first sequence is immediate from (3-4-3) since $H^1(K/F) = (1)$. The exactness of the second sequence then follows from the fact that exactness is preserved under direct limits. ▮

6-2-6. Remark. The preceding result says, in particular, that in a field formation the maps $\inf_{K/F\to L/F} : H^2(K/F) \longrightarrow H^2(L/F)$ and $\inf_{K/F\to */F} : H^2(K/F) \longrightarrow H^2(*/F)$ are monomorphisms. It is convenient to identify, and to then view these maps as inclusions; the transitivity of the various inflation maps makes these identifications permissible. We have, therefore,

$$H^2\left(\frac{*}{F}\right) = \varinjlim_{K\in\mathscr{F}} H^2\left(\frac{K}{F}\right) = \bigcup_{K\in\mathscr{F}} H^2\left(\frac{K}{F}\right)$$

The group $H^2(*/F)$ is known as the **Brauer group** at F. (The reasons for this name are historical, and arise in local class field theory from the interconnections between central simple algebras, crossed products, and 2-dimensional cohomology groups. A discussion of these topics is beyond the scope of this book, the interested reader may learn about them in [65], [67], [78].)

It is worth emphasizing how computations are done in the Brauer group—namely, in the layers. Thus, suppose that $\alpha_1, \alpha_2 \in H^2(*/F)$ Then there exist fields $K_1, K_2 \in \mathscr{F}$ such that $\alpha_1 \in H^2(K_1/F)$ and

$\alpha_2 \in H^2(K_2/F)$. Choose a field $K \in \mathscr{F}$ containing both K_1 and K_2 and consider $\inf_{K_1/F \to K/F} \alpha_1$ and $\inf_{K_2/F \to K/F} \alpha_2$ in $H^2(K/F)$. (We may note in passing that $\alpha_1 = \alpha_2$ in $H^2(*/F)$ if and only if $\inf_{K_1/F \to K/F} \alpha_1 = \inf_{K_2/F \to K/F} \alpha_2$, and also that when this relation holds for one such K then it holds for all such K.) To add α_1 and α_2 in $H^2(*/F)$ one simply takes $\inf_{K_1/F \to K/F} \alpha_1 + \inf_{K_2/F \to K/F} \alpha_2$ in $H^2(K/F)$ and views the result in $H^2(*/F)$. Of course, this procedure is independent of the choice of $K \in \mathscr{F}$.

Returning to the symbolic inflation–restriction sequence of (6-2-5), we see that the symbolic restriction $\mathrm{res}_{F \to K}$ enables us to isolate the subgroups of $H^2(*/F)$ which are associated with the normal layers; in other words, for $\alpha \in H^2(*/F)$,

$$\alpha \in H^2\left(\frac{K}{F}\right) \iff \mathrm{res}_{F \to K}\, \alpha = 0$$

It should also be noted how $\mathrm{res}_{F \to K}$ is computed in terms of the layers for an arbitrary layer E/F. Given $\alpha \in H^2(*/F)$ there exists a normal layer K/F with $\alpha \in H^2(K/F)$. Consider any field L which is normal over F and contains both K and E; then

$$\mathrm{res}_{F \to E}\, \alpha = \mathrm{res}_{L/F \to L/E}\, (\inf_{K/F \to L/F} \alpha) \in H^2\left(\frac{L}{E}\right)$$

6-3. CLASS FORMATIONS

A **class formation** is a field formation which also satisfies

Axiom II. For each field $F \in \Sigma$ there exists a monomorphism

$$\mathrm{inv}_F : H^2\left(\frac{*}{F}\right) \longrightarrow \frac{\mathbf{Q}}{\mathbf{Z}}$$

such that

(a) If K/F is a normal layer of degree n, then inv_F maps the subgroup $H^2(K/F)$ of $H^2(*/F)$ onto the subgroup $((1/n)\mathbf{Z})/\mathbf{Z}$ of $\mathbf{Q}/\mathbf{Z}$.

(b) For any layer E/F we have

$$\text{inv}_E \circ \text{res}_{F \to E} = [E : F]\, \text{inv}_F$$

For $\alpha \in H^2(*/F)$, the element $\text{inv}_F\, \alpha \in \mathbf{Q}/\mathbf{Z}$ is called the **invariant** of α.

This entire axiom is designed to give a complete description of the Brauer groups $H^2(*/F)$. Since inv_F is a monomorphism, the invariants describe the elements of $H^2(*/F)$ uniquely. It should be noted that no assertion is made about the uniqueness of inv_F. On the other hand, according to (b), the maps inv_F at the various levels are related.

From (a), it follows that if K/F is a normal layer, then $H^2(K/F)$ is cyclic of order $[K : F] = n$; in fact, inv_F provides an isomorphism of $H^2(K/F)$ onto $((1/n)\mathbf{Z})/\mathbf{Z}$. The element $(1/n)(\text{mod } \mathbf{Z})$ (we shall usually be careless and omit reference to mod $\mathbf{Z}$ in such situations) is a generator of the cyclic group $((1/n)\mathbf{Z})/\mathbf{Z}$. Thus, there exists a unique element $\alpha_{K/F}$ of $H^2(K/F)$ such that

$$\text{inv}_F\, \alpha_{K/F} = \frac{1}{n} = \frac{1}{[K : F]}$$

Thus $\alpha_{K/F}$ is a generator of the cyclic group $H^2(K/F)$, and is known as the **canonical class** or **fundamental class** of the layer K/F. If $F \subset K \subset L$ with K/F and L/F normal, then

$$\inf_{K/F \to L/F} \alpha_{K/F} = [L : K] \alpha_{L/F}$$

because both sides are elements of $H^2(L/F)$ which have the same invariant.

6-3-1. Proposition. In a class formation, let $F \subset E \subset K$ with K/F normal, then the (symbolic) restriction map

$$\text{res}_{F \to E} : H^2\left(\frac{K}{F}\right) \longrightarrow H^2\left(\frac{K}{E}\right)$$

is an epimorphism; in fact:

$$\text{res}_{F \to E}\, \alpha_{K/F} = \alpha_{K/E}$$

Furthermore, for any layer E/F

$$\operatorname{res}_{F\to E} : H^2\left(\frac{*}{F}\right) \longrightarrow H^2\left(\frac{*}{E}\right)$$

is an epimorphism.

Proof: We have

$$\operatorname{inv}_E(\operatorname{res}_{F\to E}\alpha_{K/F}) = [E:F]\operatorname{inv}_F \alpha_{K/F} = \frac{[E:F]}{[K:F]} = \frac{1}{[K:E]} = \operatorname{inv}_E \alpha_{K/E}$$

(where it is understood that everything is mod $\mathbf{Z}$), and since inv_E is a monomorphism, $\operatorname{res}_{F\to E}\alpha_{K/F} = \alpha_{K/E}$. Consequently, $\operatorname{res}_{F\to E}$ maps $H^2(K/F)$ onto $H^2(K/E)$, and since $H^2(*/F) = \bigcup H^2(K/F)$ where K runs over the normal extensions of F which contain E it follows that $\operatorname{res}_{F\to E}$ maps $H^2(*/F)$ onto $H^2(*/E)$. This says, in particular, that for any $E \supset F$ the monomorphism inv_E is determined by inv_F through the relation $\operatorname{inv}_E \circ \operatorname{res}_{F\to E} = [E:F]\operatorname{inv}_F$. ∎

6-3-2. Proposition. If E/F is an arbitrary layer in a class formation, then the symbolic corestriction

$$\operatorname{cor}_{E\to F} : H^2\left(\frac{*}{E}\right) \longrightarrow H^2\left(\frac{*}{F}\right)$$

is a monomorphism which preserves invariants; that is,

$$\operatorname{inv}_F \circ \operatorname{cor}_{E\to F} = \operatorname{inv}_E$$

Furthermore, if $F \subset E \subset K$ with K/F normal, then

$$\operatorname{cor}_{E\to F}\alpha_{K/E} = [E:F]\alpha_{K/F}$$

Proof: Since $\operatorname{res}_{F\to E}$ is onto $H^2(*/E)$, it follows from

$$\operatorname{inv}_F \circ \operatorname{cor}_{E\to F} \circ \operatorname{res}_{F\to E} = \operatorname{inv}_F \circ [E:F] = [E:F]\operatorname{inv}_F = \operatorname{inv}_E \circ \operatorname{res}_{F\to E}$$

that the symbolic corestriction preserves invariants. Combining this with the fact that both inv_E and inv_F are monomorphisms, it is immediate that $\mathrm{cor}_{E\to F}$ is a monomorphism. To verify that $\mathrm{cor}_{E\to F}\,\alpha_{K/E} = [E:F]\,\alpha_{K/F}$ one simply applies inv_F to both sides. ▌

6-3-3. Proposition. For any $F \in \Sigma$ and $\sigma \in G$ in a class formation the symbolic conjugation

$$\sigma_* = \sigma_*^F : H^2\left(\frac{*}{F}\right) \longrightarrow H^2\left(\frac{*}{F^\sigma}\right)$$

is an isomorphism onto which preserves invariants; that is,

$$\mathrm{inv}_{F^\sigma} \circ \sigma_* = \mathrm{inv}_F$$

Furthermore, if K/F is normal then

$$\sigma_*(\alpha_{K/F}) = \alpha_{K^\sigma/F^\sigma}$$

Proof: Corresponding to the full group G there exists a field F_0, which may be considered as a "base field" since it is contained in F. Thus $G = G_{F_0} = \sigma G_{F_0}\sigma^{-1}$ and $F_0^\sigma = F_0$. The conjugation map (for F_0)

$$\sigma_*^{F_0} : H^2\left(\frac{*}{F_0}\right) \longrightarrow H^2\left(\frac{*}{F_0}\right)$$

is the identity; in fact, $H^2(*/F_0) = \bigcup_K H^2(K/F_0)$ where K/F_0 is normal, and $\sigma_*^{F_0}$ is the identity map on each $H^2(K/F_0)$ (by the analog of (2-3-1) for this situation). It is clear that $\mathrm{inv}_{F_0^\sigma} \circ \sigma_* = \mathrm{inv}_{F_0}$.

As for the field F, $\sigma_*^F : H^2(*/F) \longrightarrow H^2(*/F^\sigma)$ is an isomorphism onto because $\sigma_F^* : H^2(K/F) \longrightarrow H^2(K^\sigma/F^\sigma)$ is an isomorphism onto for every normal K/F. Now, an element $\alpha \in H^2(*/F)$ is of form $\mathrm{res}_{F_0\to F}\,\beta$ for some $\beta \in H^2(*/F_0)$, and we have

$$\begin{aligned}
\mathrm{inv}_{F^\sigma}(\sigma_*^F\alpha) &= \mathrm{inv}_{F^\sigma}\,\sigma_*^F\,\mathrm{res}_{F_0\to F}\,\beta = \mathrm{inv}_{F^\sigma}\,\mathrm{res}_{F_0^\sigma\to F^\sigma}(\sigma_*^{F_0}\beta) \\
&= [F^\sigma : F_0^\sigma]\,\mathrm{inv}_{F_0^\sigma}(\sigma_*^{F_0}\beta) = [F : F_0]\,\mathrm{inv}_{F_0}\,\beta \\
&= \mathrm{inv}_F\,\mathrm{res}_{F_0\to F}\,\beta = \mathrm{inv}_F\,\alpha
\end{aligned}$$

This shows that conjugation preserves invariants, and it is immediate that conjugation carries a fundamental class to a fundamental class. ▌

In practice, we are unable to verify the axioms for a class formation directly, especially for the formation of idele classes of global theory. Instead, certain other axioms are verified. We now introduce several additional axioms in a formation and indicate how they are related to the preceding ones.

Axiom 0. For every cyclic layer K/F of prime degree, the Herbrand quotient

$$h_{2/1}\left(\frac{K}{F}\right) = [K:F]$$

This axiom says that for such layers, $h_2(K/F) = [K:F]\, h_1(K/F)$, so that in particular,

$$[K:F] \,\Big|\, h_2\left(\frac{K}{F}\right)$$

With more classical terminology in mind, we say (when this divisibility relation holds) that the **first inequality** holds for cyclic layers of prime degree.

Axiom I″. For every cyclic layer K/F of prime degree, we have the **second inequality**:

$$h_2\left(\frac{K}{F}\right) \,\Big|\, [K:F]$$

6-3-4. Proposition. In any formation,

$$\{0 + I\} \Longleftrightarrow \{0 + I'\} \Longleftrightarrow \{0 + I''\}$$

Proof: Since Axioms I and I′ are equivalent, the first equivalence is trivial. To prove the second equivalence, note that according to Axiom 0, $h_2(K/F) = [K:F]\, h_1(K/F)$ for cyclic layers of prime degree—and then, Axiom I″ holds $\Longleftrightarrow$ $h_2(K/F) = [K:F]$ $\Longleftrightarrow$ $h_1(K/F) = 1$ $\Longleftrightarrow$ Axiom I′ holds. ▮

6-3-5. Proposition. If a formation satisfies $\{0 + \text{I}''\}$, then the second inequality $h_2(K/F) \mid [K:F]$ holds for all normal layers K/F.

Proof: Since Axiom I holds we know (see (6-2-5)) that for any $F \subset K \subset L$ with K/F, L/F normal, the sequence

$$0 \longrightarrow H^2\left(\frac{K}{F}\right) \xrightarrow{\text{inf}} H^2\left(\frac{L}{F}\right) \xrightarrow{\text{res}} H^2\left(\frac{L}{K}\right)$$

is exact. It remains only to apply (6-2-2) in the case $r = 2$, $\nu = 1$. ▮

We turn next to a weaker version of Axiom II.

Axiom II′. To each field $F \in \sum$ there is associated a subgroup $\overline{H^2(*/F)}$ of the Brauer group $H^2(*/F)$ with a monomorphism

$$\overline{\text{inv}}_F : \overline{H^2\left(\frac{*}{F}\right)} \longrightarrow \frac{\mathbf{Q}}{\mathbf{Z}}$$

such that

(a) If there exists a normal layer K/F of degree n, then $\overline{H^2(*/F)}$ contains a cyclic group of order n.

(b) For any layer E/F the symbolic restriction $\text{res}_{F\to E}$ maps $\overline{H^2(*/F)}$ into $\overline{H^2(*/E)}$, and on $\overline{H^2(*/F)}$

$$\overline{\text{inv}}_E \circ \text{res}_{F\to E} = [E:F]\,\overline{\text{inv}}_F$$

6-3-6. Proposition. A formation is a class formation $\Longleftrightarrow$ it satisfies Axioms $\{0 + \text{I}'' + \text{II}'\}$.

Proof: $\Longrightarrow$: It is clear that $\{\text{I} + \text{II}\}$ implies $\{0 + \text{I}'' + \text{II}'\}$.

$\Longleftarrow$: It suffices to show that for every F, $\overline{H^2(*/F)} = H^2(*/F)$— as we may then put $\text{inv}_F = \overline{\text{inv}}_F$. Consider any normal layer K/F with $[K:F] = n$, say. By (6-3-4), Axiom I holds, so according to (6-2-5) the sequence

$$0 \longrightarrow H^2\left(\frac{K}{F}\right) \xrightarrow{\text{inf}_{K/F\to */F}} H^2\left(\frac{*}{F}\right) \xrightarrow{\text{res}_{F\to K}} H^2\left(\frac{*}{K}\right)$$

is exact. Now, II′ says that $\overline{H^2(*/F)}$ contains a cyclic group T_n of order n, and that

$$\overline{\text{inv}}_K\, \text{res}_{F\to K}\, T_n = n\, \overline{\text{inv}}_F\, T_n = \overline{\text{inv}}_F\,(nT_n) = \overline{\text{inv}}_F\,(0) = 0$$

Since $\overline{\text{inv}_K}$ is a monomorphism, we have $\text{res}_{F\to K}\, T_n = (0)$, and then (as observed in (6-2-6)) $T_n \subset H^2(K/F)$. On the other hand, by (6-3-5), $h_2(K/F) \mid [K:F] = n = \#(T_n)$. Therefore, $T_n = H^2(K/F)$, and it follows that $\overline{H^2(*/F)} = H^2(*/F)$. Putting $\text{inv}_F = \overline{\text{inv}_F}$, we note that $\text{inv}_F \{H^2(K/F)\} = ((1/n)\mathbf{Z})/\mathbf{Z}$ since $((1/n)\mathbf{Z})/\mathbf{Z}$ is the unique cyclic subgroup of order n in $\mathbf{Q}/\mathbf{Z}$. This completes the proof. ∎

The importance of Axiom II′ is then that the whole Brauer group $H^2(*/F)$ can be captured by using only certain normal layers K/F which, in practice, may be taken of a simple type. (In local theory one uses the unramified extensions; in global theory one uses the cyclotomic extensions.)

6-3-7. Exercise. A field formation is a class formation $\Leftrightarrow H^2(K/F)$ is cyclic of order $[K:F]$ for every normal layer K/F, and for $F \subset E \subset K$ $\text{res}_{K/F\to K/E}: H^2(K/F) \longrightarrow H^2(K/E)$ is an epimorphism.

Hint: Let F_0 be the base field and consider a chain of fields $F_0 \subset K_1 \subset K_2 \subset \cdots$ such that (i) K_r/F_0 is a normal layer for $r = 1, 2, \ldots$, and (ii) if E/F_0 is any layer, there exists an r for which $[E:F_0] \mid [K_r:F_0]$.

6-3-8. Exercise. Let G be an infinite cyclic multiplicative group, and consider the G-module $A = \mathbf{Z}$ with trivial action. If $\{G_F\}$ is the set of all subgroups of G (excluding (1)), then $\{G, \{G_F\}, A\}$ is a class formation.

6-4. THE MAIN THEOREM

We are now in a position to state the theorem which is the focal point of substantially all of the work in this and earlier chapters; the proof consists essentially of the collation of previous results.

6-4-1. Main Theorem. (Tate). Let K/F be any normal layer in a class formation, let $\alpha_{K/F} \in H^2(K/F) = H^2(G_{K/F}, A_K)$ be its canonical class, and let $\theta = \theta_K^{\mathrm{T}} : A_K \times \mathbf{Z} \longrightarrow A_K$ denote the natural $G_{K/F}$-pairing. Then the map

$$\Theta_{K/F}^n : H^n(G_{K/F}, \mathbf{Z}) \longrightarrow H^{n+2}(G_{K/F}, A_K) = H^{n+2}\left(\frac{K}{F}\right)$$

whose action on $\xi \in H^n(G_{K/F}, \mathbf{Z})$ is given by

$$\Theta_{K/F}^n \xi = \alpha_{K/F} \smile_\theta \xi$$

is an isomorphism onto for all integers n.

Proof: According to (6-1-5) every subgroup of $G_{K/F}$ is of form $G_{K/E}$ where $F \subset E \subset K$, and by (6-3-1),

$$\mathrm{res}_{G_{K/F} \to G_{K/E}} \alpha_{K/F} = \mathrm{res}_{K/F \to K/E} \alpha_{K/F} = \mathrm{res}_{F \to E} \alpha_{K/F} = \alpha_{K/E} \in H^2\left(\frac{K}{E}\right)$$

Therefore, it suffices to verify the hypotheses of (4-3-8) solely for $G_{K/F}$ and the fundamental class $\alpha_{K/F}$. More precisely, the proof of the main theorem is complete as soon as it has been shown that

(i) $\Theta_{K/F}^{-1}$ is an epimorphism.
(ii) $\Theta_{K/F}^{0}$ is an isomorphism onto.
(iii) $\Theta_{K/F}^{1}$ is a monomorphism.

Now, (i) is clear because $H^1(K/F) = (0)$, and (iii) is clear because (see (1-5-3)) $H^1(G_{K/F}, \mathbf{Z}) = (0)$. As for (ii), if

$$\#(G_{K/F}) = [K : F] = m$$

then $\alpha_{K/F}$ is a generator of the group $H^2(K/F)$ which is cyclic of order m. On the other hand, $H^0(G_{K/F}, \mathbf{Z}) \approx \mathbf{Z}/(m\mathbf{Z})$ (see (1-5-7)), so that $H^0(G_{K/F}, \mathbf{Z})$ is a cyclic group of order m which is generated by $\kappa(1)$. It suffices, therefore, to show that $\alpha_{K/F} \smile_\theta \kappa(1) = \alpha_{K/F}$. From (4-3-6), we know that $\alpha_{K/F} \smile_\theta \kappa(1) = (\theta^1)_* \alpha_{K/F}$, and since $\theta^1 : A_K \longrightarrow A_K$ is the identity map this equals $1_* \alpha_{K/F} = \alpha_{K/F}$. This completes the proof. ▮

6-4-2. Corollary. For any normal layer K/F of degree m in a class formation, we have isomorphisms:

$$\frac{(A_K)_N}{IA_K} \approx H^{-1}\left(\frac{K}{F}\right) \approx H^{-3}(G_{K/F}, \mathbf{Z})$$

$$\frac{A_F}{N_{K/F}A_K} \approx H^{0}\left(\frac{K}{F}\right) \approx H^{-2}(G_{K/F}, \mathbf{Z}) \approx \frac{G_{K/F}}{G^c_{K/F}}$$

$$H^{1}\left(\frac{K}{F}\right) \approx H^{-1}(G_{K/F}, \mathbf{Z}) = (0)$$

$$H^{2}\left(\frac{K}{F}\right) \approx H^{0}(G_{K/F}, \mathbf{Z}) \approx \frac{\mathbf{Z}}{m\mathbf{Z}}$$

$$H^{3}\left(\frac{K}{F}\right) \approx H^{1}(G_{K/F}, \mathbf{Z}) = (0)$$

$$H^{4}\left(\frac{K}{F}\right) \approx H^{2}(G_{K/F}, \mathbf{Z}) \approx \hat{G}_{K/F} = \operatorname{Hom}(G_{K/F}, \mathbf{Z})$$

Proof: The last isomorphism is immediate from (4-4-7). ▮

6-4-3. Proposition. The isomorphisms of the main theorem commute with restriction and corestriction; that is, if $F \subset E \subset K$ with K/F normal, then, for all n

$$\operatorname{res}_{K/F\to K/E} \circ \Theta^n_{K/F} = \Theta^n_{K/E} \circ \operatorname{res}_{K/F\to K/E}$$

$$\operatorname{cor}_{K/E\to K/F} \circ \Theta^n_{K/E} = \Theta^n_{K/F} \circ \operatorname{cor}_{K/E\to K/F}$$

Proof: We must show that the following diagram commutes:

$$\begin{array}{ccc}
H^n(G_{K/F}, \mathbf{Z}) & \xrightarrow{\Theta^n_{K/F}} & H^{n+2}\left(\frac{K}{F}\right) \\
\text{res}\downarrow\uparrow\text{cor} & & \text{res}\downarrow\uparrow\text{cor} \\
H^n(G_{K/E}, \mathbf{Z}) & \xrightarrow{\Theta^n_{K/E}} & H^{n+2}\left(\frac{K}{E}\right)
\end{array}$$

—which says that for $\xi \in H^n(G_{K/F}, \mathbf{Z})$,

$$\operatorname{res}_{K/F\to K/E}(\alpha_{K/F} \smile_\theta \xi) = \alpha_{K/E} \smile_\theta \operatorname{res}_{K/F\to K/E}\xi,$$

and for $\zeta \in H^n(G_{K/E}, \mathbf{Z})$,

$$\operatorname{cor}_{K/E\to K/F}(\alpha_{K/E} \smile_\theta \zeta) = \alpha_{K/F} \smile_\theta \operatorname{cor}_{K/E\to K/F}\zeta.$$

But, since $\alpha_{K/E}$ is the restriction of $\alpha_{K/F}$, these assertions are immediate from (4-3-7). ▮

6-4-4. Proposition. If $F \subset K \subset L$ with L/F and K/F normal, then, for all $n \geqslant 1$

$$\inf_{K/F\to L/F} \circ \Theta^n_{K/F} = [L:K]\Theta^n_{L/F} \circ \inf_{G_{K/F}\to G_{L/F}}$$

Proof: We show that the following diagram commutes:

$$\begin{array}{ccc} H^n(G_{K/F}, \mathbf{Z}) & \xrightarrow{\Theta^n_{K/F}} & H^{n+2}\left(\frac{K}{F}\right) \\ {\scriptstyle [L:K]}\downarrow{\scriptstyle \inf} & & \downarrow{\scriptstyle \inf} \\ H^n(G_{L/F}, \mathbf{Z}) & \xrightarrow{\Theta^n_{L/F}} & H^{n+2}\left(\frac{L}{F}\right) \end{array}$$

In fact, for $\xi \in H^n(G_{K/F}, \mathbf{Z})$, we have (using (4-3-9))

$$\begin{aligned} \inf_{K/F\to L/F}(\alpha_{K/F} \cup_{\theta_K^T} \xi) &= \inf_{K/F\to L/F} \alpha_{K/F} \cup_{\theta_L^T} \inf_{K/F\to L/F} \xi \\ &= [L:K]\alpha_{L/F} \cup_{\theta_L^T} \inf_{K/F\to L/F} \xi \\ &= \alpha_{L/F} \cup_{\theta_L^T} [L:K] \inf_{K/F\to L/F} \xi \end{aligned}$$ ▮

6-4-5. Corollary. Suppose that $F \subset K \subset L$ with L/F and K/F normal. If $[K:F] | [L:K]$, then, for all $n \geqslant 3$,

$$\inf_{K/F\to L/F} : H^n(K/F) \longrightarrow H^n(L/F)$$

is the trivial map.

Proof: In the preceding diagram the horizontal maps are isomorphisms onto, so it suffices to show that $[L:K] \inf_{G_{K/F}\to G_{L/F}}$ is the zero map. But this is trivial since $[K:F]\, H^n(G_{K/F}, \mathbf{Z}) = (0)$ (by (3-1-6)). ▮

This result says that $\inf_{K/F\to */K}$ is a weak map, in general. If there are sufficiently many fields—for example, if the field F has extensions of degree divisible by every integer m—then $\inf_{K/F\to */F}$ is the zero map and $H^n(*/F) = (0)$ for $n \geqslant 3$.

6-4-6. Exercise. The isomorphisms of the main theorem commute with conjugation; in other words, if K/F is a normal layer in a class formation, then the following diagram commutes for all $\sigma \in G$ and all integers n:

$$\begin{array}{ccc} H^n(G_{K/F}, \mathbf{Z}) & \xrightarrow{\Theta^n_{K/F}} & H^{n+2}\left(\frac{K}{F}\right) \\ \downarrow{\sigma_*} & & \downarrow{\sigma_*} \\ H^n(G_{K^\sigma/F^\sigma}, \mathbf{Z}) & \xrightarrow{\Theta^n_{K^\sigma/F^\sigma}} & H^{n+2}\left(\frac{K^\sigma}{F^\sigma}\right) \end{array}$$

6-4-7. Remark. For a normal layer K/F of degree m in a class formation, the main theorem asserts the existence of isomorphisms $\Theta^n_{K/F}$ of $H^n(G_{K/F}, \mathbf{Z})$ onto $H^{n+2}(K/F)$—where $\Theta^n_{K/F}$ is θ-cupping by the canonical class $\alpha_{K/F} \in H^2(K/F)$ on the left (where $\theta = \theta^{\top}_K : A_K \times \mathbf{Z} \longrightarrow A_K$ is the natural $G_{K/F}$ pairing). Given $\zeta \in H^n(G_{K/F}, \mathbf{Z})$ and $\beta \in H^{n+2}(K/F)$, we should like to decide when ζ and β correspond to each other under $\Theta^n_{K/F}$. For this we need to recall and fix some notation. There are natural $G_{K/F}$-pairings $\theta_{\mathbf{Z}} : \mathbf{Z} \times \mathbf{Z} \longrightarrow \mathbf{Z}$ and $\theta_{\mathbf{Q}/\mathbf{Z}} : \mathbf{Z} \times \mathbf{Q}/\mathbf{Z} \longrightarrow \mathbf{Q}/\mathbf{Z}$; the exact $G_{K/F}$-sequence $0 \longrightarrow \mathbf{Z} \longrightarrow \mathbf{Q} \longrightarrow \mathbf{Q}/\mathbf{Z} \longrightarrow 0$ leads to the coboundary map $\delta_* : H^{-n-1}(G_{K/F}, \mathbf{Z}) \longrightarrow H^{-n}(G_{K/F}, \mathbf{Z})$ which is an isomorphism onto; the canonical map in dimension -1, $\eta = \eta_{\mathbf{Q}/\mathbf{Z}} : (\mathbf{Q}/\mathbf{Z})_S = ((1/m)\mathbf{Z})/\mathbf{Z} \longrightarrow H^{-1}(G_{K/F}, \mathbf{Q}/\mathbf{Z})$ is an isomorphism onto.

Now,

$$\beta = \Theta^n_{K/F}\zeta \iff \beta = \alpha_{K/F} \smile_\theta \zeta \iff \beta \smile_\theta \delta_*\chi = (\alpha_{K/F} \smile_\theta \zeta) \smile_\theta \delta_*\chi$$

for all $\chi \in H^{-n-1}(G_{K/F}, \mathbf{Q}/\mathbf{Z})$ (in view of (4-4-8)), and

$$\begin{aligned} (\alpha_{K/F} \smile_\theta \zeta) \smile_\theta \delta_*\chi &= \alpha_{K/F} \smile_\theta (\zeta \smile_{\theta_{\mathbf{Z}}} \delta_*\chi) && \text{(by (4-3-3))} \\ &= (-1)^n \alpha_{K/F} \smile_\theta \delta_*(\zeta \smile_{\theta_{\mathbf{Q}/\mathbf{Z}}} \chi) && \text{(by (4-3-5))} \\ &= (-1)^n \alpha_{K/F} \smile_\theta \{\delta_*\eta[\eta^{-1}(\zeta \smile_{\theta_{\mathbf{Q}/\mathbf{Z}}} \chi)]\} \\ &= (-1)^n \alpha_{K/F} \smile_\theta \bar{\kappa}[\eta^{-1}(\zeta \smile_{\theta_{\mathbf{Q}/\mathbf{Z}}} \chi)] && \text{(by (4-5-9))} \end{aligned}$$

In the proof of the main theorem we have seen that

$$\alpha_{K/F} \smile_\theta \kappa(1) = \alpha_{K/F};$$

therefore, taking invariants, it follows that

$$\beta = \alpha_{K/F} \cup_\theta \zeta \iff \mathrm{inv}_F (\beta \cup_\theta \delta_* \chi) = (-1)^n \eta^{-1}(\zeta \cup_{\theta_{\mathbf{Q}/\mathbf{Z}}} \chi)$$

for all $\chi \in H^{-n-1}(G_{K/F}, \mathbf{Q}/\mathbf{Z})$.

6-4-8. Exercise. Suppose that K/F is a normal layer in a class formation; then according to (6-4-2), $H^4(K/F)$ is isomorphic to the character group $\hat{G}_{K/F}$—more precisely: the main theorem (6-4-1) implies that $\Theta^2_{K/F} : H^2(G_{K/F}, \mathbf{Z}) \longrightarrow H^4(K/F)$ is an isomorphism onto; furthermore, the coboundary map

$$\delta_* : H^1\left(G_{K/F}, \frac{\mathbf{Q}}{\mathbf{Z}}\right) \longrightarrow H^2(G_{K/F}, \mathbf{Z})$$

is an isomorphism onto; finally, $\hat{G}_{K/F} = \mathrm{Hom}\,(G_{K/F}, \mathbf{Q}/\mathbf{Z})$ may be identified with $H^1(G_{K/F}, \mathbf{Q}/\mathbf{Z})$ by associating with a character $\chi \in \hat{G}_{K/F}$ the standard 1-cocycle, also denoted by χ, such that $\chi([\sigma]) = \chi(\sigma)$ $\forall \sigma \in G_{K/F}$; thus, the isomorphism between $\hat{G}_{K/F}$ and $H^4(K/F)$ is given explicitly by the correspondence

$$\chi \longleftrightarrow \alpha_{K/F} \cup_\theta \delta_* \chi$$

where $\theta : A_K \times \mathbf{Z} \longrightarrow A_K$.

Consider the situation $F \subset E \subset K \subset L$ with K/F and L/F both normal, and let M/F denote the maximal abelian sublayer of K/F.

(1) The restriction map $\mathrm{res} : H^4(K/F) \longrightarrow H^4(K/E)$ corresponds to the map of $\hat{G}_{K/F} \longrightarrow \hat{G}_{K/E}$ which takes $\chi \longrightarrow \chi \mid G_{K/E}$. If $M \subset E$, then res is the trivial map.

(2) The corestriction map $\mathrm{cor} : H^4(K/E) \longrightarrow H^4(K/F)$ corresponds to the map of

$$\hat{G}_{K/E} = \widehat{\left(\frac{G_{K/E}}{G^c_{K/E}}\right)} \longrightarrow \hat{G}_{K/F} = \widehat{\left(\frac{G_{K/F}}{G^c_{K/F}}\right)}$$

which takes $\chi \longrightarrow \chi \circ \overline{V}$ where

$$\overline{V} : \frac{G_{K/F}}{G^c_{K/F}} \longrightarrow \frac{G_{K/E}}{G^c_{K/E}}$$

is the reduced group theoretical transfer (see (3-5-4)). If $M \subset E$, then cor is the trivial map.

(3) The inflation map $\inf : H^4(K/F) \longrightarrow H^4(L/F)$ corresponds to the map of $\hat{G}_{K/F} \longrightarrow \hat{G}_{L/F}$ given by $\chi \longrightarrow \bar{\chi}^{[L:K]}$, where $\bar{\chi} \in \hat{G}_{L/F}$ is defined as follows: let $\sigma \longrightarrow \bar{\sigma}$ be the natural map of $G_{L/F} \longrightarrow G_{L/F}/G_{L/K} \approx G_{K/F}$, then $\bar{\chi}(\sigma) = \chi(\bar{\sigma})$.

(4) Suppose that $\chi \in \hat{G}_{K/F}$ and $\beta \in H^4(K/F)$; if χ and β correspond, then (see (4-5-4))

$$\chi(\sigma) = \mathrm{inv}_F\,((\sigma^\#)_*\beta) \qquad \forall \sigma \in G$$

Hint: It is convenient to use (6-5-1) and (6-5-2).

6-5. RECIPROCITY LAW AND NORM RESIDUE SYMBOL

In this section we derive the key consequences of the main theorem; these are the basic results of class field theory.

Consider the normal layer K/F of degree m. (It is understood here, and henceforth, that we are dealing with a class formation.) The main theorem, in dimension -2, provides an isomorphism onto:

$$\Theta_{K/F} = \Theta_{K/F}^{-2} : H^{-2}(G_{K/F}, \mathbf{Z}) \longrightarrow H^0\left(\frac{K}{F}\right) = H^0(G_{K/F}, A_K)$$

which is given by a cup product. Since

$$H^{-2}(G_{K/F}, \mathbf{Z}) \approx \frac{G_{K/F}}{G_{K/F}^c} \qquad \text{and} \qquad H^0(G_{K/F}, A_K) \approx \frac{A_F}{N_{K/F}A_K}$$

(in order to conform to the applications, we write the formation module A multiplicatively) we have

$$\frac{G_{K/F}}{G_{K/F}^c} \approx \frac{A_F}{N_{K/F}A_K} \qquad (*)$$

Moreover, this isomorphism can be given explicitly.

To do this, recall first that $\kappa : A_F \longrightarrow H^0(K/F)$ is a homomorphism onto with kernel $N_{K/F}A_K$ which induces the isomorphism $\bar{\kappa}$ of $A_F/(N_{K/F}A_K)$ onto $H^0(K/F)$. On the other hand, the map $\sigma \longrightarrow \zeta_\sigma = \delta_*^{-1}\eta(\sigma - 1)$ is a homomorphism of $G_{K/F}$ onto $H^{-2}(G_{K/F}, \mathbf{Z})$ with kernel $G^c_{K/F}$ (see (3-5-2)) which induces an isomorphism of $G_{K/F}/G^c_{K/F}$ onto $H^{-2}(G_{K/F}, \mathbf{Z})$. (Here η is the canonical map of I onto $H^{-1}(G_{F/K}, I)$, which is an isomorphism; $\delta_* : H^{-2}(G_{K/F}, \mathbf{Z}) \longrightarrow H^{-1}(G_{K/F}, I)$ is the coboundary arising from the exact sequence $0 \longrightarrow I \longrightarrow \Gamma \longrightarrow \mathbf{Z} \longrightarrow 0$—it is an isomorphism onto.) Combining these isomorphisms with $\Theta_{K/F}$ gives the desired isomorphism in (*). If, for $a \in A_F$ and $\sigma \in G_{K/F}$, the elements $aN_{K/F}A_K$ and $\sigma G^c_{K/F}$ correspond to each other under this isomorphism, we write $aN_{K/F}A_K \longleftrightarrow \sigma G^c_{K/F}$ and also $a \longleftrightarrow \sigma$ (even though the latter is not a 1–1 correspondence). We have proved:

6-5-1. Theorem. (Reciprocity Law). For any normal layer K/F we have

$$\frac{A_F}{N_{K/F}A_K} \approx \frac{G_{K/F}}{G^c_{K/F}}$$

In fact, if $a \in A_F$ and $\sigma \in G_{K/F}$, then

$$a \longleftrightarrow \sigma \iff \kappa a = \alpha_{K/F} \cup_\theta \zeta_\sigma$$

where $\theta = \theta^{\mathsf{T}}_K : A_K \times \mathbf{Z} \longrightarrow A_K$.

It is also possible to give a dual description of the reciprocity law isomorphism.

6-5-2. Corollary. If $a \in A_F$ and $\sigma \in G_{K/F}$, then

$$a \longleftrightarrow \sigma \iff \operatorname{inv}_F(\kappa a \cup_\theta \delta_*\chi) = \chi(\sigma) \text{ for all } \chi \in \hat{G}_{K/F}.$$

Proof: (Here $\delta_* : H^1(G_{K/F}, \mathbf{Q}/\mathbf{Z}) \longrightarrow H^2(G_{K/F}, \mathbf{Z})$, and by the standard identification $\chi \in \hat{G}_{K/F} = \operatorname{Hom}(G_{K/F}, \mathbf{Q}/\mathbf{Z})$ is viewed also as an element of $H^1(G_{K/F}, \mathbf{Q}/\mathbf{Z})$.) Using the case $n = -2$ of (6-4-7) we have:

$$a \longleftrightarrow \sigma \iff \kappa a = \alpha_{K/F} \cup_\theta \zeta_\sigma \iff \operatorname{inv}_F(\kappa a \cup_\theta \delta_* \chi) = \eta^{-1}(\zeta_\sigma \cup_{\theta_{\mathbf{Q}/\mathbf{Z}}} \chi)$$

for all $\chi \in H^1(G_{K/F}, \mathbf{Q}/\mathbf{Z})$, and according to (4-5-8)

$$\zeta_\sigma \cup_{\theta_{\mathbf{Q}/\mathbf{Z}}} \chi = \eta(\chi(\sigma)). \quad ▮$$

6-5-3. Remark. The isomorphism of the reciprocity law provides maps in two directions. One way there is a homomorphism of $G_{K/F}$ onto $A_F/(N_{K/F}A_K)$ with kernel $G^c_{K/F}$. If $u = u_{K/F}$ is a standard 2-cocycle belonging to the fundamental class $\alpha_{K/F}$, then (since $\alpha_{K/F} \cup_\theta \zeta_\sigma = \zeta_\sigma \cup_{\theta_K} \alpha_{K/F}$) it follows from (4-5-7) that this map is given by

$$\sigma \longrightarrow \prod_{\rho \in G_{K/F}} u[\rho, \sigma] (\operatorname{mod} N_{K/F}A_K) \qquad \sigma \in G_{K/F}$$

The other way there is a homomorphism of A_F onto $G_{K/F}/G_c^{K/F}$ with kernel $N_{K/F}A_K$; we denote it by

$$a \longrightarrow \left(a, \frac{K}{F}\right) \qquad a \in A_F$$

and call it the **norm residue map** and $(a, K/F)$ the **norm residue symbol.**

Henceforth, it will be convenient to denote the factor commutator group $G_{K/F}/G^c_{K/F}$ by $G^{ab}_{K/F}$.

6-5-4. Proposition. If K/F is a normal layer and $a \in A_F$, then the element $(a, K/F)$ of $G^{ab}_{K/F}$ is characterized by the formula

$$\operatorname{inv}_F(\kappa a \cup_\theta \delta_* \chi) = \chi\left(a, \frac{K}{F}\right)$$

for all characters χ of $G_{K/F}$.

Proof: Since $\hat{G}_{K/F}$ is the same as $\widehat{G^{ab}_{K/F}}$, this is immediate from (6-5-2). ▮

6-5-5. Theorem. Suppose that $F \subset E \subset K \subset L$ with K/F and L/F normal. The functorial properties of the norm residue map are expressed by the commutativity of the following diagrams in which the vertical maps are the appropriate norm residue maps.

$$\begin{array}{ccc} A_F & \xrightarrow{i} & A_E \\ \downarrow & & \downarrow \\ G_{K/F}^{ab} & \xrightarrow{\bar{V}} & G_{K/E}^{ab} \end{array} \tag{1}$$

where $\bar{V}$ is the reduced group theoretical transfer; in other words,

$$\bar{V}_{G_{K/F} \to G_{K/E}}\left(a, \frac{K}{F}\right) = \left(a, \frac{K}{E}\right) \qquad a \in A_F$$

$$\begin{array}{ccc} A_E & \xrightarrow{NE/F} & A_F \\ \downarrow & & \downarrow \\ G_{K/E}^{ab} & \xrightarrow{\bar{i}} & G_{K/F}^{ab} \end{array} \tag{2}$$

where $\bar{i}$ is the map induced by the inclusion $i : G_{K/E} \longrightarrow G_{K/F}$; in other words,

$$\left(N_{E/F}a, \frac{K}{F}\right) = \left(a, \frac{K}{E}\right) G_{K/F}^{c} \qquad a \in A_E$$

$$\begin{array}{ccc} A_F & \xrightarrow{\sigma} & A_{F\sigma} \\ \downarrow & & \downarrow \\ G_{K/F}^{ab} & \xrightarrow{\bar{\sigma}} & G_{K\sigma/F\sigma}^{ab} \end{array} \tag{3}$$

where $\bar{\sigma}$ is the map induced by the conjugation map

$$\sigma : G_{K/F} \longrightarrow G_{K\sigma/F\sigma}$$

in other words,

$$\left(a^{\sigma}, \frac{K^{\sigma}}{F^{\sigma}}\right) = \left(a, \frac{K}{F}\right)^{\sigma} \qquad a \in Ae$$

$$\begin{array}{ccc} A_F & \xrightarrow{1} & A_F \\ \downarrow & & \downarrow \\ G_{L/F}^{ab} & \xrightarrow{\bar{\pi}} & G_{K/F}^{ab} \end{array} \tag{4}$$

where $\bar{\pi}$ is induced by the projection map

$$\pi : G_{L/F} \longrightarrow \frac{G_{L/F}}{G_{L/K}} \approx G_{K/F}$$

in other words,

$$\left(a, \frac{K}{F}\right) = \left(a, \frac{L}{F}\right) G_{L/K} \qquad a \in A_F$$

Proof: To prove (1) consider the diagram

$$\begin{array}{ccc}
A_F & \xrightarrow{\quad i \quad} & A_E \\
\kappa \downarrow & & \downarrow \kappa \\
H^0(G_{K/F}, A_K) & \xrightarrow{\text{res}} & H^0(G_{K/E}, A_K) \\
\Theta_{K/F} \uparrow & & \uparrow \Theta_{K/E} \\
H^{-2}(G_{K/F}, \mathbf{Z}) & \xrightarrow{\text{res}} & H^{-2}(G_{K/F}, \mathbf{Z}) \\
\updownarrow & & \updownarrow \\
G^{ab}_{K/F} & \xrightarrow{\quad \bar{V} \quad} & G^{ab}_{K/E}
\end{array}$$

The top, middle, and bottom squares commute—according to (2-5-5), (6-4-3), and (3-5-5), respectively.

Since $N_{K/F}A_K \subset N_{K/E}A_K$ we also have commutativity when the top row of (1) is replaced by the induced map

$$\bar{i} : \frac{A_F}{N_{K/F}A_K} \longrightarrow \frac{A_E}{N_{K/E}A_K}$$

The proof of (2) depends on substantially the same diagram, and on (2-5-5), (6-4-3), and (3-5-3). Since $N_{E/F}(N_{K/E}A_K) \subset N_{K/F}A_K$ we have commutativity when the top row is replaced by the induced map $\overline{N}_{E/F} : A_F/(N_{K/F}A_K) \longrightarrow A_E/(N_{K/E}A_K)$.

As for (3), the proof follows from (2-5-5), (6-4-6), and (3-5-7). Note that the conjugation map σ takes $G_{K/F} = G_F/G_K$ isomorphically onto $(\sigma G_F \sigma^{-1})/(\sigma G_K \sigma^{-1}) = G_F^\sigma/G_K^\sigma = G_{F^\sigma}/G_{K^\sigma} = G_{K^\sigma/F^\sigma}$, and that σ takes $G^c_{K/F}$ onto $G^c_{K^\sigma/F^\sigma}$, so that $\bar{\sigma} : G^{ab}_{K/F} \longrightarrow G^{ab}_{K^\sigma/F^\sigma}$ is an isomorphism onto. Also, $\sigma : A_F \longrightarrow \sigma A_F = A_F^\sigma = A_{F^\sigma}$ is 1–1 onto, and so is $\sigma : N_{K/F}A_K \longrightarrow N_{K^\sigma/F^\sigma}A_{K^\sigma}$ (since for $a \in A_K$, $\sigma : \prod_{\tau \in G_{K/F}} (\tau a) \longrightarrow \prod_{\tau \in G_{K/F}} (\sigma\tau\sigma^{-1})(\sigma a))$. We then have commutativity when the top row of (3) is replaced by the induced map $\bar{\sigma} : A_F/(N_{K/F}A_K) \longrightarrow A_{F^\sigma}/(N_{K^\sigma/F^\sigma}A_{K^\sigma})$.

With regard to (4), note first that the projection π maps $G_{L/F}^c \longrightarrow G_{K/F}^c = (G_{L/F}^c G_{L/K})/G_{L/K}$. Since $G_{L/F}^c$ and $G_{L/K}$ are both normal in $G_{L/F}$, so is $G_{L/F}^c G_{L/F}$. Therefore,

$$G_{K/F}^{ab} \approx \frac{G_{L/F}}{G_{L/F}^c G_{L/K}}$$

and we see that $\bar{\pi}(a, L/F) = (a, L/F)\, G_{L/K}$. The proof of (4) cannot proceed like the others because we have not discussed cohomological maps corresponding to 1 and $\bar{\pi}$. Instead we make use of (6-5-4). It suffices to show that for $a \in A_F$

$$\chi\left(\left(a, \frac{L}{F}\right) G_{L/K}\right) = \mathrm{inv}_F\,(\kappa_K a \cup_\theta \delta_* \chi)$$

for every character χ of $G_{K/F} \approx G_{L/F}/G_{L/K}$. Viewing

$$\chi \in H^1\left(G_{K/F}, \frac{\mathbf{Q}}{\mathbf{Z}}\right)$$

we know that $\psi = \inf_{G_{K/F} \to G_{L/F}} \chi \in H^1(G_{L/F}, \mathbf{Q}/\mathbf{Z}) \approx \hat{G}_{L/F} \approx \widehat{G_{L/F}^{ab}}$ and $\psi((a, L/F)) = \chi((a, L/F)\, G_{L/K})$ (see (3-5-4)). Let

$$\theta' = \theta_L^\top : A_L \times \mathbf{Z} \longrightarrow \mathbf{Z}$$

be the natural pairing; so $\theta = \theta_K^\top$ is θ' restricted to $A_K \times \mathbf{Z}$. Now, compute:

$$\begin{aligned}
\chi\left(\left(a, \frac{L}{F}\right) G_{L/K}\right) = \psi\left(\left(a, \frac{L}{F}\right)\right) &= \mathrm{inv}_F\,(\kappa_L a \cup_{\theta'} \delta_* \psi) \\
&= \mathrm{inv}_F\,(\kappa_L a \cup_{\theta'} \inf \delta_* \chi) \\
&= \mathrm{inv}_F \inf\,(\kappa_K a \cup_\theta \delta_* \chi) \qquad \text{(by (4-3-9))} \\
&= \mathrm{inv}_F\,(\kappa_K a \cup_\theta \delta_* \chi)
\end{aligned}$$

This completes the proof of (4). Of course, the top row of (4) may be replaced by $\bar{1} : A_F/(N_{L/F}A_L) \longrightarrow A_F/(N_{K/F}A_K)$. ∎

6-5-6. Proposition. Suppose that $F \subset E \subset K$ with K/F normal; then for $a \in A_F$

$$a \in N_{E/F}A_E \iff \left(a, \frac{K}{F}\right) \in \frac{G_{K/E}G^c_{K/F}}{G^c_{K/F}} = \{\tau G^c_{K/F} \mid \tau \in G_{K/E}\}$$

Proof: $\Longrightarrow$: If $a = N_{E/F}b$, $b \in A_E$, then (since $(b, K/E) \in G_{K/E}/G^c_{K/E}$ and $G^c_{K/E} \subset G^c_{K/F}$) part (2) of (6-5-5) says that

$$\left(a, \frac{K}{F}\right) = \left(b, \frac{K}{E}\right) G^c_{K/F} \in \frac{G_{K/E}G^c_{K/F}}{G^c_{K/F}}$$

$\Longleftarrow$: Suppose $(a, K/F) \in (G_{K/E}G^c_{K/F})/G^c_{K/F}$, so there exists $\rho \in G_{K/E}$ such that $(a, K/F) = \rho G^c_{K/F}$. Because $b \longrightarrow (b, K/E)$ maps A_E onto $G_{K/E}/G^c_{K/E}$, there exists $b \in A_E$ such that $(a, K/F) = (b, K/E)\, G^c_{K/F} = (N_{E/F}b, K/F)$. Since the homomorphism $c \longrightarrow (c, K/F)$ of $A_F \longrightarrow G^{ab}_{K/F}$ has kernel $N_{K/F}A_K$, there exists $d \in A_K$ such that

$$a = (N_{E/F}b)(N_{K/F}d) = N_{E/F}(b \cdot N_{K/E}d) \in N_{E/F}A_E \,. \quad ∎$$

A subgroup of A_F of form $N_{E/F}A_E$, for any extension E/F, is said to be a **norm group** and $(A_F : N_{E/F}A_E)$ is called the **norm index** of E/F. The fundamental properties of the norm subgroups for a given field are now accessible.

6-5-7. Theorem. (1) The norm group of any extension E/F is the same as that of its maximal abelian subextension M/F; that is

$$N_{E/F}A_E = N_{M/F}A_M$$

(2) For each F, the correspondence

$$M \longleftrightarrow N_{M/F}A_M$$

is 1–1 between the set of all abelian extensions M of F and the set of all norm subgroups of A_F. Moreover, this correspondence has the following properties:

(a) $M_1 \subset M_2 \Longleftrightarrow N_{M_1/F}A_{M_1} \supset N_{M_2/F}A_{M_2}$
(b) $N_{(M_1M_2)/F}A_{M_1M_2} = (N_{M_1/F}A_{M_1}) \cap (N_{M_2/F}A_{M_2})$
(c) $N_{(M_1 \cap M_2)/F}A_{M_1 \cap M_2} = (N_{M_1/F}A_{M_1}) \cdot (N_{M_2/F}A_{M_2})$
(d) $[M : F] = (A_F : N_{M/F}A_M)$

(3) Every subgroup of A_F which contains a norm subgroup is itself a norm subgroup.

Proof: (1) Imbed E/F in a normal layer K/F, and apply (6-5-6) to both $F \subset E \subset K$ and $F \subset M \subset K$. Consequently, for $a \in A_F$

$$a \in N_{E/F}A_E \Longleftrightarrow \left(a, \frac{K}{F}\right) \in \frac{G_{K/E}G^c_{K/F}}{G^c_{K/F}}$$

$$a \in N_{M/F}A_M \Longleftrightarrow \left(a, \frac{K}{F}\right) \in \frac{G_{K/M}G^c_{K/F}}{G^c_{K/F}}$$

and it suffices to show $G_{K/E}G^c_{K/F} = G_{K/M}G^c_{K/F}$. Since

$$G_M = [G_E \cup G^c_F] = G_E \cdot G^c_F$$

we have

$$G_{K/M} = \frac{G_M}{G_K} = \frac{G_E G^c_F}{G_K} = \frac{G_E G^c_F G_K}{G_K} = \left(\frac{G_E}{G_K}\right)\left(\frac{G^c_F G_K}{G_K}\right) = G_{K/E}G^c_{K/F}$$

—which proves (1). In particular, this shows that every norm subgroup of A_F is the norm group of some abelian extension.

(2) In virtue of (1) the correspondence $M \longleftrightarrow N_{M/F}A_M$ is onto in each direction; we must show that it is 1–1. Consider any abelian extension M/F. Since $a \longrightarrow (a, M/F)$ is a homomorphism of A_F onto $G_{M/F}/G^c_{M/F} = G_{M/F}$ with kernel $N_{M/F}A_M$, there is a 1–1 correspondence between the set of all subgroups of A_F which contain $N_{M/F}A_M$ and the set of all subgroups (which are of form $G_{M/M'}$, where $F \subset M' \subset M$ and M'/F is automatically abelian) of the Galois group $G_{M/F}$. Moreover, in this correspondence

$N_{M'/F}A_{M'}$ corresponds to $G_{M/M'}$; in fact, from (6-5-6) applied to $F \subset M' \subset M$, we have for $a \in A_F$

$$a \in N_{M'/F}A_{M'} \iff \left(a, \frac{M}{F}\right) \in \frac{G_{M/M'}G^c_{M/F}}{G^c_{M/F}} = G_{M/M'}$$

Of course, $G_{M/M'} \longleftrightarrow M'$ is a 1–1 correspondence between the set of subgroups of $G_{M/F}$ and the set of intermediate fields.

Suppose then that M_1/F and M_2/F are any abelian extensions. Putting $M = M_1M_2$ the extension M/F is abelian, and from the foregoing we have $M_1 = M_2 \iff N_{M_1/F}A_{M_1} = N_{M_2/F}A_{M_2}$ and also (2a).

Furthermore, $M \longrightarrow N_{M/F}A_M$ is a 1–1 order-inverting map of the lattice of all abelian extensions of F onto the lattice of all norm subgroups of A_F; therefore, it is a "lattice isomorphism"—that is, (b) and (c) hold. As for (d), it is immediate from the reciprocity law isomorphism $A_F/(N_{M/F}A_M) \approx G_{M/F}$.

Finally, (3) is obvious; for if H is a subgroup of A_F containing a norm subgroup, which must be of form $N_{M/F}A_M$, M/F abelian, then H corresponds to a subgroup $G_{M/M'}$ of $G_{M/F}$ and $H = N_{M'/F}A_{M'}$. This completes the proof. ∎

6-5-8. Corollary. The norm index of an arbitrary extension E/F divides its degree, and equality holds if and only if the extension is abelian.

The top field of an abelian extension M/F is said to be a **class field** over F. If B is a norm subgroup of A_F, then, by (6-5-7), there is a unique class field M over F corresponding to it (namely, the one with $N_{M/F}A_M = B$), and M is called the **class field belonging to** B.

6-5-9. Corollary. Let B be a norm subgroup of A_F with M the class field over F belonging to B; then for $\sigma \in G$, M^σ is the class field over F^σ belonging to B^σ.

Proof: It is clear that M/F abelian implies M^σ/F^σ abelian, and then

$$N_{M^\sigma/F^\sigma}(A_{M^\sigma}) = N_{M^\sigma/F^\sigma}(\sigma A_M) = \sigma(N_{M/F}A_M) = B^\sigma$$ ∎

6-5-10. Corollary. (Translation Theorem). Let B be a norm subgroup of A_F with M the class field over F belonging to B. If E/F is any extension, then $C = N_{E/F}^{-1}(B)$ is a norm subgroup of A_E and the class field over E belonging to it is the compositum ME.

Proof: Let E' be any extension of E, and let M'/F be the maximal abelian subextension of E/F. Then

$$\begin{aligned} M \subset E' &\Longleftrightarrow M \subset M' \Longleftrightarrow N_{M/F}A_M \supset N_{M'/F}A_{M'} \\ &\Longleftrightarrow B \supset N_{E'/F}A_{E'} = N_{E/F}(N_{E'/E}A_{E'}) \\ &\Longleftrightarrow C = N_{E/F}^{-1}(B) \supset N_{E'/E}A_{E'} \end{aligned}$$

In particular, C contains the norm subgroup of $E' = ME$, so it is a norm subgroup of A_E. Since ME is the smallest extension of E whose norm subgroup is contained in C, it follows that ME is the class field over E belonging to C. Note that it is a by-product of the proof that $(ME)/E$ is abelian. ▮

We can now use the norm residue symbol and the (group theoretic) principal ideal theorem (5-4-8) to prove what may be considered an abstract version of the principal ideal theorem of arithmetic.

6-5-11. Theorem. Let K/F be a normal extension in a class formation, and let M/F be the maximal abelian subextension; then

$$A_F \subset N_{K/M}A_K$$

Proof: The statement that M/F is the maximal abelian subextension of K/F is equivalent to saying that $G_{K/M} = G_{K/F}^C$, the commutator subgroup of $G_{K/F}$. According to (5-4-8), the reduced transfer from $G_{K/F}$ to $G_{K/M}$ is the trivial map. From (6-5-5), we see that for all $a \in A_F$,

$$\left(a, \frac{K}{M}\right) = \bar{V}_{G_{K/F} \to G_{K/M}}\left(a, \frac{K}{M}\right) = 1$$

Since the kernel of the norm residue map $a \longrightarrow (a, K/M)$ is $N_{K/M}A_K$ the desired result follows. ∎

6-5-12. Exercise. Given any normal layer K/F of a class formation, we have the reciprocity law isomorphism

$$\frac{A_F}{N_{K/F}A_K} \approx H^0\left(\frac{K}{F}\right) \approx H^{-2}(G_{K/F}, \mathbf{Z}) \approx \frac{G_{K/F}}{G^c_{K/F}}$$

and the concomitant "norm residue map" $a \longrightarrow (a, K/F)$ of the ground level A_F onto the factor commutator group $G_{K/F}/G^c_{K/F}$ of the Galois group $G_{K/F}$.

Suppose that the Galois group G of our formation $\{G, \{G_F\}, A\}$ is complete and hence compact also. We shall piece together the various finite norm residue maps $a \longrightarrow (a, K/F)$ and construct an infinite norm residue homomorphism

$$a \longrightarrow \left(a, \frac{*}{F}\right)$$

of A_F into the factor commutator group $G_F/\overline{G^c_F}$ of the topological group G_F. Here, of course, G^c_F is the algebraic commutator group of G_F, and $\overline{G^c_F}$ denotes its topological closure in G_F. Note that

$$\overline{G^c_F} = \bigcap_K G_K \cdot G^c_F$$

the intersection being over all K which are finite normal over F; this follows from the fact that such G_K constitute a fundamental system of neighborhoods of the identity in G_F.

Because $G_{K/F} = G_F/G_K$, its commutator group is

$$G^c_{K/F} = \frac{G_K \cdot G^c_F}{G_K}$$

so the factor commutator group $G^{ab}_{K/F} = G_{K/F}/G^c_{K/F}$ may be identified with $G_F/(G_K \cdot G^c_F)$ and the norm residue symbol $(a, K/F)$ may be viewed as a coset of $G_K G^c_F$ in G_F. Each $(a, K/F)$ is then a closed subset of the compact group G_F, because it is a coset of the closed subgroup $G_K G_F^c$. Therefore, in order to show that

$$\bigcap_K \left(a, \frac{K}{F}\right) \neq \varnothing \qquad (*)$$

it suffices to show that the norm residue symbols have the finite intersection property—namely,

$$\bigcap_{i=1}^{n} \left(a, \frac{K_i}{F}\right) \neq \varnothing$$

To prove this, take a normal extension M of F such that $F \subset K_i \subset M$ for $i = 1, \ldots, n$; then by a standard property of the finite norm residue symbol (see (6-5-5), part (4)) we have $(a, K_i/F) \supset (a, M/F)$ for all i.

Now that $(*)$ helds, consider $\sigma \in \bigcap_K (a, K/F)$; then

$$\left(a, \frac{K}{F}\right) = \sigma G_K G_F^c \qquad \text{for all} \quad K$$

and consequently

$$\bigcap_K \left(a, \frac{K}{F}\right) = \bigcap_K \sigma G_K G_F^c = \sigma \left(\bigcap_K G_K G_F^c\right) = \sigma \overline{G_F^c}$$

We define

$$\left(a, \frac{*}{F}\right) = \bigcap_K \left(a, \frac{K}{F}\right)$$

which is indeed an element of $G_F/\overline{G_F^c}$. Clearly,

$$\left(a, \frac{*}{F}\right) \subset \left(a, \frac{K}{F}\right) \qquad \text{for all} \quad K$$

and because $\overline{G_F^c} \cdot G_K = G_F^c \cdot G_K = G_K \cdot G_F^c$ it follows that

$$\left(a, \frac{*}{F}\right) G_K = \left(a, \frac{K}{F}\right)$$

Moreover, because each finite norm residue symbol is a homomorphism, it is easy to see that the infinite norm residue symbol is a homomorphism—that is,

$$\left(ab, \frac{*}{F}\right) = \left(a, \frac{*}{F}\right)\left(b, \frac{*}{F}\right)$$

Furthermore, using the fact that each finite norm residue symbol $(a, K/F)$ maps A_F onto $G_F/(G_K G_F^c)$, we may show that the infinite norm residue symbol $(a, */F)$ maps A_F onto an everywhere dense subgroup of $G_F/\overline{G_F^c}$. In fact, a fundamental neighborhood of an arbitrary element $\sigma\overline{G_F^c}$ of $G_F/\overline{G_F^c}$ is of form $(\sigma G)_F^c G_K = \sigma G_K G_F^c$; for this K, there exists $a \in A_F$ such that $(a, K/F) = \sigma G_K G_F^c$, so that

$$\left(a, \frac{*}{F}\right) \subset \left(a, \frac{*}{F}\right) G_K = \sigma G_F^c G_K = (\sigma\overline{G_F^c})\, G_K$$

We have proved the existence of a homomorphism $a \longrightarrow (a, */F)$ which takes A_F onto a dense subgroup of $G_F/\overline{G_F^c}$ and such that for every normal layer K/F

$$\left(a, \frac{K}{F}\right) = \left(a, \frac{*}{F}\right) \cdot G_K \qquad a \in A_F$$

In addition, it may be noted that a homomorphism with these properties is unique.

The formal properties of the infinite norm residue symbol are as follows:

(1) If $F \subset E$ and $a \in A_F$, then

$$\overline{V}_{G_F \to G_E}\left(a, \frac{*}{F}\right) = \left(a, \frac{*}{F}\right)$$

(2) If $F \subset E$ and $a \in A_E$, then

$$\left(N_{E\to F}a, \frac{*}{F}\right) = \left(a, \frac{*}{E}\right)\overline{G_F^c}$$

(3) If $\sigma \in G$ and $a \in A_F$, then

$$\left(\sigma a, \frac{*}{\sigma F}\right) = \sigma \left(a, \frac{*}{F}\right) \sigma^{-1}$$

In connection with (1), it is necessary to clarify the meaning of $\overline{V}_{G_F \to G_E}$. Consider any K finite normal over F with $F \subset E \subset K$. The reduced group theoretical transfer $\overline{V}_{G_F \to G_E}$ from G_F to G_E induces the reduced group theoretical transfer from G_F/G_K to G_E/G_K (see (3-5-6)), and it also maps $(G_K G_F^c)/G_F^c$ into $(G_K G_E^c)/G_E^c$. Taking the intersection over all such K, it follows that $\overline{V}_{G_F \to G_E}$ maps $\overline{G_F^c}/G_F^c$ into $\overline{G_E^c}/G_E^c$; consequently, there is induced a map of $G_F/\overline{G_F^c} \longrightarrow G_E/\overline{G_E^c}$ which, by an abuse of notation, we call $\overline{V}_{G_F \to G_E}$. Then, to prove (1), one simply shows that for all such K

$$\left(\overline{V}_{G_F \to G_E}\left(a, \frac{*}{F}\right)\right) G_K = \left(a, \frac{*}{E}\right) G_K$$

6-6. THE EXISTENCE THEOREM

In virtue of the results of the preceding section, it is desirable to be able to characterize the norm subgroups of a ground level A_F, as this provides information about the abelian extensions of F. An elegant way to give such a characterization makes use of some simple topological considerations.

6-6-1. Definition. A **topological formation** is a formation in which:

(1) Each level A_F is a topological group (this includes Hausdorff).

(2) For each layer E/F, the topology of the ground level A_F is the topology induced from the top level A_E; in particular, $i : A_F \longrightarrow A_E$ is continuous.

(3) The Galois group G acts continuously on every level; that is, for any $\sigma \in G$ and field F the map $\sigma : A_F \longrightarrow (A_F)^\sigma = A_{F^\sigma}$ is continuous.

Among the immediate consequences of the definition, we note the following:

(i) Because $\sigma^{-1} : A_{F^\sigma} \longrightarrow A_F$ (the inverse map of σ) is continuous, it follows that $\sigma \in G$ acts as a homeomorphism on the levels.

(ii) Any given layer E/F may be imbedded in some normal layer K/F, and we have a coset decomposition $G_{K/F} = \bigcup_i \sigma_i G_{K/E}$. Each map $a \longrightarrow a^{\sigma_i}$ is a continuous map of $A_K \longrightarrow A_K$; hence, the norm map $a \longrightarrow N_{E/F}a = \prod_i a^{\sigma_i}$ is a continuous map of $A_K \longrightarrow A_K$. Because A_E and A_F have topologies induced from A_K, and since $N_{E/F}A_E \subset A_F$, we see that $N_{E/F}$ is a continuous map of $A_E \longrightarrow A_F$.

(iii) For F, E, K as above, the map $(\sigma - 1) : A_K \longrightarrow A_K$ given by $a \longrightarrow a^{\sigma-1}$ is continuous for each $\sigma \in G_{K/F}$. Since A_K is Hausdorff, $\ker(\sigma - 1)$ is closed in A_K. Thus,

$$A_F = \bigcap_{\sigma \in G_{K/F}} \{\ker(\sigma - 1)\}$$

is closed in A_K, and in similar fashion A_E is closed in A. Therefore, A_F is closed in A_E.

For any field F, let us write D_F for the group of **universal norms**—that is, the elements of A_F which are norms from every extension of F; symbolically, we have

$$D_F = \bigcap_{E \supset F} N_{E/F}A_E$$

The key to the characterization of the norm subgroups is given by the following axiom; it arises from a careful analysis of the proofs of the existence theorem in the concrete situations of local and global class field theory.

Axiom III. (a) For each layer E/F, the norm map $N_{E/F} : A_E \longrightarrow A_F$ has compact kernel and closed image—that is $N_{E/F}^{-1}(1)$ is a compact subgroup of A_E, and $N_{E/F}A_E$ is a closed subgroup of A_F.

(b) For each prime p there exists a field F_p such that for all $E \supset F_p$ the map $a \longrightarrow a^p$ of $A_E \longrightarrow A_E$ has compact kernel, and image containing D_E; in other words, for sufficiently large E, $\{1^{1/p}\}$ is compact in A_E and $D_E \subset A_E^p$.

(c) For each field F there exists a compact subgroup U_F of A_F such that every open subgroup of finite index in A_F which contains U_F is a norm subgroup.

Of course, any open subgroup is also closed. Conversely, it is easy to see that a closed subgroup of finite index is open. Thus, part (c) could also be stated in terms of closed subgroups of finite index.

6-6-2. Proposition. Suppose that E/F is any layer in a topological formation satisfying Axiom III(a); then

$$D_F = N_{E/F} D_E$$

Proof: Suppose $a \in D_E$. For any $E' \supset F$ we have $EE' \supset E$ and there exists $b \in A_{EE'}$ with $N_{EE'/E} b = a$. Consequently, $N_{E/F} a = N_{E'/F}(N_{EE'/E'} b) \in N_{E'/F} A_{E'}$. Because E' containing F is arbitrary, this implies that $N_{E/F} D_E \subset D_F$. (This part of the proof requires no axioms.)

To show that $D_F \subset N_{E/F} D_E$, consider any $a \in D_F$. For each field $K \supset E$ (we do not require that K/E be normal) form the set

$$X_K = \{N_{K/E} A_K\} \cap \{N_{E/F}^{-1}(a)\}$$

which consists of the elements of A_E which are norms from A_K and whose norm to A_F is a. There exists $b \in A_K$ such that $a = N_{K/F}(b) = N_{E/F}(N_{K/E} b)$, so $X_K \neq \varnothing$. According to III(a), $N_{K/E} A_K$ is closed in A_E and $N_{E/F}^{-1}(a)$ is a compact subset of A_E; hence X_K is a closed subset of the compact set $N_{E/F}^{-1} a$. Consequently, to show that $\bigcap_{K \supset E} X_K \neq \varnothing$, it suffices to verify the finite intersection property—but this is trivial, because from $K \subset K' \Longrightarrow X_K \supset X_{K'}$ it follows that $X_{K_1} \cap X_{K_2} \supset X_{K_1 K_2} \neq \varnothing$. If we take $b \in \bigcap_{K \supset E} X_K$ then $b \in \bigcap_{K \supset E} N_{K/E} A_K = D_E$ and $N_{E/F} b = a$. ∎

6-6-3. Proposition. For every natural number m and every field F in a topological formation satisfying Axiom III(a), (b) we have

$$D_F = D_F^m = \bigcap_{n=1}^{\infty} A_F^n$$

Proof: To show that $D_F = D_F^m$ (i.e., that the group D_F is infinitely divisible) it suffices to prove that $D_F = D_F^p$ for every prime p. Consider any $a \in D_F$. For sufficiently large E—that is, for $E \supset FF_p$—we have (using III(a) and (b))

$$D_F = N_{E/F}D_E \subset N_{E/F}(A_E^p) = (N_{E/F}A_E)^p$$

so that

$$X_E = \{N_{E/F}A_E\} \cap \{a^{1/p}\} \neq \varnothing$$

By $\{a^{1/p}\}$ we mean $\{b \in A_F \mid b^p = a\}$; it is a closed set in A_F, and a subset of $\{b \in A_E \mid b^p = a\}$ which is compact in A_E (in virtue of Axiom III(b)). Thus, $\{a^{1/p}\}$ is compact in A_F and $N_{E/F}A_E$ is closed in A_F—so that X_E is compact and closed in A_F, and in $\{a^{1/p}\}$. Now, the X_E have the finite intersection property since $X_{E_1} \cap X_{E_2} \supset X_{E_1E_2} \neq \varnothing$, and by compactness $X = \bigcap_{E \supset FF_p} X_E \neq \varnothing$. For $b \in X$ we have $b^p = a \in D_F$ and

$$b \in \bigcap_{E \supset FF_p} N_{E/F}A_E = \bigcap_{E \supset F} N_{E/F}A_E = D_F$$

This implies that $D_F \subset D_F^p$, and then $D_F = D_F^p$.

Turning to the remaining part of the proof, we have $D_F = D_F^n \subset A_F^n$ for $n = 1, 2, \ldots$, so $D_F \subset \bigcap_{n=1}^{\infty} A_F^n$. On the other hand, for each $E \supset F$ we have

$$\bigcap_{n=1}^{\infty} A_F^n \subset A_F^{[E:F]} = N_{E/F}A_F \subset N_{E/F}A_E$$

Hence, $\bigcap_{n=1}^{\infty} A_F^n \subset D_F$ and the proof is complete. ▮

6-6-4. Abstract Existence Theorem. In a topological class formation satisfying Axiom III, a subgroup of any level A_F is a norm subgroup $\Longleftrightarrow$ it is an open subgroup of finite index in A_F.

Proof: $\Longrightarrow$: A norm subgroup $N_{E/F}A_E$ of A_F is closed (by Axiom III(a)) and of finite index (by (6-5-8)), so it is also open in A_F.

$\Longleftarrow$: Let C be an open subgroup of finite index in A_F, and let $\mathscr{B} = \{B\}$ denote the set of all norm subgroups of A_F. From (6-6-3), we have $D_F = D_F^{(A_F:C)} \subset A_F^{(A_F:C)} \subset C$—so $\bigcap_{B \in \mathscr{B}} B \subset C$ and then

$(A_F - C) \subset \bigcup_B (A_F - B)$. Since $A_F - C$ is closed in A_F, $\{(A_F - B) | B \in \mathscr{B}\}$ is an open covering of the compact set $(A_F - C) \cap U_F$. Extract a finite subcovering

$$U_F \cap (A_F - C) \subset \bigcup_1^r (A_F - B_i).$$

By (6-5-7), $B' = \bigcap_1^r B_i$ is an element of $\mathscr{B}$, and it is clear that $(A_F - C) \cap U_F \subset A_F - B'$. Therefore, $(A_F - C) \cap U_F \cap B' = \varnothing$ and $U_F \cap B' \subset C$.

Because B' and C are open subgroups of finite index, so is $B' \cap C$, and then $(B' \cap C)\, U_F$ is an open subgroup of finite index containing U_F. By Axiom III(c), $(B' \cap C)\, U_F$ is a norm subgroup. Hence $B' \cap (B' \cap C)\, U_F$ is a norm subgroup. Since $B' \cap U_F \subset C$ it is easy to see that

$$B' \cap (B' \cap C) U_F \subset (B' \cap C)(B' \cap U_F) \subset C$$

Thus, C contains a norm subgroup, and is therefore itself a norm subgroup.

References

1. I. T. Adamson, Cohomology theory for non-normal subgroups and non-normal fields. *Proc. Glasgow Math. Assoc.* **2** (1954), 66–76.
2. E. Artin, Ideal Klassen in Oberkörpern und allgemeines Reziprozitatsgesetz, *Abh. Math. Sem. Univ. Hamburg* **7** (1930), 46–51.
3. E. Artin, "Galois Theory" (Notre Dame Math. Lectures, No. 2). Notre Dame, Indiana, 1946.
4. E. Artin, C. J. Nesbitt, and R. M. Thrall, "Rings with Minimum Condition." Univ. of Michigan Press, Ann Arbor, Michigan, 1948.
5. E. Artin, "Algebraic Numbers and Algebraic Functions." N.Y.U., New York, 1951.
6. E. Artin and J. T. Tate, "Class Field Theory." Harvard Univ. Press, Cambridge, Massachusetts, 1961.
7. M. Auslander and D. Goldman, The Brauer group of a commutative ring. *Trans. Amer. Math. Soc.* **97** (1960), 367–409.
8. R. Baer, Erweiterung von Gruppen und ihren Isomorphismen. *Math. Z.* **38** (1934), 375–416.
9. R. Baer, Abelian groups that are direct summands of every containing abelian group. *Bull. Amer. Math. Soc.* **46** (1940), 800–806.
10. N. Bourbaki, "Topologie Generale" (*Actualités Sci. Indust.*, No. 916). Hermann, Paris, 1942.
11. N. Bourbaki, "Algèbre Multilinéaire" (*Actualités Sci. Indust.*, No. 1044). Hermann, Paris, 1948.
12. R. Brauer, Uber Systeme hyperkomplexer Zahlen. *Math. Z.* **30** (1929), 79–107.
13. R. Brauer, Uber die algebraische Struktur von Schiefkorpern. *J. Reine Angew. Math.* **166** (1932), 241–252.
14. D. A. Buchsbaum, Exact categories and duality. *Trans. Amer. Math. Soc.* **80** (1955), 1–34.
15. H. Cartan and S. Eilenberg, "Homological Algebra." Princeton Univ. Press, Princeton, New Jersey, 1956.
16. C. Chevalley, La theorie du corps de classes dans les corps finis et les corps locaux (These). *J. Fac. Sci. Univ. Tokyo Sect. I* (1933), 365–474.
17. C. Chevalley, "Class Field Theory." Nagoya Univ., Nagoya, Japan, 1954.

18. C. W. Curtis and I. Reiner "Representation Theory of Finite Groups and Associative Algebras." Wiley (Interscience), New York, 1962.
19. A Douady, Cohomologie des groupes compacts totalement discontinus. *Sem. Bourbaki, 1959/1960*, exp. 189.
20. B. Eckmann, Cohomology of groups and transfer. *Ann. of Math.* **58** (1953), 481–493.
21. B. Eckmann and A. Schopf, Uber injektive Moduln. *Arch Math. (Basel)* **4** (1953), 75–78.
22. S. Eilenberg and S. MacLane, General theory of natural equivalences. *Trans. Amer. Math. Soc.* **58** (1945), 231–294.
23. S. Eilenberg and S. MacLane, Cohomology theory in abstract groups. *Ann. of Math.* **48** (1947), 51–78.
24. S. Eilenberg and N. Steenrod, "Foundations of Algebraic Topology." Princeton Univ. Press, Princeton, New Jersey, 1952.
25. L. Evens, The cohomology ring of a finite group. *Trans. Amer. Math. Soc.* **101** (1961), 224–239.
26. L. Evens, An extension of Tate's theorem on cohomology triviality. *Proc. Amer. Math. Soc.* **00** (1965), 289–291.
27. P. Furtwängler, Beweis der Hauptidealsatzes fur Klassenkorper algebraischer Zahlkorper. *Abh. Math. Sem. Univ. Hamburg* **7** (1930), 14–36.
28. K. Grant and G. Whaples, Abstract class formations. *Bull. Amer. Math. Soc.* **78** (1961), 393–395.
29. A. Grothendieck, Sur quelques points d'algèbre homologique. *Tohoku Math. J.* **9** (1957), 119–221.
30. M. Hall, Group rings and extensions. *Ann. of Math.* **39** (1938), 220–234.
31. M. Hall, "The Theory of Groups." Macmillan, New York, 1959.
32. G. Hochschild, On the cohomology groups of an associative algebra. *Ann. of Math.* **46** (1945), 58–67.
33. G. Hochschild, On the cohomology theory for associative algebras. *Ann. of Math.* **47** (1946), 568–579.
34. G. Hochschild, Local class field theory. *Ann. of Math.* **51** (1950), 331–347.
35. G. Hochschild, Note on Artin's reciprocity law. *Ann. of Math.* **52** (1950), 694–701.
36. G. Hochschild, Relative homological algebra. *Trans. Amer. Math. Soc.* **82** (1956), 246–269.
37. G. Hochschild and T. Nakayama, Cohomology in class field theory. *Ann. of Math.* **55** (1952), 348–366.
38. G. Hochschild and J. P. Serre, Cohomology of group extensions. *Trans. Amer. Math. Soc.* **74** (1953), 110–134.
39. S. Iyanaga, Zum beweis des Hauptidealsatzes. *Abh. Math. Sem. Univ. Hamburg* **10** (1934), 349–357.
40. S. Iyanaga, "Class Field Theory Notes." Univ. of Chicago, Chicago, Illinois, 1961.
41. N. Jacobson, "Lectures in Abstract Algebra," Vol. III: Theory of Fields and Galois Theory. Van Nostrand, Princeton, New Jersey, 1964.
42. I. Kaplansky, "Infinite Abelian Groups." Univ. of Michigan Press, Ann Arbor, Michigan, 1954.

43. Y. Kawada, Class formations. *Duke Math. J.* **22** (1955), 165–177.
44. Y. Kawada, Cohomology of group extensions. *J. Fac. Sci. Univ. Tokyo Sect. I.* **9** (1963), 417–431.
45. A. G. Kurosh, "The Theory of Groups." Chelsea, New York, 1960.
46. S. Lang, "Rapport sur la Cohomologie des Groupes." Benjamin, New York, 1966.
47. R. C. Lyndon, The cohomology theory of group extensions. *Duke Math. J.* **15** (1948), 271–292.
48. S. MacLane, "Homology." Springer, New York, 1963.
49. W. Magnus, Über den Beweis des Hauptidealsatzes. *J. Reine Angew. Math.* **70** (1934), 235–240.
50. W. S. Massey, Some new algebraic methods in topology. *Bull. Amer. Math. Soc.* **00** (1954), 111–123.
51. T. Nakayama, Über die Beziehungen zwischen den Faktorensystemen und der Normklassengruppe eines galaisschen Erweiterungskörpers. *Math. Ann.* **112** (1935), 85–91.
52. T. Nakayama, Note on 3-factor sets. *Kodai Math. Sem. Rep.* **3** (1949), 11–14.
53. T. Nakayama, Factor system approach to the isomorphism and reciprocity theorems. *J. Math. Soc. Japan* **3** (1951), 52–58.
54. T. Nakayama, Idèle-class factor sets and class field theory. *Ann. of Math.* **55** (1952), 73–84.
55. T. Nakayama, On modules of trivial cohomology over a finite group, I. *Illinois J. Math.* **1** (1957), 36–43.
56. T. Nakayama, On modules of trivial cohomology over a finite group, II. *Nagoya Math. J.* **12** (1957), 171–176.
57. T. Nakayama, Cohomology of class field theory and tensor product modules. *Ann. of Math.* **65** (1957), 255–267.
58. D. G. Northcott, "An Introduction to Homological Algebra." Cambridge Univ. Press, London and New York, 1962.
59. L. Pontrjagin, "Topological Groups." Princeton Univ. Press, Princeton, New Jersey, 1946.
60. D. S. Rim, Modules over finite groups. *Ann. of Math.* **69** (1959), 700–712.
61. O. Schreier, Über die Erweiterung von Gruppen. *Monatsh. Math. Phys.* **37** (1926), 165–180.
62. O. Schreier, Über die Erweiterung von Gruppen, II. *Abh. Math. Sem. Univ. Hamburg* **4** (1926), 321–346.
63. I. Schur, Über die Darstellung der endlichen Gruppe durch gebrochene lineare Substitutionen. *J. Reine Angew. Math.* **27** (1904), 20–50.
64. W. R. Scott, "Group Theory." Prentice-Hall, Englewood Cliffs, New Jersey, 1964.
65. J. P. Serre, Applications algébriques de la cohomologie des groupes. *Sem. Cartan, 1950/1951.*
66. J. P. Serre, Groupes finis à cohomologie périodique (d'après R. Swan). *Sem. Bourbaki, 1960/1961*, exp. 209.
67. J. P. Serre, "Corps Locaux." Hermann, Paris, 1962.
68. J. P. Serre, "Cohomologie Galoisienne" (Lecture Notes in Math. No. 5). Springer, New York, 1964.

69. E. Snapper, Cohomology of permutation representations, I. Spectral sequences. *J. Math. Mech.* **13** (1964), 133–162.
70. E. Snapper, Cohomology of permutation representations, II. Cup product. *J. Math. Mech.* **13** (1964), 1047–1064.
71. R. Swan, Induced representations and projective modules. *Ann. of Math.* **71** (1960), 552–578.
72. R. Swan, Periodic resolutions for finite groups. *Ann. of Math.* **72** (1960), 267–291.
73. R. Swan, The non-triviality of the restriction map in cohomology of groups. *Proc. Amer. Math. Soc.* **11** (1960), 885–887.
74. J. Tate, The higher dimensional cohomology groups of class field theory. *Ann. of Math.* **56** (1952), 294–297.
75. J. T. Tate, The cohomology groups of algebraic number fields. *Proc. Internat. Congr. Math., Amsterdam, 1954.*
76. J. T. Tate, Duality theorems in galois cohomology over number fields. *Proc. Internat. Congr., Stockholm, 1962*, pp. 288–295.
77. A. Weil, Sur la théorie du corps de classes. *J. Math. Soc. Japan.* **3** (1951), 1–35.
78. A. Weil, "Basic Number Theory." Springer, New York, 1967.
79. E. Weiss, A deflation map. *J. Math. Mech.* **8** (1959), 309–330.
80. E. Witt, Bemerkungen zum Beweiss des Hauptidealsatzes von S. Iyanaga. *Abh. Math. Sem. Univ. Hamburg* **11** (1936), 221.
81. E. Witt, Verlegerung von Gruppen un Hauptidealsatz. *Proc. Internat. Math. Congr. 2nd, Amsterdam, 1954*, pp. 71–73.
82. N. Yoneda, On the homology theory of modules. *J. Fac. Sci. Univ. Tokyo Sect. I* **7** (1954), 193–227.
83. H. Zassenhaus, "The Theory of Groups." Chelsea, New York, 1949.

Subject Index

6
B 7
C 8
D 9
E 0
F 1
G 2
H 3
I 4
J 5